Designing with Linear Integrated Circuits

Contributors

Alan C. Bahr, Amelco Semiconductor, Mountain View, California

Randy Brandt, Raytheon Company, Mountain View, California

J. F. Bogusz, Philco-Ford Corporation, Blue Bell, Pennsylvania

J. E. Byerly, General Electric Company, Syracuse, New York

Jack Gifford, Fairchild Semiconductor, Mountain View, California

Alfredo Gomez, ITT Semiconductors, West Palm Beach, Florida

H. B. Grutchfield, Signetics Corporation, Sunnyvale, California

P. E. Hermann, Philco-Ford Corporation, Blue Bell, Pennsylvania

L. H. Hoke, Jr., Philco-Ford Corporation, Blue Bell, Pennsylvania

Don Jones, ITT Semiconductors, West Palm Beach, Florida

J. A. Livezey, Raytheon Company, Mountain View, California

M. A. Maidique, Transitron Electronic Corporation, Wakefield, Massachusetts

L. A. Pope, Sylvania Electric Products, Woburn, Massachusetts

Ed Renschler, Motorola Semiconductor Products, Inc., Phoenix, Arizona

William S. Routh, Radiation, Inc., Melbourne, Florida

V. Sarkissian, Signetics Corporation, Sunnyvale, California

D. R. Sullivan, Transitron Electronic Corporation, Wakefield, Massachusetts

Robert J. Widlar, National Semiconductor, Santa Clara, California

W. H. Williams, Westinghouse Electric Corporation, Elkridge, Maryland

H. A. Wittlinger, RCA Electronic Components, Somerville, New Jersey

Charles C. Wu, Raytheon Company, Mountain View, California

Designing With Linear Integrated Circuits

Jerry Eimbinder, Editor

John Wiley & Sons, Inc.

New York • London • Sydney • Toronto

Library of Congress Catalog Card Number: 68-56161
SBN 471 23455 9
Printed in the United States of America

Preface

In a recent article[1] in *EEE* Magazine Motorola's Wally Raisanen criticized major conferences and engineering magazines for letting users of integrated circuits be misled. Instead of devoting unwarranted attention to glamorous subjects, years away from practical application, Raisanen called for more emphasis on papers with immediate bearing.

It was with much the same sentiment that *EEE* launched its program of annual semiconductor-device clinics in 1966. The subject selected for the first clinic[2] was field-effect transistors, and all the speakers were applications engineers from FET manufacturers. The papers were concerned with helping engineers who use FETs to do their jobs; the speakers drew from experiences with customers in preparing their material.

The second clinic,[3] held in 1967, was patterned after the first, except that the topic picked was linear integrated circuits instead of FETs. The switch was prompted by the rapid rise in popularity of linear ICs and the lack of available application information.

The sudden ascendancy of the linear IC surprised even the experts. It had been generally accepted throughout the industry in 1965 and early 1966 that linear ICs would gain acceptance slowly because their prices, it was believed, could not possibly taildive as digital ICs had. Many good reasons were advanced to support this theory, the key one being that demand for any one particular type would undoubtedly be limited. As it turned out, the price decline for linear ICs was even more rapid than it had been for digital ICs.

Again, in 1968, linear integrated circuits were chosen as the topic for the clinic[4]; again the prime motivation was the gap between advancing technology and application information readily accessible to the IC user.

Dropping prices continued to play an important part in opening new areas to the linear IC during 1967 and early 1968. Many operational-amplifier ICs now sell for less than four dollars in large quantities, and a number of linear ICs are available for less than one dollar.

Of course, low cost is not the only reason for the present popularity of

[1]Wally Raisanen of Motorola Speaks Out On The "Real" World of ICs, *EEE* Magazine, 38–41 (December 1967).
[2]First Annual EEE Semiconductor-Device Clinic, March 25, 1966.
[3]Second Annual EEE Semiconductor-Device Clinic, March 24, 1967.
[4]Third Annual EEE Semiconductor-Device Clinic, March 21, 1968.

the linear IC. As many of the chapters in this book show, linear IC performance is equal to and sometimes considerably superior to that of discrete component counterparts.

In addition, improvements in existing ICs and announcements of new ones have been made and still are coming along at a rate that makes it difficult for the user to keep up; but, then, that is why this book was written.

The chapters in this book represent the latest thinking in the application of linear ICs and are based on the papers given at the 1968 clinic. The authors, as applications engineers at IC manufacturers, interface between their companies and its customers. They are therefore uniquely qualified to write about the use of linear ICs.

JERRY EIMBINDER

New York, New York
June 1968

Contents

Designing with Linear Integrated Circuits

1

The Linear Integrated Circuit: Past, Present and Future

by Jack Gifford
Fairchild Semiconductor

1.1 INTRODUCTION

The analog phase of the integrated-circuit evolution began with the introduction and acceptance of the solid-state differential dc amplifier. This device, which used matched discrete differential pairs, clearly demonstrated temperature drift and stability advantages and allowed the design of stable, high-accuracy analog feedback control systems.

This resulted in a new class of electronic equipment for guidance control, process control, analog computation, and instrumentation. From both the standpoints of performance and economics this approach was superior to the vacuum tube and transistor differential-amplifier approaches. Equipment design breakthroughs could be traced directly to the availability of the matched-transistor pair; as a result, users sought for further improvements in transistor matching and pricing.

Monolithic differential transistor pairs offered a solution to many problems. In 1964 it was possible to match a pair of monolithic transistors to 10-mV offset voltage with high yields. Matched discrete transistors could also achieve this, but the assembly costs and yields for discrete chip matching made this approach less attractive than that of using monolithic matched transistor pairs. The first monolithic amplifiers, designed in 1964, did accomplish a major goal, in that they provided low-cost, high-performance matched transistors. In 1968 the matched monolithic pair remained the most salient linear integrated-circuit performance and marketing advantage.

Although the monolithic linear integrated circuits of 1964 did offer matched transistors, they could not provide adequate low-current h_{fe} ($I_C < 100\ \mu A$), high breakdown voltages ($BV_{CEO} > 20$ V), low noise, high power dissipation, resistance magnitude and tolerance, or *npn-pnp* flexibility. The analog circuit designer in 1964 could not use the linear integrated circuit as a building-block replacement for many applications developed earlier for the discrete transistor amplifier.

Figure 1.1 indicates the level of linear integrated-circuit activity and interest in 1965; approximately 500,000 circuits were sold at an average price well above $30. These early ICs did not offer sufficient advantages, other than size, to initiate the growth spiral shown in Figure 1.1. They did, however, develop interest and curiosity in engineering circles for what was to come; the stage was set for the high-gain, high-impedance monolithic operational amplifier, the μA709.

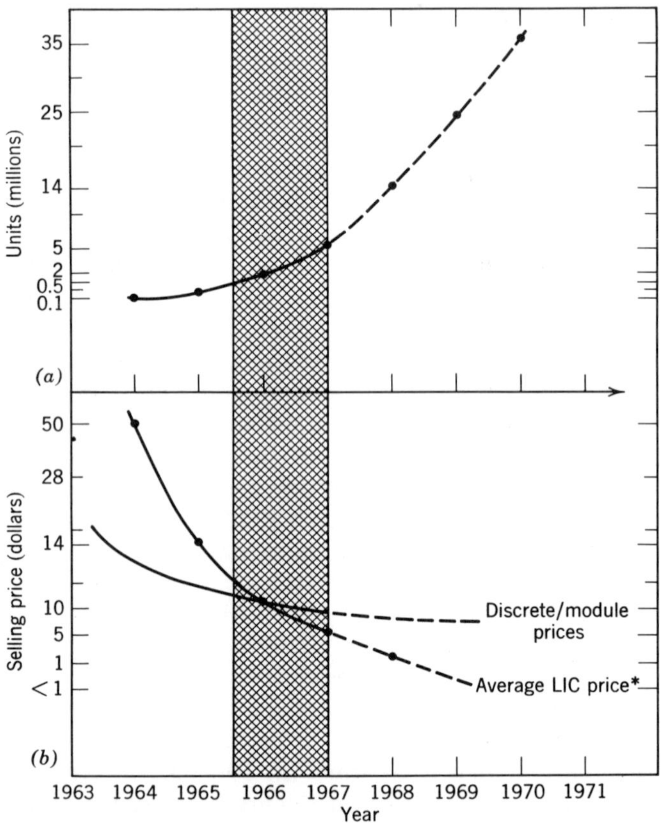

*1968 average price is affected considerably by increasing volume of low-cost consumer CKTS.

Figure 1.1. In (*a*) unit sales are plotted from 1964 through 1970 (1968 through 1970 are projected figures). In (*b*) average selling price is plotted.

1.2 THE 709

The μA709 or 709 was close to what was needed by the user. It provided adequate output voltage (exceeding ± 10 V) and adequate input characteristics ($R_{in} > 100$ kΩ; $V_{in} > \pm 8$ V), and stressed the advantages of linear monolithic construction by featuring offset voltages and currents of 2 mV and 100 nA, with proportionally reduced temperature coefficients. Semiconductor manufacturers provided the final ingredient, volume price projections, thereby showing that the use of linear ICs would provide a significant cost advantage.

The μA709 was introduced in the summer of 1965. Figure 1.1*a* shows that linear IC volume grew from 500,000 circuits in 1965 to two million in 1966. Significant in this growth was the availability of linear integrated circuits at prices less than their discrete-amplifier counterparts. This is also indicated in Figure 1.1. The analog circuit designer working in the aerospace and defense fields could actually buy a monolithic amplifier for less than he could design and manufacture his own amplifier.

In 1966 80% of the linear IC volume was in aerospace and defense applications. The industrial and consumer circuit designer had developed low-cost manufacturing techniques that kept the linear IC beyond his economic reach, but in 1967 these economic constraints were removed. The industrial equipment manufacturer was offered linear ICs at prices that were competitive with or less than the cost of discrete-component counterparts.

Market growth to date is embryonic. It has been estimated that less than 25% of the aerospace and defense electronic systems in production in 1968 use monolithic amplifiers, although this was the first market to recognize and take advantage of the performance and cost savings of linear ICs. The linear IC renaissance in the industrial and computer markets followed aerospace and defense acceptance by about two years. The consumer market is incorporating linear ICs in significant equipment designs, which should result in 1970 and 1971 production quantities in excess of 30 million units.

Why was the linear-circuit market growth so spontaneous? The answer lies in the obvious need for the first linear IC developed. Sophisticated market research and product definition was hardly necessary to determine that operational amplifiers and voltage comparators were marketable products. Parameter descriptions were quickly and accurately developed even by the novice analog-design engineer. Operational and comparator amplifiers were restricted from market success only by monolithic-process limitations or by the inability of circuit designers to circumvent these process limitations through unique design. Accurate product planning and subtle definition of sophisticated circuits for complex systems was not a requirement for these first successful circuits. Processing uniqueness, reliability, and clever circuit design were the key needs.

The next phase of linear-IC market development was initiated by most major IC manufacturers during 1967. This phase included examination of potential major linear integrated-circuit applications for other possible circuit configurations through systematic product and market planning. Certainly product areas with significant potential did exist, but the problem of defining them and pinpointing the market size was complex and difficult and required well-defined manufacturer-user communications.

The linear circuit manufacturer is required to understand how his product will be applied. On the other hand, user participation in circuit definition is also needed in today's product planning cycle. Technology advances resulted in increased complexity for 1968 devices. In 1965 the linear-circuit die size that could provide good yields was about 35 by 35 mils. In 1967 high yields were obtained with 55 by 55 mil dies. In the first quarter of 1968 linear integrated circuits with 60 by 150 mil complexity were in off-the-shelf supply.

Electronic equipment performance and availability is being affected to an increasing degree by the semiconductor manufacturer. The semiconductor manufacturer has become in many cases a dominant factor influencing the profitability and competitiveness of the final equipment design. Many equipment manufacturers have created survey teams to keep abreast of new semiconductor product developments. Integrated circuit developments can provide competitive advantage, reduced cost, higher performance, and even new product feasibility to the equipment manufacturer.

1.3 EVALUATING THE IC SUPPLIER

Equipment and systems manufacturers, whose product lines are affected by linear integrated-circuit development and production capabilities, attempt to minimize equipment and delivery risks by thorough evaluation of potential suppliers. Criteria commonly used for evaluation are the following:

1. Linear integrated-circuit production capability. Past experience with suppliers' product line is weighed with respect to volume delivery capability, quality, and reliability.
2. Major supplier with broad lines. For most equipment both linear integrated circuits and digital integrated circuits are required in addition to diodes and discrete transistors. Selecting suppliers with total capability sometimes reduces procurement costs and develops a good communicative relationship.

3. Organizational structure and resources to provide adequate customer applications and field sales service. Both linear and digital integrated circuits are increasing in complexity, application, and definition; therefore, efficient use of these circuits requires dependable field sales and factory service.

4. High-reliability 100% processing capability for aerospace and defense requirements. Major manufacturers of equipment for military and space applications must be assured of dependable circuit processing so that products will meet rigid high-reliability requirements. The vital nature of equipment and usage demands supplier integrity and competence.

5. A standard, complete, industry-accepted linear integrated-circuit product line. Multiplicity of source has become a byword in today's electronics industry. Most linear circuit users will not commit production equipment to new products without a firm commitment that the circuit will be second sourced by other suppliers. Two or three of today's major circuit suppliers are able to offer this advantage either through cross-licensing agreements or a past history of industry standardization on their product line. New circuits (providing 5 to 10% performance improvement over existing standard circuits) offered and manufactured by a small semiconductor supplier have less chance of market success due to a lack of dependable second sourcing. The equipment manufacturer usually will not jeopardize a multimillion dollar equipment product line for a 5% performance or economic advantage from an untested, unqualified, single-source supplier of a new semiconductor circuit. In the past users have delayed converting to an improved linear integrated circuit until a similar product is available from a major supplier. In the semiconductor industry this delay is typically less than six months. The small supplier will usually find it profitable to second source the major manufacturers.

1.4 TYPES OF LINEAR ICs

Today's available linear integrated circuits can be grouped into the following product-type categories: operational amplifiers, comparators, audio amplifiers, high-frequency and communications amplifiers, voltage regulators, and analog-digital interface circuits.

Eighty percent of the eight million linear integrated circuits manufactured in 1968 can be grouped into the first two categories. Going into 1968 twenty linear integrated-circuit manufacturers supplied more than 55 operational amplifier and comparator configurations. The electronics industry, however, has standardized on a single operational amplifier and one voltage comparator. More than 90% of the operational amplifiers and comparators shipped in 1967 were 709 and 710 linear integrated circuits.

Attempts to supply an acceptable audio amplifier to the consumer market have been generally unsuccessful due to economics requiring that the circuit must be manufactured and sold profitably in the 80c to $1.50 price range. At present, audio circuits offer neither price nor performance inducements. Both communications circuits and regulators require high-frequency and matched-transistor performance. Both product areas provide a profitable market when circuit types are designed consistent with an understanding of customer performance and economic problems and requirements.

Unfortunately, a lack of communications in the past between user and manufacturer, which caused application ignorance or misunderstanding, has resulted in market rejection for many new functionally complex linear integrated circuits.

New linear integrated circuits, however, will be available from all the major suppliers. Major manufacturers will continue to second source one another on standard circuit types. It is expected that major suppliers will not second source new products offered by less significant manufacturers. Market forces of mass media advertising, established customer followings, and large product distribution and sales channels are not available to the small supplier to speed product acceptance and standardization.

Several new products and technologies are being made available by the three or four major linear circuit suppliers. It is also expected that two circuits such as the μA709 and μA710 can never again completely dominate customer acceptance and standardization as they did in 1965 and 1966. It is reasoned that these products enjoyed this standardization through a protective competitive umbrella, which provided more than two years of competition-free user design experience.

How did this occur? There are many reasons, but some of the better understood are the following:

1. Lack of confidence and knowledge in the size and availability of the potential linear integrated-circuit market.
2. An industry deficiency in high-voltage processing know-how and monolithic linear integrated-circuit design technology.
3. A not-invented-here attitude toward these accepted circuits by some companies. This resulted in more than 40 other circuit configurations competing for the established μA709 and μA710 market.
4. Errors in forecasting the potential low cost nature of linear integrated circuits.

1.5 NEW DEVELOPMENTS

The linear integrated-circuit industry will announce products to extend existing product groups and will create new linear integrated-circuit product areas. Each new product will extend linear integrated-circuit processing and circuit complexity technology. Operational amplifier improvements will result in low-drift/low-noise instrumentation amplifiers, easier-to-use high-performance general-purpose operational amplifiers, high input-impedance operational amplifiers, high slew-rate, wide-band operational amplifiers, and micropower operational amplifiers.

New low-level linear integrated circuits will be targeted at instrumentation applications that require chopper-stabilized circuits at present. Low current, low noise, and improved temperature-stabilized substrate technologies will make these products possible.

A typical low-noise instrumentation amplifier will feature low offset currents coupled with adjustable offset voltage. Input-matched transistor development will provide current and voltage tracking approaching the optimum practical performance that can be expccted from a differential transistor circuit. These circuits will approach drift performance of chopper-stabilized amplifiers with advantages of smaller size and lower cost.

Amplifiers will offer performance described by these parameters:

Open-loop gain	100 k
Input offset voltage	0.5 mV
Input offset current	10 nA
Input bias current	30 nA
Common-mode range	±13 V
CMRR	120 dB
PSRR	100 dB
I_{os} drift	0.05 nA/°C
V_{os} drift	1.0 μV/°C
Noise voltage (f_o= 10 Hz, f= 1 Hz)	8 nV
Noise current (f_o= 10 Hz, f= 1 Hz)	0.15 pA

Low-level technology will also be improved by thermal feedback control of monolithic substrates. This technology was in use in 1967 in the μA726 matched-differential transistor pair. Technology advances will improve temperature drift characteristics, stability, and complexity. A low level, temperature-stable preamplifier for operational amplifiers would be a logical extension of this technology.

Such ICs could replace chopper-stable amplifiers in low-level instrumentation applications such as thermocouple and strain gauge transducers with signal levels as low as 5 μV.

A typical preamplifier might have the following specifications:

Open-loop gain	100 V/V
Input offset voltage	2 mV
Input offset current	2 nA
Input impedance	100 MΩ
Common mode range	±13 V
CMRR	100 dB
V_{os} drift	0.3 μV/°C
I_{os} drift	2 pA/°C
Noise (10 Hz–500 Hz)	3 μVrms

Typical applications for these low-level instrumentation amplifiers could be for precision transducer amplifiers for process control systems, ultra-stable control systems, and measurement and data amplifiers.

A successor circuit to the μA709 has been developed. Each major manufacturer will probably supply a version of this circuit, incorporating his individual perception of the ideal configuration. In general successor circuits will be easier to apply with respect to frequency compensation, over-voltage, over-current, offset voltage adjustment, and other conveniences. Most will be compatible with the standardized μA709 pin configuration and will be plug-in replacements. Standardization will not follow to the extent of μA709 standardization.

A typical specification for such a general-purpose operational amplifier could resemble the following:

Open-loop gain	100
Input offset voltage	1 mV
Offset voltage drift	3 μV/°C
Input offset current	35 nA
Input bias current	150 nA
Common-mode range	±13 V
CMRR	100 dB
Output swing ($R_L = 2$ kΩ)	±13 V
Power consumption	50 mW
Offset null potentiometer	5 kΩ

A high input-impedance operational amplifier could find applications in active filters, high impedance transducers, and integrators and voltage followers requiring high input impedance.

Typical performance requirements would be the following:

Input offset voltage	20 mV
Input offset current	50 pA
Input bias current	200 pA
Input impedance	10^{11} Ω

Common-mode range	±12 V
PSRR	65 dB
V_{os} drift	15 μV/°C
Bandwidth	1 MHz

Existing module and monolithic operational amplifiers are not adequate for large signal applications in which high slew rate and fast settling times are required. These applications include high-speed A-to-D and D-to-A comparator and phase locked loops, sample and hold, multiplexed analog gates, pulse width modulators and video amplifiers.

Typical circuit requirements are the following:

Open-loop gain	96 dB
Input offset voltage	1 mV
Input offset current	50 nA
Input bias current	500 nA
Input impedance	10 MΩ
Common-mode range	±10 V
CMRR	100 dB
PSRR	90 dB
V_{os} drift	3 μV/°C
Slew rate (unity gain)	25 V/μs
Open-loop bandwidth (small signal)	20 MHz
Open-loop bandwidth (large signal)	500 kHz
Output swing	±12 V

A micropower amplifier could be used in commercial, military, and space equipment where available power, weight, and/or system heating is critical. Maximum amplifier power dissipation should be less than 100 μW with ±3 V supplies.

Micropower operational amplifier

Open-loop gain	25 k	75 k
Input offset voltage	1 mV	1 mV
Input offset current	2 nA	4 nA
Input bias current	5 nA	10 nA
Common-mode range	±0.5 V	±12 V
V_{os} Drift	3 μV/°C	3 μV/°C
I_{os} Drift	0.05 nA/°C	0.1 nA/°C
Unity-gain bandwidth (small signal)	10 kHz	
Unity-gain bandwidth (large signal)	2.5 kHz	
Power consumption	0.1 mW	2.5 mW
Supply voltage	±3 V	±15 V

High frequency and communications circuits will receive more development emphasis from most linear integrated-circuit manufacturers in the coming years. Linear product areas of interest are rf and microwave receiver and transmitter i-f for both military and industrial applications, microwave solid-state sources, television and hi-fi receiving and small-signal processing circuits, and video and sense amplifiers for tape and film memory systems.

Recent monolithic high-frequency transistor technology has provided devices of 800 MHz to more than 3 GHz, thus paving the way for increased IC effort in these frequency regions.

Linear integrated-circuit manufacturers became aware of high-speed computer manufacturers' core sense-amplifier needs during 1967. Interest and attention given to providing a monolithic circuit solution to this costly computer performance problem area has resulted in the development of a dual-channel sense-amplifier approach. Such a dual-channel (4 k bits/channel) sense amplifier is designed for core memories operating at cycle times down to 300 ns. Unique circuit design enables a worst-case threshold accuracy of ± 2 mV to be achieved, thus eliminating the need for threshold adjustment of individual sense amplifiers.

System timing is facilitated by a very tight specification on strobe release time variation coupled with a provision to monitor the core signal after thresholding. Machine logic flexibility is assured by output either as a memory data register or as the drive to the MDR. In either case multiple amplifiers may be OR'd together to handle large memory systems with no loss in speed, logic flexibility, threshold accuracy, or component economy. A major feature of circuit output logic is the ease in terms of external logic with which data may be placed into or out of the memory data register.

A brief typical specification is listed.

Input threshold accuracy, temperature, unit-to-unit (V_T 0 − 20 mV)	2.0 mV
Propagation delay	30 ns
Strobe release time	10 ns
Common-mode recovery time (2 V)	40 ns
Differential-mode recovery time (100 mW)	40 ns
Power consumption	200 mW

Digital data-processing equipment requires low-cost devices for analog data acquisition, display, and control to complement the digital integrated circuits already being used. Linear integrated circuits can provide a way to make digital-to-analog and analog-to-digital converters that meet performance specifications that satisfy many users; the integrated circuits can be produced in quantity to assure low manufacturing cost.

An A-D converter design contains a digital programmer, a holding register, a D-A converter reference, a current-setting resistor array, and two standard linear integrated circuits—an operational amplifier and a comparator. It can be projected that an A-D converter with resolution to 10 bits can be assembled in production quantities using linear integrated circuits, as shown in Figure 1.2, for less than $100.

The buffer amplifier and comparator performance require the high-speed operational amplifier described earlier. The logic register can be obtained using three four-bit bipolar registers (such as the Fairchild 9300). The resistor network would be selected depending upon coding, accuracy, and full-scale requirements. A high-speed, medium accuracy, voltage digital-to-current analog converter is required to complete the 10-bit A-D converter. The unit has been constructed on a single monolithic chip with associated switches. The full-scale voltage range as well as the coding format (BCD or binary up to 10 bits) is variable through an external resistor array. The array resistors may be preselected and fixed for normal usage. If greater accuracy or extended temperature range

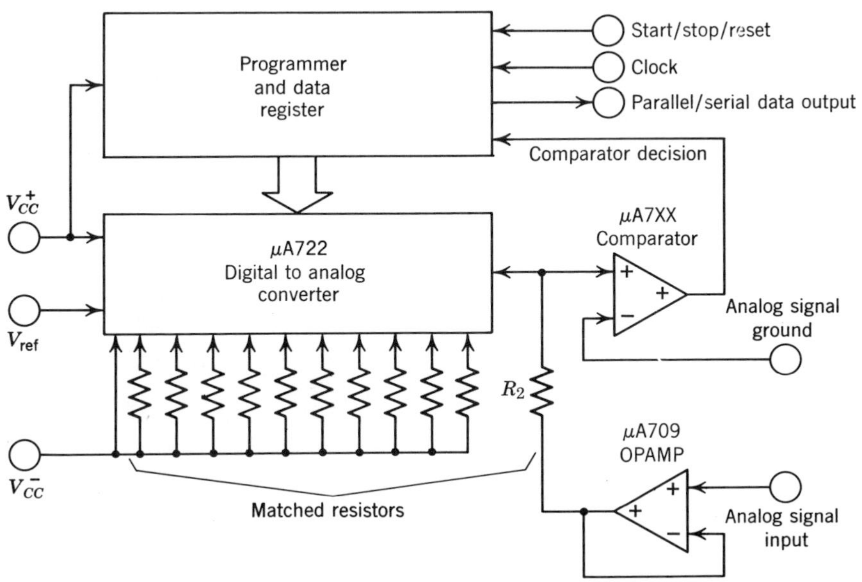

Figure 1.2. Block diagram of analog-to-digital converter using representative IC devices.

for a given accuracy is desired, the individual bit currents may be trimmed by adjusting the external array resistors. This circuit can be used directly as a digital-to-analog converter.

10-bit A-D/D-A Programmable Switched Current Source

Digital input, source "on"	0.5 V
Digital input, source "off"	2.0 V
Resolution	0.1%
Range (all inputs "1")	2940 μA
Range (all inputs "0")	100 nA
Speed (from 90% of logic find to	1 μs/wd parallel
within ±10 ±0.1% of its final value)	1 μs/bit serial
Accuracy* ($T_A = 25°C$)	0.01%
($T_A = 0-70°$)	0.50%
($T_A = -55$ to $125°C$)	0.50%
Nonlinearity* ($T_A = 25°C$)	0.01%
($T_A = 0-70°$)	0.02%
($T_A = -55$ to $125°C$)	0.08%
Power consumption	150 mW

*May be improved by trimming individual resistors.

2

Monolithic Operational Amplifiers—The Universal Linear Component

by Robert J. Widlar
National Semiconductor

2.1 INTRODUCTION

Operational amplifiers undoubtedly provide the easiest and best way to perform a wide range of linear functions from simple amplification to complex analog computation. In addition, the low cost of monolithic operational amplifiers now makes it attractive to design them into circuits in which, only a short time ago, they would not have been considered. Yet low cost is not the only attractive feature of monolithic operational amplifiers. Since all of the components are simultaneously fabricated on one chip, it is economically feasible to use much higher order circuit complexities than would have been used with discrete amplifiers. This freedom to include additional components permits considerably improved performance. Furthermore, there are no insurmountable technical difficulties to temperature stabilizing the amplifier chip; this makes possible chopper-stabilized performance with little added cost.

Operational-amplifier integrated circuits are designed for high gain, low offset voltage, and low input current. As a result, dc biasing is considerably simplified in most applications, and because many potential error terms can be neglected, fairly simple design rules apply.

In the following paragraphs several examples of the use of operational-amplifier ICs are presented. The circuits shown cover a wide variety of applications. The examples, however, are certainly not all-inclusive and should stimulate even more ideas from the reader. A few practical hints on preventing oscillations in operational amplifiers are also given, since this is probably the largest single problem area for many engineers.

The designs presented use the LM101 operational amplifier [1] and the LM102 voltage follower (described in chapter 7), both of which are products of National Semiconductor. Most of the considerations mentioned, however, are generally applicable to all monolithic operational-amplifier ICs, if the manufacturer's recommended frequency compensation is used and the differences in maximum ratings are taken into account. The LM101 differs from most other monolithic amplifiers, such as the LM709[2], in that it has a ±30 V differential input voltage range and a +15 V, −12 V, common mode range with ±15 V supplies, and it can be compensated with a single 30-pF capacitor. The LM102 has been designed specifically as a voltage follower and features a maximum input current of 10 nA and a 10 V/μs slew rate.

2.2 OPERATIONAL-AMPLIFIER OSCILLATOR

The free-running multivibrator shown in Figure 2.1 is an excellent example of an application for which the use of an operational amplifier is not normally considered. This circuit, however, requires only relatively small capacitor values because the threshold point of the operational amplifier is well determined, thus allowing the use of a longer portion of the capacitor time constant. In addition the circuit operates at low frequencies and it has a completely-symmetrical output waveform along with a buffered output.

Another advantage of the circuit in Figure 2.1 is that it will always be self-starting and cannot hang up, since there is more dc negative feedback

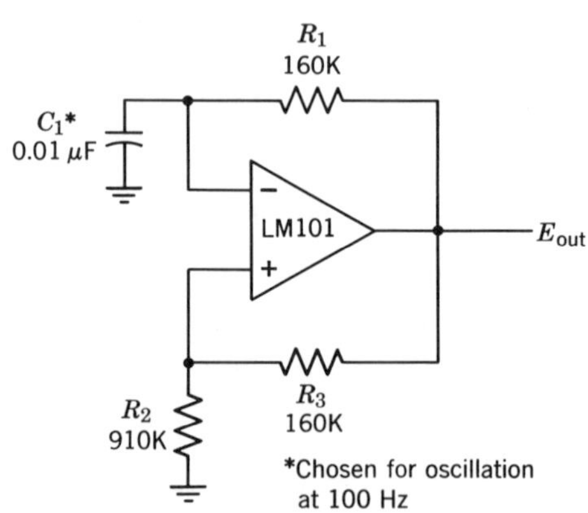

Figure 2.1. Free-running multivibrator.

than positive feedback. This can be a problem with many "textbook" multivibrators.

Since the operational-amplifier IC is used open-loop, the usual frequency compensation components are not required and would only slow it down. But even without the 30-pF capacitor normally used for compensation, the LM101 has speed limitations which restrict the use of this circuit to frequencies below about 2 kHz.

The large input voltage range of the LM101 (both differential and single-ended) permits large voltage swings on the input so that several time constants of the timing capacitor C_1 can be used. With most other amplifiers, R_2 must be reduced to keep from exceeding these ratings, which requires that C_1 be increased. Nonetheless, even when large values are needed for C_1, smaller polarized capacitors may be used by returning them to the positive supply voltage instead of ground.

2.3 LEVEL SHIFTING AMPLIFIER

In the design of linear equipment, it is frequently necessary to take a voltage that is referred to some dc level and produce an amplified output that is referred to ground. The most straightforward way of doing this is to use a differential amplifier as shown in Figure 2.2*a*. This circuit, however, has the disadvantage that the signal source is loaded by current from the input divider R_3 and R_4; also, the feedback resistors must be well matched to prevent erroneous outputs from the common-mode input signal.

A circuit that lacks these problems is shown in Figure 2.2*b*. Here the use of a FET transistor at the output of the operational amplifier produces a voltage drop across the feedback resistor R_1 that is equal to the input voltage. The voltage across R_2 is equal to the input voltage multiplied by the ratio R_2/R_1; and the common-mode rejection is as good as the basic rejection of the amplifier, independent of the resistor tolerances. This voltage is buffered by an LM102 voltage follower to give a low-impedance output.

One advantage of using an LM101 in this circuit is that it will work with input voltages up to its positive supply voltages as long as the supplies are less than ± 15 V.

2.4 VOLTAGE COMPARATORS

The LM101 is well suited for comparator applications because it has a large differential input voltage range and its output is easily clamped to make it compatible with various driver and logic circuits. It does not have the speed of the LM710 [3] (10 μs versus 40 ns, under equivalent

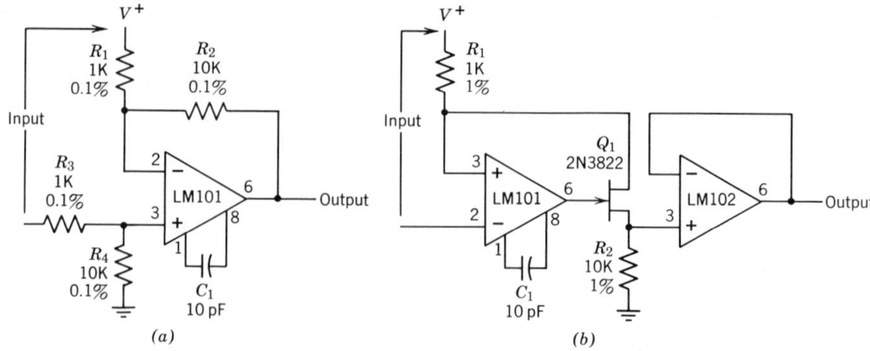

Figure 2.2. Level-shifting amplifiers: (*a*) standard differential amplifier; (*b*) level-isolation amplifier.

conditions); however, in many linear applications speed is not a problem and the lower input currents along with the higher voltage capability of the LM101 are important advantages.

Two comparator circuits using the LM101 are shown in Figure 2.3. The circuit in Figure 2.3*a* illustrates a clamping scheme which makes the output signal directly compatible with DTL or TTL integrated circuits. An LM103 breakdown diode clamps the output at 0 V or 4 V in

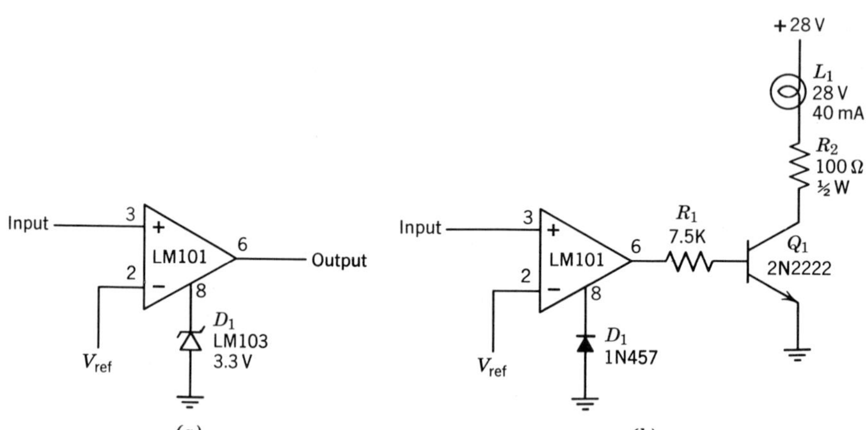

Figure 2.3. Voltage comparator circuits: (*a*) comparator for driving DTL and TTL integrated circuits; (*b*) comparator and lamp driver.

the low or high states, respectively. This particular diode was chosen because it has a sharp breakdown and low equivalent capacitance. When working as a comparator, the amplifier operates open loop so that normally no frequency compensation is needed. Nonetheless, the stray capacitance between pins 5 and 6 of the amplifier should be minimized to prevent low-level oscillations when the comparator is in the active region. If this capacitance becomes a problem, a 3-pF capacitor connected to the normal compensation terminals will eliminate it.

Figure 2.3*b* shows the connection of the LM101 as a comparator and lamp driver; Q_1 switches the lamp off and on, R_2 limits the current surge that results from turning on a cold lamp, R_1 determines the base drive to Q_1, and D_1 keeps the amplifier from putting excessive reverse bias on the emitter-base junction of Q_1 when it turns off.

2.5 MORE OUTPUT CURRENT SWING

Nearly all monolithic amplifiers use class-B output stages, and as a result they have good loaded output-voltage swings, delivering in the order of ± 10 V at 5 mA with ± 15 V supplies. To obtain higher current levels from the ICs would require that the output transistors be made considerably larger. In addition, the increased dissipation that would result could give rise to troublesome thermal gradients on the chip as well as excessive package heating in high-temperature applications. It is therefore advisable to use an external buffer when large output currents are needed.

A simple way of doing this is shown in Figure 2.4. A pair of complementary transistors are used at the output of the LM101 to get the increased current swing. Although this circuit does have a dead zone, it is not important at frequencies below 100 Hz when the amplifier is operated closed loop because of the high open-loop gain. Resistor R_1 is included to eliminate parasitic oscillations in the output transistors. In addition adequate bypassing should be used at the collectors of the output transistors to insure that the output signal is not coupled back into the amplifier. This circuit does not have current limiting, but this can be added by putting 50-Ω resistors in series with the collectors of Q_1 and Q_2.

2.6 AN FET AMPLIFIER

For ambient temperatures less than about 70°C the use of junction field-effect transistors at the input stage of an operational amplifier can provide exceptionally low input currents. Monolithic FET amplifiers, however, are not now available, for it is no simple matter to diffuse high

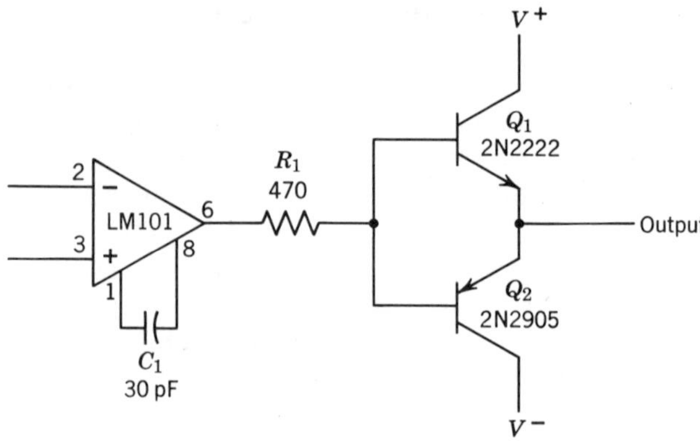

Figure 2.4. High-current output buffer.

quality FETs on the same chip as the amplifier. It is possible, however, to make a good FET amplifier using a pair of discrete FETs in conjunction with a monolithic circuit. Such a circuit is illustrated in Figure 2.5.

In this circuit, a matched FET pair, connected as source followers, is put in front of an integrated operational amplifier. The composite circuit provides roughly the same gain as the integrated circuit by itself; the IC is compensated for unity gain with a 30-pF capacitor as shown. Although the circuit works well as a summing amplifier, it leaves something to be desired in applications requiring high common-mode rejection. This happens both because resistors are used for current sources and because the FETs by themselves do not have common-mode rejection.

2.7 STORAGE CIRCUITS

A sample-and-hold circuit that combines the low input current of FETs with the low offset voltage of monolithic amplifiers is shown in Figure 2.6. The circuit is a unity-gain amplifier employing an operational amplifier and an FET source follower. In operation, when the sample switch Q_2 is turned on, it closes the feedback loop to make the output equal to the input, differing only by the offset voltage of the LM101. When the switch is opened, the charge stored on C_1 holds the output at a level equal to the last value of the input voltage.

Some care must be taken in the selection of the holding capacitor. Certain capacitor types, including paper and mylar, exhibit a polarization phenomenon which causes the sampled voltage to drop off by about 50 mV and then stabilize, when the capacitor is exercised over a 5-V range during the sample interval. This dropoff has a time constant in the order of seconds. The effect, however, can be minimized by using capacitors with teflon, polyethylene, glass, or polycarbonate dielectrics.

Although this circuit does not have a particularly low output resistance, fixed loads do not upset the accuracy because the loading is automatically compensated for during the sample interval. If the load is expected to change after sampling, however, a buffer such as the LM102 must be added between the FET and the output.

A second pole is introduced into the loop response of the amplifier by the switch resistance and the holding capacitor C_2. This can cause problems with overshoot or oscillation if it is not compensated for by adding a resistor R_2 in series with the LM101 compensation capacitor such that the breakpoint of the R_1C_1 combination is roughly equal to that of the switch and the holding capacitor.

It is possible to use an MOS transistor for Q_1 without worrying about the threshold stability. The threshold voltage is balanced out during every sample interval so that only the short-term threshold stability is important. When MOS transistors are used along with mechanical switches, drift rates less than 10 mV/min can be realized.

Additional features of the circuit are that the amplifier acts as a buffer so that the circuit does not load the input signal. Furthermore, gain can also be provided by feeding back to the inverting input of the LM101 through a resistive divider instead of going back directly.

The peak detector in Figure 2.7 is similar in many respects to the sample-and-hold circuit. A diode is used in place of the sampling switch. Connected as shown, it will conduct whenever the input is greater than

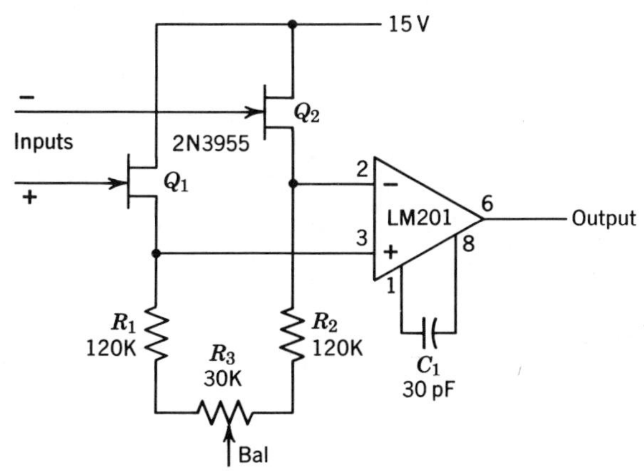

Figure 2.5. FET operational amplifier.

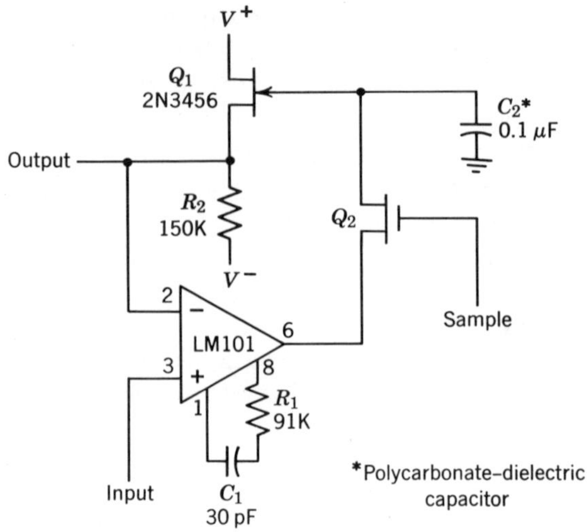

Figure 2.6. Low-drift sample-and-hold circuit.

the output, so that the output will be equal to the peak value of the input voltage. In this case an LM102 is used as a buffer for the storage capacitor which gives low drift along with a low output resistance.

As with the sample-and-hold, the differential input voltage range of the LM101 permits large differences between the input and output voltages when the circuit is holding.

2.8 NONLINEAR AMPLIFIERS

If a nonlinear transfer function is needed from an operational amplifier, there are many methods that can be employed to obtain it. These methods usually require diodes, however, and are therefore difficult to temperature compensate for accurate breakpoints. One way of getting around this is to make the output swing so large that the diode threshold is negligible by comparison, but this is not always practical.

A method of producing very sharp, temperature-stable breakpoints in the transfer function of an operational amplifier is shown in Figure 2.8. For small input signals, the gain is determined by R_1 and R_2. Both Q_2 and Q_3 are conducting to some degree, but they do not affect the gain because their current gain is high and they do not feed any appreciable current back into the summing mode. When the output voltage rises to 2 V (determined by R_3, R_4 and V^-), Q_3 draws enough current to saturate, thus connecting R_4 in parallel with R_2. This cuts the gain in half. Similarly, when the output voltage rises to 4 V, Q_2 will saturate, again halving the gain.

Temperature compensation is achieved in this circuit by including Q_1 and Q_4. Q_4 compensates the emitter-to-base voltage of Q_2 and Q_3 to keep the voltage across the feedback resistors R_4 and R_6 very nearly equal to the output voltage; Q_1 compensates for the emitter-to-base voltage of these transistors as they go into saturation, making the voltage across R_3 and R_5 equal to the negative supply voltage. A detrimental effect of Q_4 is that it causes the output resistance of the amplifier to increase at high output levels. It may therefore be necessary to use an output buffer if the circuit must drive an appreciable load.

2.9 SERVO PREAMPLIFIER

In certain servo systems it is desirable to get the rate signal required for loop stability from some sort of electrical lead network. This can be accomplished, for example, with reactive elements in the feedback network of the servo preamplifier.

Many saturating servo amplifiers operate over an extremely wide dynamic range. For example, the maximum error signal could easily be 1000 times the signal required to saturate the system. Cases like this create problems with electrical rate networks because they cannot be

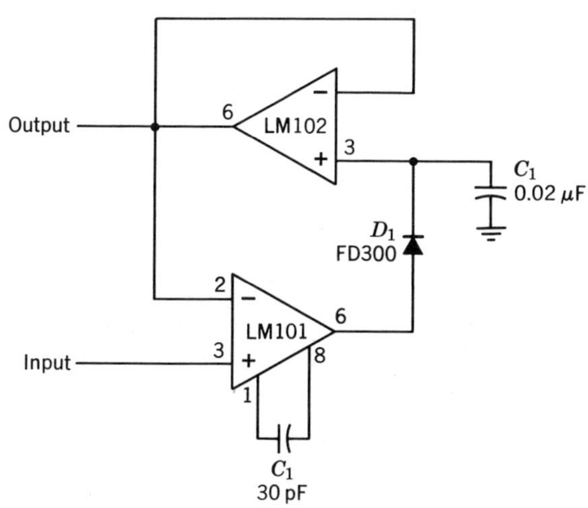

Figure 2.7. Positive peak detector with beffered output.

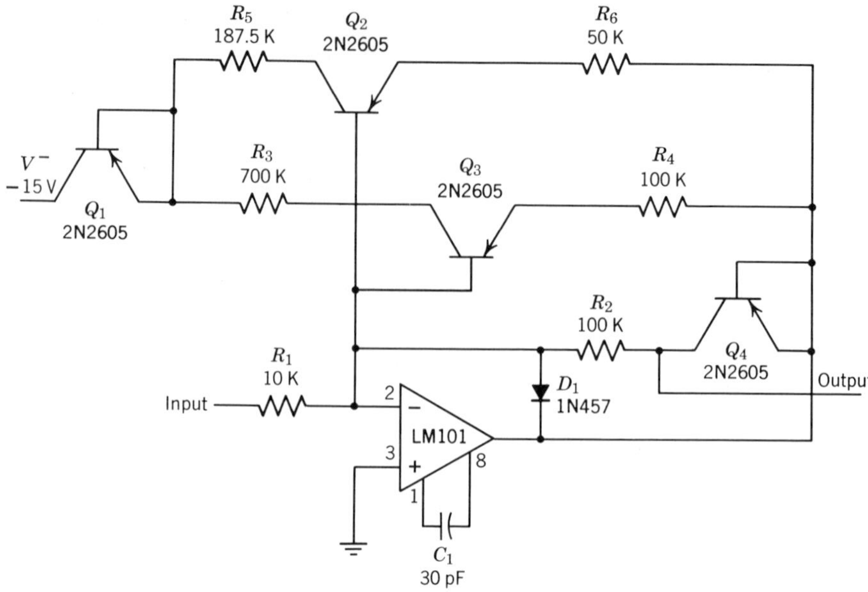

Figure 2.8. Non-linear operational amplifier with temperature-compensated breakpoints.

placed in any part of the system which saturates. If the signal into the rate network saturates, a rate signal will be developed only over a narrow range of system operation, and instability will result when the error becomes large. Attempts to place the rate networks in front of the error amplifier or make the error amplifier linear over the entire range of error signals frequently gives rise to excessive dc error from signal attenuation.

These problems can be largely overcome by using the kind of circuit shown in Figure 2.9. This amplifier operates in the linear mode until the output voltage reaches approximately 3 V with 30 μA output current from the solar cell sensors. At this point the breakdown diodes in the feedback loop begin to conduct, drastically reducing the gain. A rate signal will still be developed, however, because current is being fed back into the rate network (R_1, R_2, and C_1) just as it would if the amplifier had remained in the linear operating region. In fact the amplifier will not actually saturate until the error current reaches 6 mA, which would be the same as having a linear amplifier with a $\pm$600 V output swing.

2.10 COMPUTING CIRCUITS

In analog computation it is a relatively simple matter to perform such operations as addition, subtraction, integration, and differentiation by incorporating the proper resistors and capacitors in the feedback circuit of an amplifier. Many of these circuits are described in the handbook referred to in [4]. Multiplication and division, however, are a bit more difficult. These operations are usually performed by taking the logarithms of the quantities, adding or subtracting as required, and then taking the antilog.

At first glance it might appear that obtaining the log of a voltage is difficult; but it has been shown [5] that the emitter-to-base voltage of a silicon transistor follows the log of its collector current over as many as nine decades. This means that common transistors can be used to perform the log and antilog operations.

A circuit which performs both multiplication and division in this fashion is shown in Figure 2.10. It gives an output that is proportional to the product of two inputs divided by a third, and it has about the same complexity as a divider alone. The circuit consists of three log converters and an antilog generator. Log converters similar to these have been described elsewhere [6], but a brief description follows. Taking amplifier A_1, a logging transistor Q_1 is inserted in the feedback loop such that its collector current is equal to the input voltage divided by the input resistor R_1. Hence the emitter-base voltage of Q_1 will vary as the log of the input voltage E_1.

A_2 is a similar amplifier operating with logging transistor Q_2. The emitter-to-base junctions of Q_1 and Q_2 are connected in series, thus adding the log voltages. The third log converter produces the log of E_3. This is series-connected with the antilog transistor Q_4, and the combina-

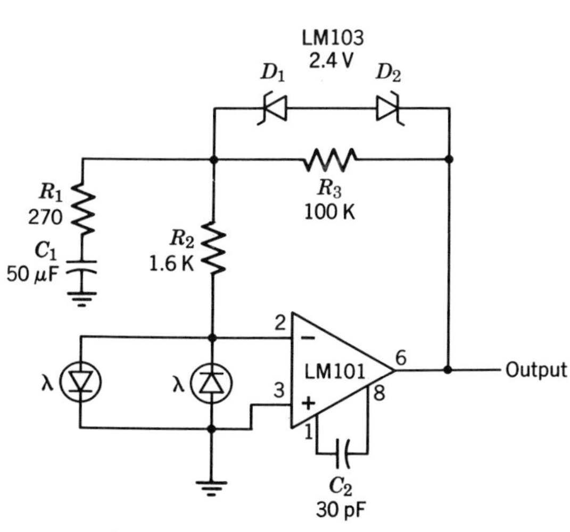

Figure 2.9. Saturating servo preamplifier with rate feedback.

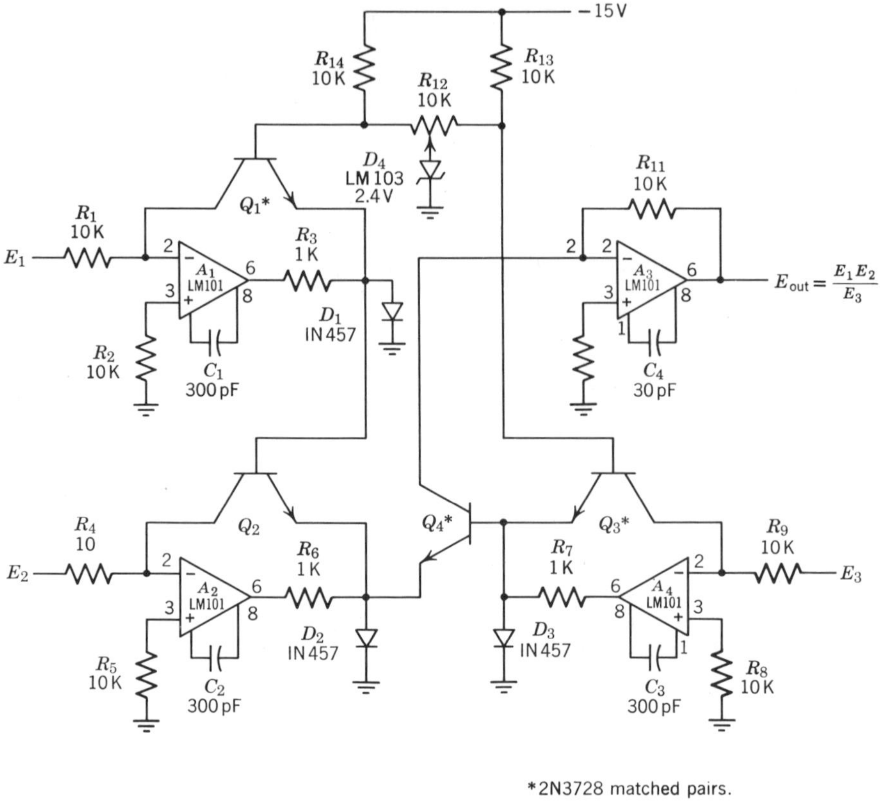

Figure 2.10. Analog multiplier/divider.

tion is hooked in parallel with the output of the other two log converters. Therefore the emitter-to-base of Q_4 will see the log of E_3 subtracted from the sum of the logs of E_1 and E_2. Since the collector current of a transistor varies as the exponent of the emitter-to-base voltage, the collector current of Q_4 will be proportional to the product of E_1 and E_2 divided by E_3. This current is fed to the summing amplifier A_4, thus giving the desired output.

This circuit can provide 1% accuracy with input voltages of 500 mV to 50 V. To get this precision at lower input voltages, the offset of the amplifiers handling them must be individually balanced out. The zener diode D_4 increases the collector-to-base voltage across the logging transistors to improve high-current operation. The diode is not needed and is in fact undesirable when these transistors are running at currents less than 0.3 mA. At currents above 0.3 mA the lead resistances of the transistors can become important (0.25 Ω is 1% at 1 mA), so that the transistors should be installed with short leads and no sockets.

An important feature of this circuit is that its operation is independent of temperature. It is only required that Q_1, Q_2, Q_3 and Q_4 be at the same temperature. Dual transistors should be used and arranged as shown in the figure so that thermal mismatches between cans appear as inaccuracies in scale factor (0.3%/°C) rather than as a balance error (8%/°C). R_{12} is a balance potentiometer which nulls out the offset voltages of all the logging transistors. It is adjusted by setting all input voltages equal to 2 V and adjusting for a 2 V output voltage.

The logging transistors provide a gain that is dependent on their operating level, which complicates frequency compensation. Resistors (R_3, R_6, and R_7) are put in the amplifier output to limit the maximum loop gain, the compensation capacitor being chosen to correspond with this gain. As a result the amplifiers are not especially designed for speed, but techniques for optimizing this parameter are given in a paper by Gibbons and Horn [5].

Finally, clamp diodes D_1 through D_3 prevent exceeding the maximum reverse emitter-to-base voltage of the logging transistors with negative inputs.

2.11 ROOT EXTRACTOR*

Taking the root of a number using log converters is a fairly simple matter. All that is needed is to take the log of a voltage, divide it by, say, one-half for the square root, and then take the antilog. A circuit which does this is shown in Figure 2.11. A_1 and Q_1 form the log converter for the input signal. This feeds Q_2, which in turn produces a level shift to give zero voltage into the R_4, R_5 divider for a 1 V input. This divider reduces the log voltage by the ratio for the root desired and drives the buffer amplifier A_2. A_2 has a second level shifting diode Q_3 in its feedback network that gives the output voltage needed to get a 1 V output from the antilog generator, consisting of A_3 and Q_4, with a unity input. The offset voltages of the transistors are nulled out by imbalancing R_6 and R_8 to give 1 V output for 1 V input, since any root of one is one.

Q_2 and Q_3 are connected as diodes to simplify the circuitry. This does not introduce problems because both operate over a very limited current

*The "root extraction" referred to here doubtless has origin in the dental operation that most of us would find less frightening than having to find even a square root without tables or other aids.

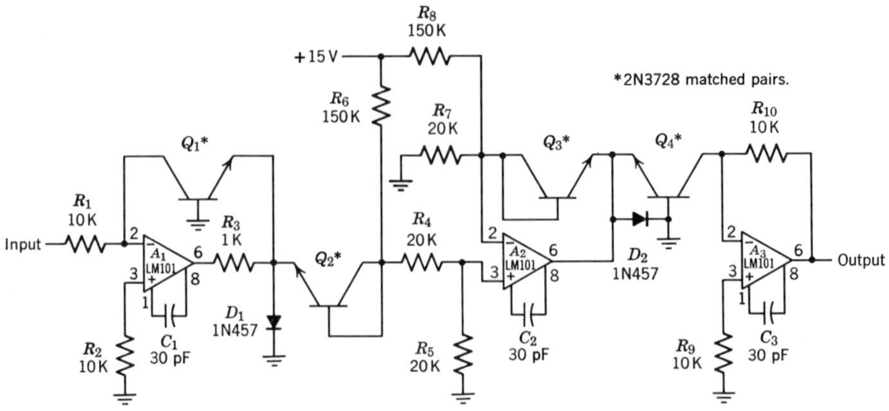

Figure 2.11. Root extractor.

range, and it is really only required that they match; R_7 is a gain-compensating resistor that keeps the currents in Q_2 and Q_3 equal with changes in signal level. As in the case of the multiplier/divider, the circuit is insensitive to temperature as long as all the transistors are at the same temperature. Using transistor pairs and matching them as shown minimizes the effects of gradients.

The circuit has 1% accuracy for input voltages between 0.5 and 50 V. For lower input voltages, A_1 and A_3 must have their offsets balanced out individually.

2.12 FREQUENCY COMPENSATION HINTS

The ease of designing with operational amplifiers sometimes obscures some of the rules which must be followed when using any feedback amplifier to keep it from oscillating. In general these problems stem from stray capacitance, excessive capacitive loading, inadequate supply bypassing or improper frequency compensation.

In frequency compensating an operational amplifier, it is best to follow the manufacturer's recommendations. If operating speed and frequency response is not a consideration, however, a greater stability margin can usually be obtained by increasing the size of the compensation capacitors. For example, replacing the 30-pF compensation capacitor suggested for the LM101 with a 300-pF capacitor would make it ten times less susceptible to oscillation problems in the unity-gain connection. Similarly, for the LM709, use of 0.05 μF, 1.5 kΩ, 200 pF, and 51 Ω will give 20 dB more stability margin. Capacitor values less than those specified by the manufacturer for a particular gain connection should not be used because they will make the amplifier more sensitive to strays and capacitive loading; the circuit can even oscillate with worst-case units.

The basic requirement for frequency compensating a feedback amplifier is to keep the frequency rolloff of the loop gain from exceeding 12 dB/octave when it goes through unity gain. Figure 2.12*a* shows what is meant by loop gain. The feedback loop is broken at the output, and the input sources are replaced by their equivalent impedance. Then the response is measured such that the feedback network is included.

Figure 2.12*b* gives typical responses for both uncompensated and compensated amplifiers. An uncompensated amplifier generally rolls off at 6 dB/octave, then 12 dB/octave, and even 18 dB/octave as various frequency-limiting effects within the amplifier come into play. If a loop with this kind of response were closed, it would oscillate. Frequency compensation causes the gain to roll off at a uniform 6 dB/octave right down through unity gain. This allows some margin for excess rolloff in the external circuitry.

Some of the external influences which can affect the stability of an operational amplifier are shown in Figure 2.13. One is the load capacitance that can come from wiring, cables, or an actual capacitor on the output. This capacitance works against the output impedance of the amplifier to attenuate high frequencies. If this added rolloff occurs before the loop gain goes through zero, it can cause instability. It should be remembered that this single rolloff point can give more than 6-dB/octave rolloff, since the output impedance of the amplifier can be increasing with frequency.

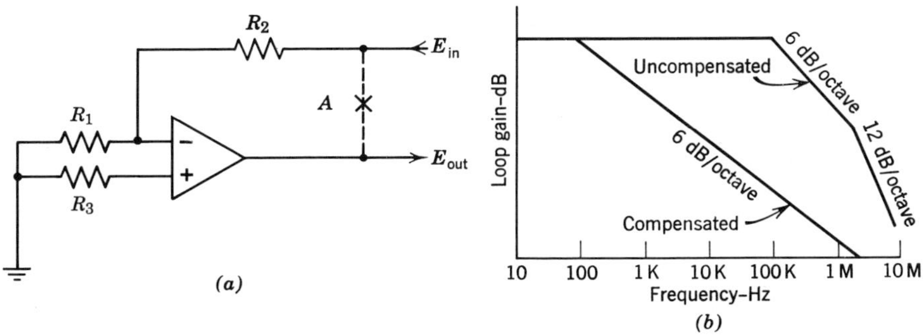

Figure 2.12. Loop gain measurement and response: (*a*) measuring loop gain; (*b*) typical response.

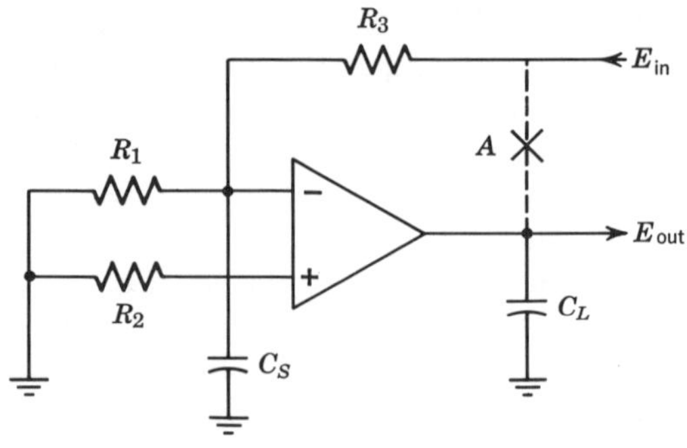

Figure 2.13. External capacitances that affect stability.

A second source of excess rolloff is stray capacitance on the inverting input. This becomes extremely important with large feedback resistors that might be used with an FET-input amplifier. A relatively simple method of compensating for this stray capacitance is shown in Figure 2.14: a lead capacitor C_1 is put across the feedback resistor. Ideally the ratio of the stray capacitance to the lead capacitor should be equal to the closed-loop gain of the amplifier. The lead capacitor can be made larger, however, as long as the amplifier is compensated for unity gain. The only disadvantage of doing this is that it will reduce the bandwidth of the amplifier. Oscillations can also result if there is a large resistance on the noninverting input of the amplifier. The differential input impedance of the amplifier falls off at high frequencies (especially with bipolar input transistors) so that this resistor can produce troublesome rolloff

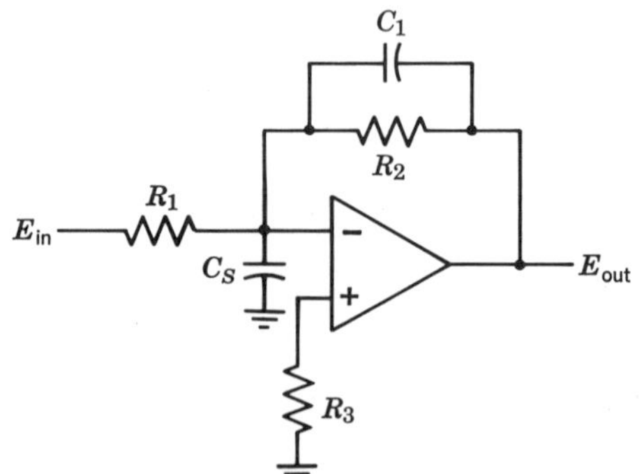

Figure 2.14. Compensating for stray input capacitance.

with most amplifiers if it is much greater than 10 kΩ. This is easily corrected by bypassing the resistor to ground.

When the capacitive load on an integrated amplifier is much greater than 100 pF, some consideration must be given to its effect on stability. Even though the amplifier does not oscillate readily, there may be a worst-case set of conditions under which it will. The amplifier can be stabilized, however, for any value of capacitive loading by using the circuit shown in Figure 2.15. The capacitive load is isolated from the output of the amplifier with R_4 which has a value of 50 to 100 Ω for both the LM101 and the LM709. At high frequencies the feedback path is through the lead capacitor C_1, so that the lag produced by the load capacitance does not cause instability. If one is to use this circuit, the amplifier must be compensated for unity gain, regardless of the closed loop dc gain. The value of C_1 is not too important, but at a minimum its capacitive reactance should be one-tenth the resistance of R_2 at the unity-gain crossover frequency of the amplifier.

Problems encountered with supply bypassing are insidious in that they will hardly ever show up in a Nyquist plot. This problem has not really been thoroughly investigated, probably because one sure cure is known: bypass the positive and negative supply terminals of each amplifier to ground with at least a 0.01 μF capacitor; for example, a LM101 can handle over 1 mH inductance in either supply lead without oscillation. This should not suggest that the ICs be run without bypass capacitors. It has been established that 100 LM101s on a single printed circuit board with common supply busses will oscillate if the supplies are not bypassed at about every fifth device. This happens even though the inputs and outputs are completely isolated. The LM709, on the other hand, will oscillate under many load conditions with as little as 18 in. of wire between the negative supply lead and a bypass capacitor. Therefore it is almost essential to have a set of bypass capacitors for every device.

When high-current buffers are used in conjunction with operational amplifiers, supply bypassing and decoupling are even more important since they can feed a considerable amount of signal back into the supply lines. For reference, bypass capacitors of at least 0.1 μF are required for a 50-mA buffer.

When emitter followers are used to drive long cables, additional precautions are required. An emitter follower by itself, not contained in a feedback loop, will frequently oscillate when connected to a long length of cable. When an emitter follower is connected to the output of an operational amplifier, it can produce oscillations that will persist no matter how the loop gain is compensated. An analysis to determine why this happens is not very enlightening, so suffice it to say that these oscilla-

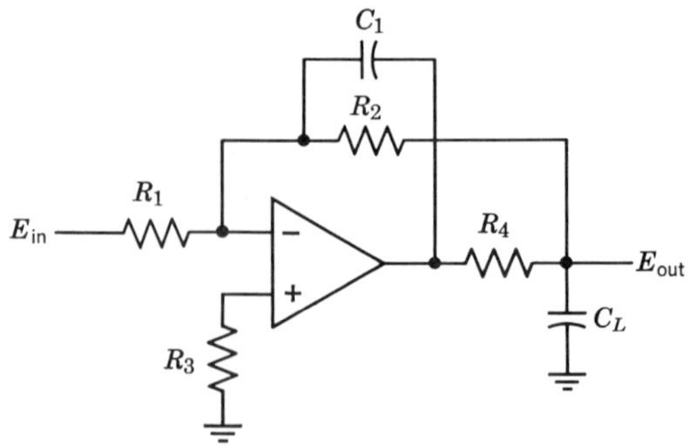

Figure 2.15. Compensating for very large capacitive loads.

tions can usually be eliminated by putting a ferrite bead [7] between the emitter follower and the cable.

Looking at the loop gain of an amplifier can be valuable in understanding the influence of various factors on the stability of feedback amplifiers. But it is not too helpful in determining if the amplifier is indeed stable. The reason is that most problems in a well-designed system are caused by secondary effects that occur only under certain conditions of output voltage, load current, capacitive loading, temperature, etc. Making frequency-phase plots under all these conditions would require unreasonable amounts of time, so it is invariably not done.

A better check on stability is the small-signal transient response. It can be shown mathematically that the transient response of a network has a one-for-one correspondence with the frequency domain response.* The advantage of transient response tests is that they are displayed instantaneously on an oscilloscope, so that it is reasonable to test a circuit under a wide range of conditions.

Exact methods of analysis using transient response will not be presented here. This is not because these methods are difficult, although they are. Instead, it is because it is very easy to determine which conditions are unfavorable from the overshoot and ringing on the step response. The stability margin can be determined much more easily by seeing how much greater the aggravating condition can be made before the circuit oscillates than by analysis of the response under given conditions. A little practice with this technique can quickly yield much better results than classical methods, even for the inexperienced engineer.

*The frequency-domain characteristics can be determined from the impulse response of a network and this is directly related to the step response through the convolution integral.

2.13 SUMMARY

A number of circuits using operational amplifiers have been proposed to show their versatility in circuit design. These have ranged from low-frequency oscillators through circuits for complex analog computation. Because of the low cost of monolithic amplifiers, it is almost foolish to design dc amplifiers without integrated circuits. Moreover, the price makes it practical to take advantage of operational-amplifier performance in a variety of circuits in which they are not normally used.

Many of the potential oscillation problems that can be encountered in both discrete and integrated operational amplifiers have been described, and some conservative solutions to these problems presented. The areas discussed included stray capacitance, capacitive loading, and supply bypassing. Finally, a simplified method of quickly testing the stability of amplifier circuits over a wide range of operating conditions was suggested.

REFERENCES

[1] Widlar, R. J., "Monolithic Op Amp with Simplified Frequency Compensation," *EEE*, **15**, No. 7, 58-63 (July 1967).

[2] Widlar, R. J., "A Unique Circuit Design for a High Performance Operational Amplifier Especially Suited to Monolithic Construction," *Proc. NEC*, **XXI**, 85-89 (October 1965).

[3] Widlar, R. J., "The Operation and Use of a Fast Integrated Circuit Comparator," *Fairchild Semiconductor APP-116*, February 1966.

[4] "Handbook of Operational Amplifier Applications," Burr-Brown Research Corporation, Tucson, Arizona.

[5] Gibbons, J. F. and H. S. Horn, "A Circuit with Logarithmic Transfer Response Over Nine Decades," *IEEE Trans. on Circuit Theory*, **CT-11**, 378-384 (September 1964).

[6] Widlar, R. J. and J. N. Giles, "Designing with Off-the-Shelf Linear Microcircuits," *Fairchild Semiconductor APP-124*, January 1966.

[7] Solomon, Leslie, "Ferrite Beads," *Electronics World*, 42-43 (October 1966).

3

Linear ICs in Industrial Applications

by Alfredo Gomez and Don Jones
ITT Semiconductors

3.1 INTRODUCTION

Integrated circuits have wide application in the field of industrial control systems. Two typical applications are illustrated in this chapter. One is a servo amplifier circuit, containing an ITT 709 IC. This circuit may be used in either ac or dc control systems. The description includes power supply information. The second application shows a method for using a differential comparator, such as the ITT 710, with input isolated from the system ground. This allows the use of the comparator in situations where floating inputs are required, such as in wide range voltage monitors, analog-to-digital converters, and bridge balance monitors.

3.2 BOOSTER AMPLIFIER

A diagram of a self-contained booster amplifier is shown in Figure 3.1. The complementary emitter followers Q_1 and Q_2 effectively isolate the MIC 709 IC from the load. Transistor Q_1 handles positive-going signals and Q_2 negative-going ones. The 100-Ω resistors connected to the transistor base protect the circuit against output short circuits and divide the load more uniformly between the two power transistors. The value of the transistor collector supply voltage is not critical; however, it must supply the current taken by the load. As shown in Figure 3.1, this voltage is taken from the unregulated source. The isolation emitter followers affect the output-voltage swing as shown in Figure 3.2. Depending on the transistors used, full output swing can be obtained with load impedances as low as 10 Ω. The amplifier's response to an input square wave is depicted in Figure 3.3. The output swings ±13 V with an 82 Ω load impedance. The time scale is 0.5 ms/division.

Resistor R_{10} and potentiometer R_{11} can be adjusted to obtain zero-output offset voltage, which will eliminate any dc currents flowing in the load during standby. Diodes D_7 and D_8 afford input protection against overvoltage. The selection of Q_1 and Q_2 is determined by the maximum current that must be supplied to the load as well as the average power being dissipated. Their $V_{BE(ON)}$ values should match as closely as possible. The frequency response of the amplifier connected for a gain of 100 is shown in Figure 3.4.

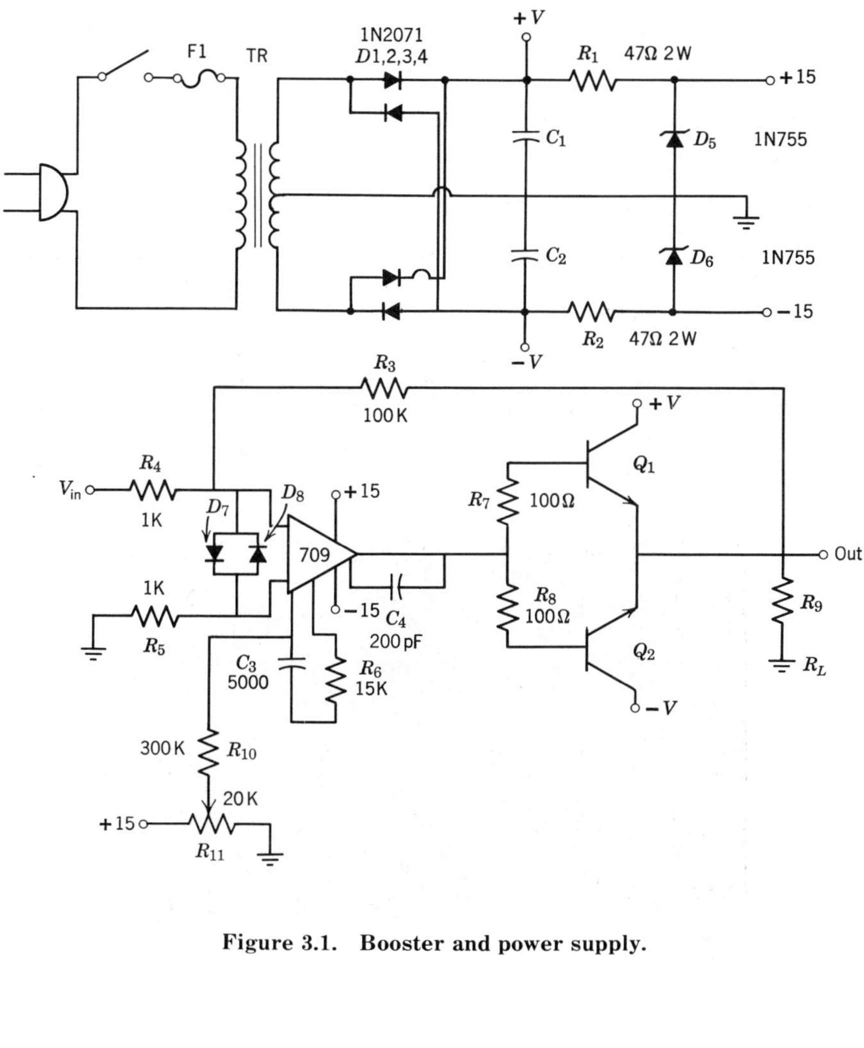

Figure 3.1. Booster and power supply.

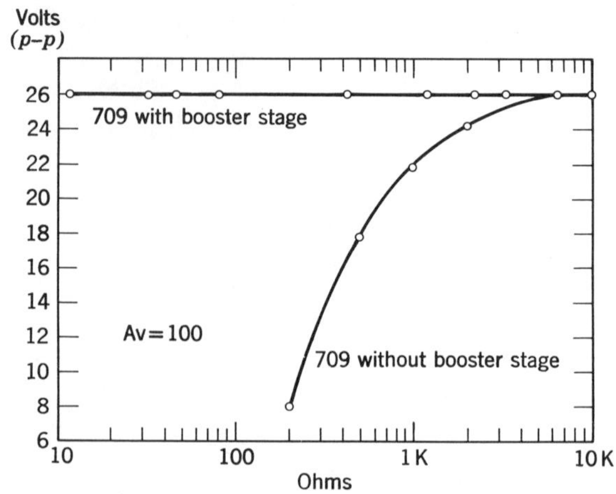

Figure 3.2. Effect of isolation emitter followers on output voltage swing.

3.3 TYPICAL APPLICATION

A typical setup in which the IC power amplifier can be used is shown in Figure 3.5. In this application the physical position of the load is controlled by an ac motor.

The error signal is equal to the difference in amplitude and phase between the command pot and the load pot. The amplified error signal drives the motor that in turn changes the position of the load. A mechanical feedback signal then attempts to reduce the error signal back to zero. A dc input signal can also be used to control a dc motor.

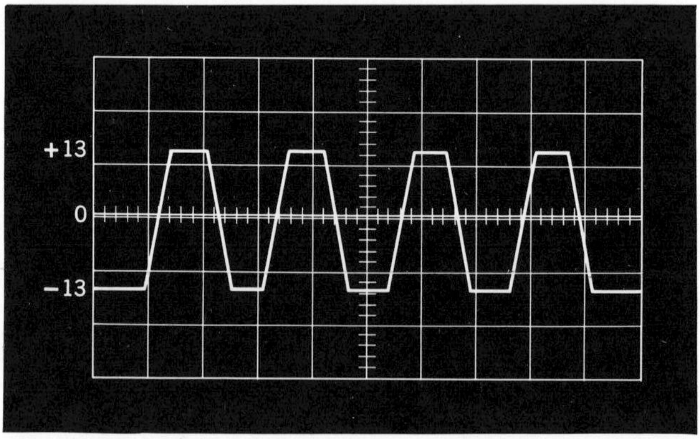

Figure 3.3. Booster output.

3.4 DESIGN OF AN ISOLATED DIFFERENTIAL COMPARATOR

There would be many applications for a four-terminal device having a differential input and an isolated output, operating in the same manner as a polarized relay, except that switching would occur in less than a microsecond at input differentials of a few millivolts.

An integrated differential comparator circuit, such as the popular 710, meets the speed and sensitivity requirements but, by itself, does not have the isolation necessary for a truly universal device. Input voltages must be related to the power supply and must remain within ± 5 V of the power-supply ground. The digital output ground is the power-supply ground; thus the comparator circuit by itself is limited to monitoring small signals related to the output digital-system ground. In order to float the differential comparator, it is necessary not only to furnish floating power supplies for the circuit, but also to include a dc isolator circuit in the output line to make the input independent of the output digital-system ground. The design described below is suitable for construction in modular or printed-circuit card form as a general-purpose building block in industrial measurement systems.

The schematic of the isolated comparator circuit is shown in Figure 3.6. The circuit consists of a power supply, the comparator, and a switched high-frequency oscillator transformer coupled to a detector with digital output. A simple zener-regulated power supply is used. For best isolation a power transformer with electrostatic shielding between windings may be desirable. The comparator circuit is a type 710 integrated circuit with diode protection between the inputs. R_2 and R_3 should be equal and as small as possible; however, if high differential voltage is encountered, it will be necessary to increase the value of R_2 and R_3 or to compute the power rating necessary for the resistors and diodes. Capacitance may be added across CR_7 and CR_8 to reduce ac sensitivity.

In Figure 3.6 the first MIC 932 dual power gate is wired as a free-running multivibrator. When the 710 output is high level, the multivibrator oscillates at approximately 5 MHz; when the 710 output is low level, oscillation stops. R_4 is selected so that the voltage at pin 14 of the MIC 932 is 5 V$\pm$10% when the circuit oscillates. This voltage drops when oscillation is cut off, thereby preventing the gate from drawing excessive current.

Isolation is achieved by coupling the multivibrator to a detector through a pulse transformer. A MIC 932 dual gate is used as the detector. In the absence of oscillation the first gate input is grounded through the transformer secondary so that its output is high and the second gate output is low. When the multivibrator is oscillating, the first gate turns on and off at the multivibrator frequency. The second gate, which has a capacitor

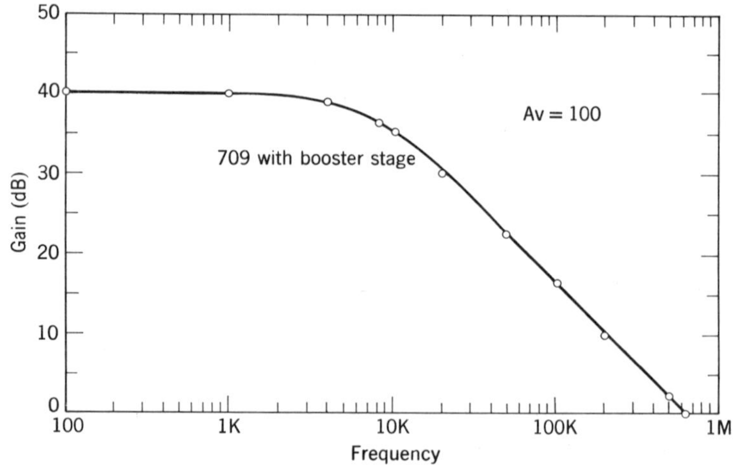

Figure 3.4. Amplifier response.

at the extender input, has a continuous high output. This occurs because the capacitor is discharged each time the first gate output is low and there is not sufficient time between pulses for the capacitor to charge up to the point at which the second gate turns on. The output is sufficient to drive 12 TTL gates or 25 DTL gates in parallel.

The response of the circuit to a short input pulse is shown in Figure 3.7. There is about 0.4 μs delay from input to output. The leading edge delay is due to the starting time of the multivibrator. The trailing edge delay is caused by the time interval required to charge the capacitor in the detector circuit sufficiently for the output gate to conduct. The multivibrator frequency increases as its supply voltage builds up. The circuit will respond to input frequencies from direct current up to 1 MHz.

Figures 3.8 through 3.14 show some of the uses for such a device. Figure 3.8 depicts a conventional hookup for an integrated-circuit differential comparator as a dc voltage monitor. The line to be monitored

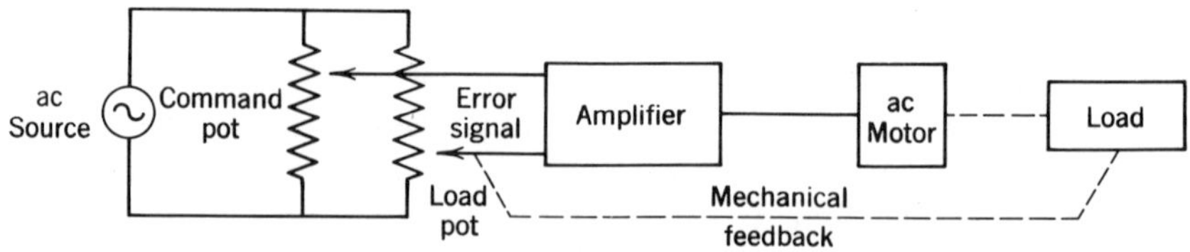

Figure 3.5. A motor control application.

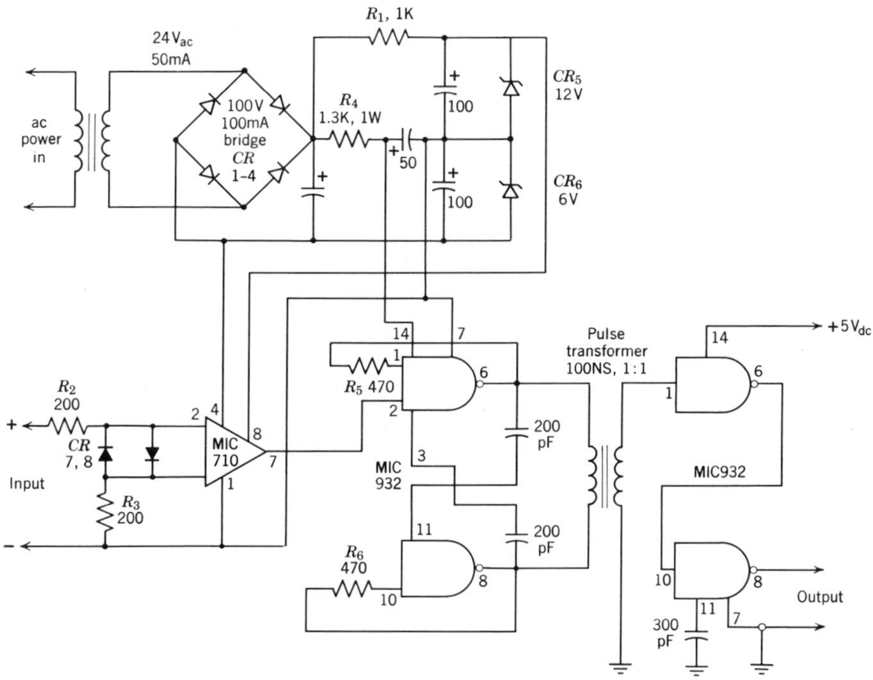

Figure 3.6. Circuit diagram of isolated differential comparator.

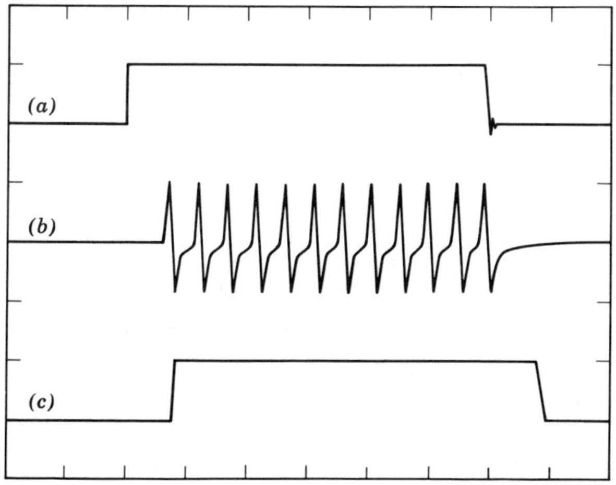

Figure 3.7. Typical waveforms of isolated differential comparator: (*a*) input pulse 0.1 V/div.; (*b*) pulse transformer secondary 2 V/div.; (*c*) detector output 5V/div.; each horizontal box is 0.5 μs/div.

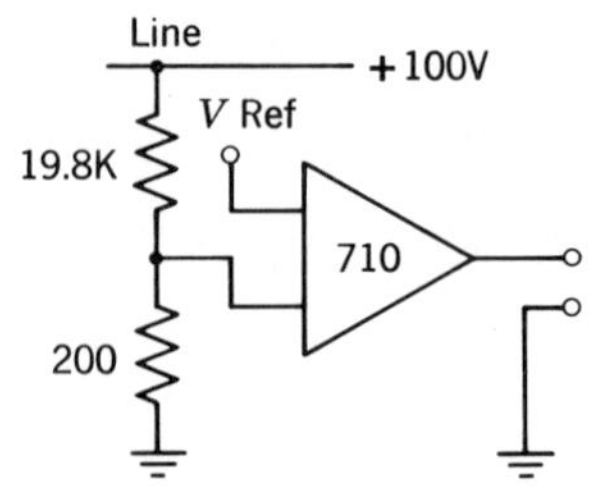

Figure 3.8. Conventional voltage monitor.

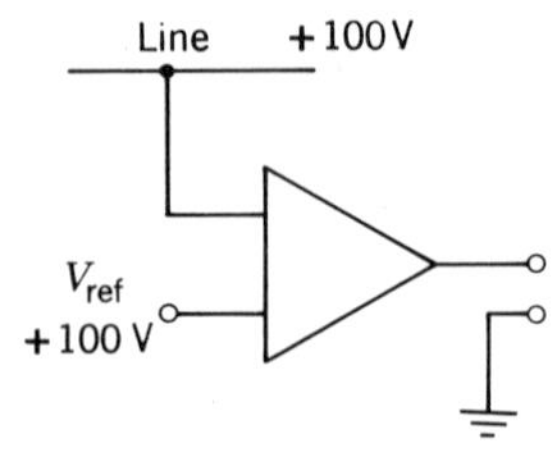

Figure 3.9. Voltage monitor using isolated comparator.

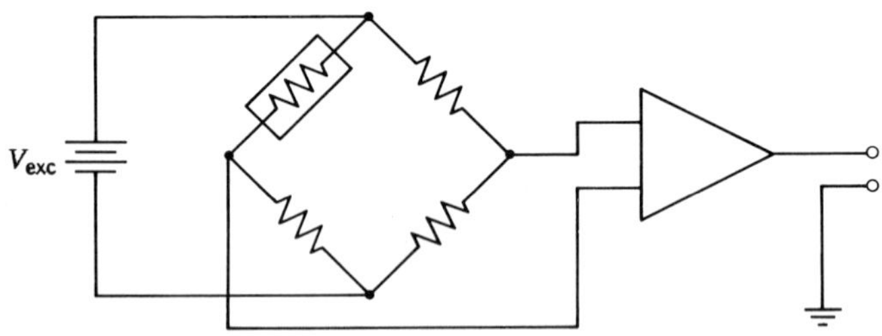

Figure 3.10. Bridge monitor.

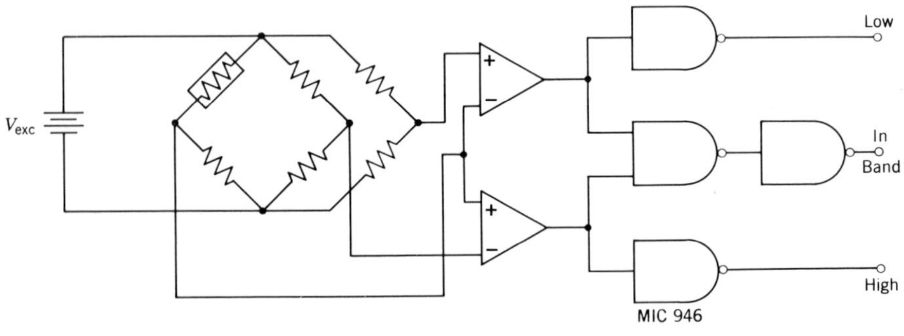

Figure 3.11. Double-limit bridge monitor.

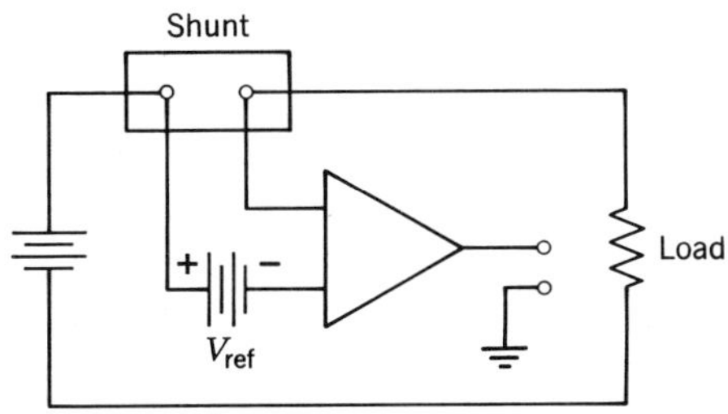

Figure 3.12. Current monitor.

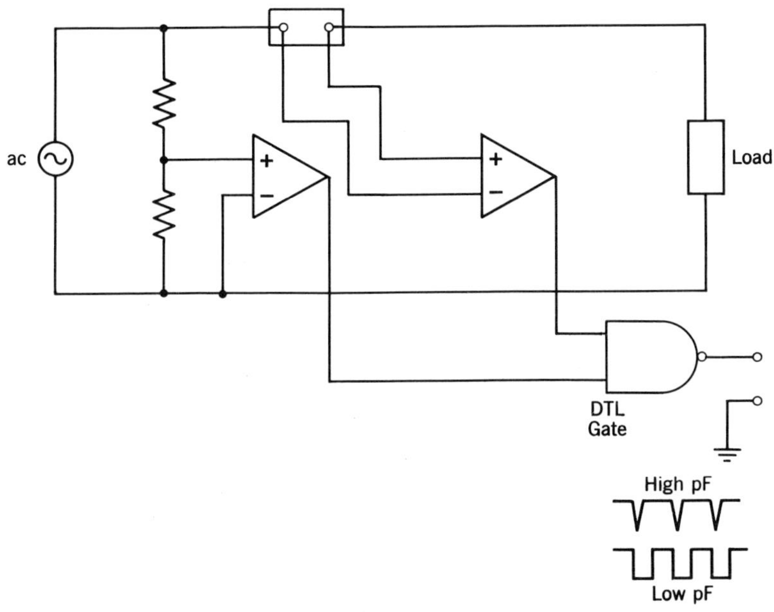

Figure 3.13. Power factor monitor.

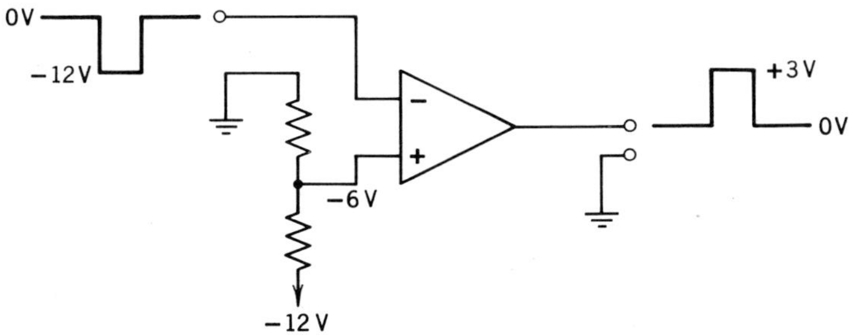

Figure 3.14. Digital system translator.

must have a common ground or a fixed dc relationship with the digital output ground. In the example shown offset and drift at the differential comparator input of 5 mV would produce an uncertainty at the 100-V line of 0.5 V, or 0.5%, but, if the same differential comparator could be made to float off ground, as in Figure 3.9, the uncertainty would be reduced to 0.005%. Two separate monitor circuits may be used to determine if the line voltage is within a specified band.

Figure 3.10 shows an isolated comparator circuit used to monitor the balance of a resistive bridge, independent of the bridge grounding scheme. In Figure 3.11 the addition of another comparator circuit with another pair of bridge legs has been made so that the circuit will indicate if the transducer resistance is within a specified band.

Figure 3.12 illustrates the use of a floating comparator as a current monitor in which the output trips when a predetermined current is reached. A similar hookup with the comparator across a current shunt in an ac line could be used to measure power factor, as shown in Figure 3.13.

In Figure 3.14 the comparator is used as a universal translator for coupling two different digital systems together when different signal levels, polarities, or ground references are encountered; DTL or TTL systems can be driven from virtually any other type of digital system including discrete component, RTL, ECL, MOS, vacuum tube or relay logic with logic level differences ranging from a few millivolts to hundreds of volts.

4

Field-Effect-Transistor Input Operational Amplifiers

by Allan C. Bahr
Amelco Semiconductor, Division of Teledyne

4.1 INTRODUCTION

When operational-amplifier applications require high input impedance or very low bias and offset currents, conventional bipolar monolithic circuits may not be able to supply the desired performance. Instead, field-effect-transistor circuitry must be used. There are many different types of amplifiers employing FET inputs; however, this chapter deals primarily with the application of a hybrid configuration having a FET differential source follower and a monolithic amplifier fabricated in a single package.

4.2 INPUT OFFSET VOLTAGE DRIFT

An ideal amplifier will have exactly zero output for zero input, infinite input impedance, zero output impedance, infinite open-loop gain, and infinite bandwidth. Since such an ideal amplifier cannot be built using today's technology; compromises have to be made in the parameters.

A practical amplifier will exhibit some output for zero input. This output may include random ac signals (noise) and dc signals; the dc signals are called offset. It is customary to refer both signals to the input so that they are independent of the amplifier's gain. Input offset can then be defined as the input signal required to zero the dc component of the output with zero input signal applied.

A fixed input offset is usually not a problem, since biasing circuits can be devised to cancel this signal, but changes in input offset due to variations in temperature, supply voltage, or aging of component values with time introduce additional errors. These errors cannot readily be distinguished from input signals, and therefore compensating for them is no easy task.*

The voltage offset e_{os} is defined as the voltage required at the input to zero the amplifier's output. The current offset i_{os} is defined as the difference between the input currents required to zero the amplifier's output. Using a simple inverting amplifier as shown in Figure 4.1, we find that the total error due to voltage and current offset when referred to the source voltage is

$$V_s' = i_{os} R_1 + \frac{e_{os}(R_1 + R_2)}{R_2}, \tag{4.1}$$

where R_1 is equal to the series value of R_3 and R_s, R_s is the resistance of the source generator V_s, and R_4 is the parallel combination of R_1 and R_2.

For low values of circuit impedance in the external feedback network (R_1 and R_2) the magnitude of the input offset is nearly independent of the amount of current offset. For large values of impedance, however, the input offset increases almost proportionally to the magnitude of the impedance used, and the magnitude of input offset is nearly independent of the amount of voltage offset.

The offset voltage of (4.1) can be referred to the output by multiplying by the closed-loop gain (taking into consideration that the amplifier is not ideal):

$$V_o = \left[\frac{(-R_2}{R_1)} V_s - (i_{os} R_2) - \frac{e_{os}(R_1 + R_2)}{R_1} \right] \left[\frac{1}{1 + 1/A \left[\dfrac{1 + R_2 R_1 R_{in}}{(R_1 + R_{in})} \right]} \right], \tag{4.2}$$

where A is the open loop voltage gain, $i_{os}R_2$ is the error due to current offset, e_{os} [$(R_1 + R_2)/R_1$] is the error due to voltage offset, $R_2R_1R_{in}/(R_1 + R_{in})$ is the error due to finite input impedance, and $1/[1 + 1/A\ (1 + R_2R_1R_{in}/(R_1 + R_{in})]$ is the error due to finite open-loop gain.

As can be seen, finite open-loop input impedance R_{in} has no effect on voltage and current offset when referred to the source voltage regardless of how large R_1 and R_2 may be compared to R_{in}, and when referred to the output R_{in} has only a second-order effect inasmuch as it contributes to the gain error.

The effect of current offset referred to the source voltage can be determined by assuming that all of the offset current is supplied by the source voltage; the voltage drop generated is $i_{os}R_1$. In practice the offset current

*Temperature compensation for an individual unit can be performed but this is expensive and time consuming.

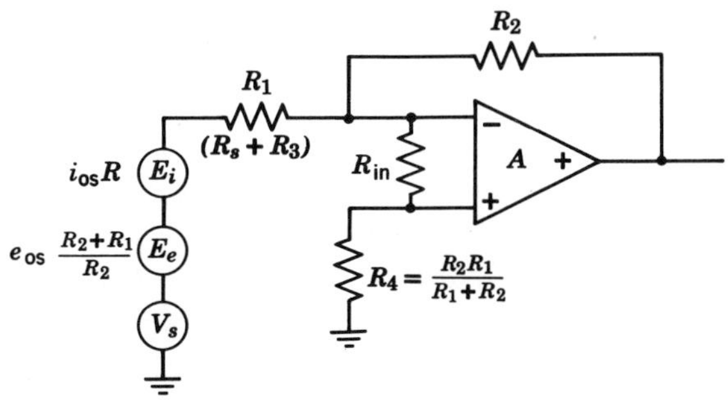

Figure 4.1. Simple inverting amplifier.

is a function of both the output and the source voltage, in ratios depending on R_1 and R_2. The portion supplied by the output must be divided by the closed-loop gain to be referred to the input; this yields the same results as if all the offset current were supplied by the input.

The input voltage is

$$V_i = \underbrace{\left[\frac{i_{os}R_2}{R_1+R_2}\right]}_{\uparrow \text{ supplied by the input}} R_1 + \underbrace{\left[\frac{i_{os}R_1}{R_1+R_2}\right]}_{\uparrow \text{ supplied by the output}} R_2 \frac{R_1}{R_2} = i_{os}R_1. \tag{4.3}$$

As mentioned before, the relative importance of voltage and current offset depend upon the magnitude of R_1. In Table 4.1 offset drift due to temperature changes is shown for a typical monolithic amplifier and a typical FET input amplifier.

TABLE 4.1

	Monolithic IC (809)	IC with FET input (2809)
Offset voltage drift	10 μV/°C	20 μV/°C
Offset current drift	3 nA/°C	0.1 nA/°C*
Input resistance	400 kΩ	10^{10} Ω
Input voltage drift		
$R_1 = 1$ kΩ	13 μV/°C	20 μV/°C
$R_1 = 10$ kΩ	40 μV/°C	21 μV/°C
$R_1 = 100$ kΩ	310 μV/°C	30 μV/°C
$R_1 = 1$ MΩ	3000 μV/°C	120 μV/°C

* 10 pA at 25°C. Offset current doubles approximately every 10°C. For the calculation of input voltage drift, a temperature delta of +100°C was used, or an average temperature coefficient of 0.1 nA/°C.

Curves of offset drift vs input impedance are plotted for both the monolithic and the FET input devices in Figure 4.2 (the input impedance shown in Fig. 4.2 is the combined source resistance and resistance of the inverting input resistor). At an input impedance of 3 kΩ, the contributions due to voltage drift and current drift of the monolithic amplifier and the FET amplifier are equal, and for input impedance greater than the 3 kΩ, the FET device has superior thermal drift characteristics.

When very large input impedances are required for an inverting amplifier (where the closed-loop input impedance is equal to R_1), they can only be obtained at the expense of offset errors. The offset errors of a non-inverting amplifier (where the input impedance is the product of the amplifier's open-loop input impedance and the amount of A − Beta, negative feedback) are shown in Figure 4.3. Again as in Figure 4.1, the offsets are referred to the source. R_s is the source impedance.

Unlike the inverting amplifier, the voltage offset referred to the source is independent of the closed-loop gain. Also, as in the case of the inverting amplifier, the voltage and current offsets referred to the source voltage are unaffected by the value of R_{in}.

The effect of current offset referred to the source voltage is directly proportional to the source impedance R_s. Many devices, such as photocells and photomultipliers, produce an output current from a relatively high source impedance. These devices can be characterized by a current source as shown in Figure 4.4.

In this circuit configuration the operational amplifier converts the input current to an output voltage with a low output impedance. The offset errors referred to the signal source are

$$I_s = e_{\text{os}} \frac{R_s + R_2}{R_s R_2} + i_{\text{os}}. \tag{4.4}$$

To refer the signal and offset errors to the output, we multiply by the closed-loop gain:

$$V_o = \left[-I_s R_2 - e_{\text{os}} \frac{R_s + R_2}{R_s} - i_{\text{os}} R_2 \right] \left[\frac{1}{1 + 1/A\,[\,1 + R_2 R_{\text{in}} R_s/(R_{\text{in}} + R_s)\,]} \right]. \tag{4.5}$$

For a value of R_s much greater than R_2, which is usually the case for a high-impedance current source, the voltage offset is not amplified when referred to the output and therefore can often be ignored.

So far we have been discussing the offset of differential input amplifiers. The differential amplifier has individual current bias sources and depends upon the symmetry of the input circuitry for its proper operation. Therefore it is reasonable to expect that the bias currents at each input would

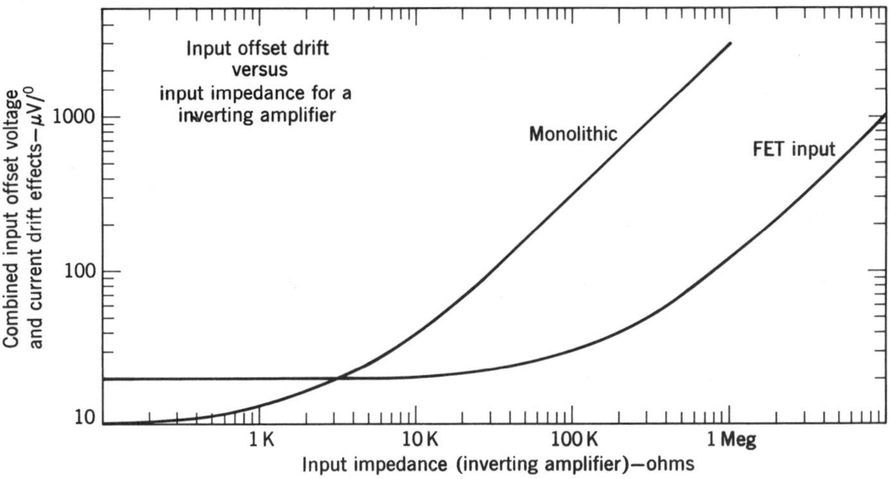

Figure 4.2. Offset drift vs input impedance for a conventional monolithic IC and a hybrid IC with an FET input.

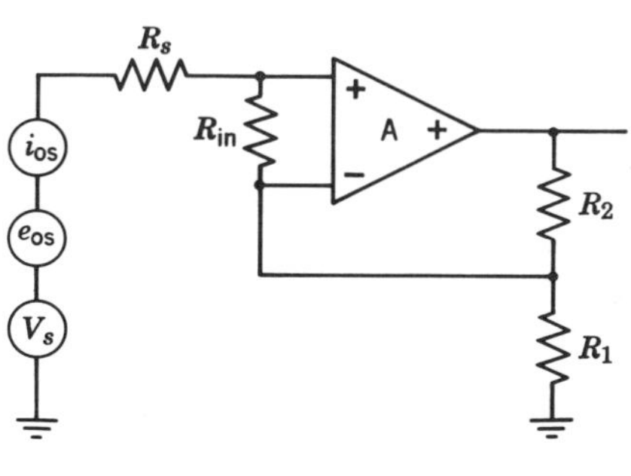

Figure 4.3. Noninverting amplifier.

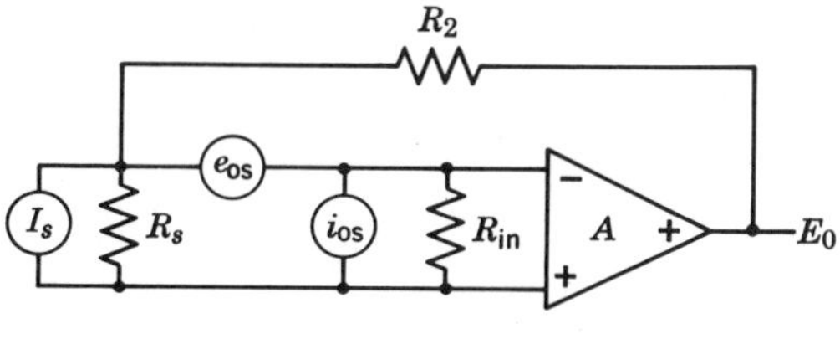

Figure 4.4. Operational amplifier circuit used to convert input current to output voltage with low output impedance.

be about equal and that changes in bias current would tend to track with changes in ambient temperature and supply voltages. For this reason errors due to offset current drift are minimized with a differential pair as compared to a single-ended amplifier.

Two differential amplifiers that use FETs appear in Figure 4.5. The circuit in Figure 4.5*a* is a pair of matched source followers intended for impedance matching. In Figure 4.5*b* a true source-coupled differential amplifier that provides voltage gain is shown. In either circuit one gate may be grounded for single-ended input operation. In Figure 4.5*a* the FET with the grounded input is required for temperature drift compensation. In Figure 4.5*b* the FET with the grounded input continues to act as an amplifier, but it must also exhibit equal drift with temperature changes.

For good linearity the transconductance of the two FETs should be nearly equal at the operating point, and if the gate-to-source voltages of the two FETs (offset voltage) are identical at equal drain currents, no offset adjustment is needed in either case. Since this condition is rarely met, offset adjustment is required to equalize the V_{GS} of the FETs for zero output with zero input.

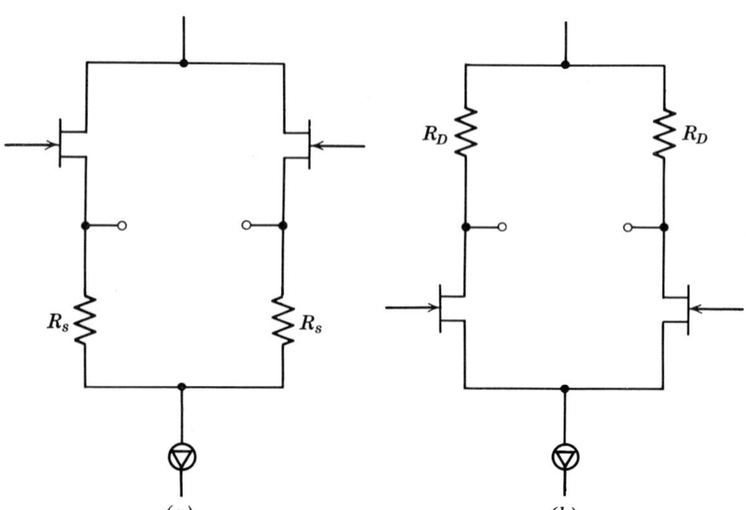

Figure 4.5. Two differential amplifier circuits using FETs. In (*a*) a pair of matched source followers are used for impedance matching; the circuit in (*b*) is a true source-coupled differential amplifier.

4.3 INPUT IMPEDANCE

An amplifier having a high input impedance is by definition one that requires very little input current. Conventional transistors are current operated so that their use as input devices for very-high-input-impedance circuits is limited. On the other hand FETs are voltage operated like vacuum tubes and have been known principally for their high-input-impedance characteristics. In addition the FET is the lowest-noise device available for use in amplifiers driven from high source impedances, such as a vidicon or photomultiplier.

The gate input characteristic is that of a reverse biased diode, and therefore the device exhibits an input impedance which is a function of the physical size and quality of the gate junction. The dc leakage current I_{GSS} of the junction FET may be in the one-to-five pA range at 10 V; this is equivalent to $10^{13}\ \Omega$ input impedance. The input impedance is given by

$$Z_{in} = \frac{V_{GS}}{I_{GSS}}. \tag{4.6}$$

To obtain very low leakage, gain must be sacrificed. The device mentioned (Amelco's U1714) still exhibits a transconductance of 400 μmho. It is important to note that the gate current is a nonlinear function of both gate-to-source voltage and temperature.

4.4 TYPICAL APPLICATIONS FOR FET-INPUT AMPLIFIERS

The following are a few of the applications for which a FET-input amplifier is ideally suited: potentiometer unloading, sample-and-hold circuits, A-to-D and D-to-A converters, PH meters, long-term integrators, picoampere-current measurements, high Z_{in} adders-subtractors, electrometers, high-Z_{in} comparators, demodulator-amplifier-filter networks, voltage-controlled oscillators, low-noise amplifiers, low-drift amplifiers, and charge-sensitive amplifiers.

Figure 4.6 shows the circuit diagram for the Amelco 2809 hybrid operational amplifier. This amplifier combines the advantages of a FET differential source-follower input stage with those of the monolithic 809 operational amplifier. The 2709 hybrid operational amplifier uses the same FET differential source-follower input stage, but uses a 709 monolithic operational amplifier in place of the 809 (this device is not shown). The combination of the FET input stage and the monolithic operational amplifier results in a low-cost high-performance amplifier. The typical electrical characteristics of both amplifiers (the 2709 and the 2809) are shown in Table 4.2.

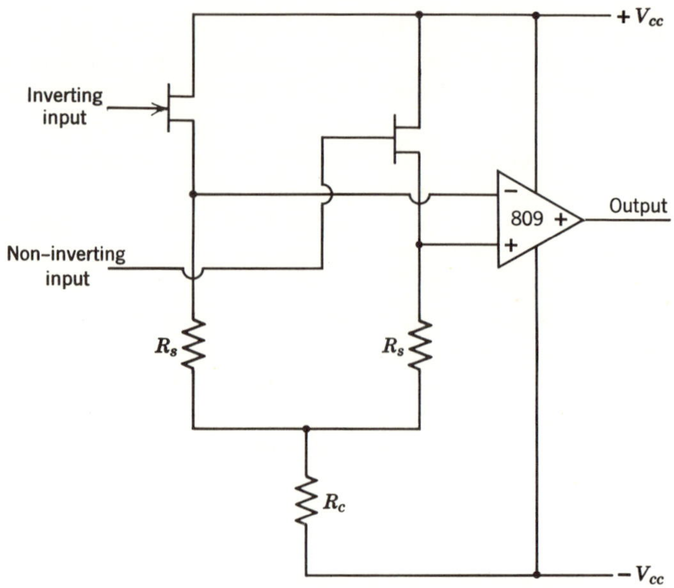

Figure 4.6. Circuit diagram for a commercially available hybrid IC with an FET input (Amelco's 2809).

Figure 4.7. Photo of the 2709 hybrid IC amplifier.

TABLE 4.2 Hybrid IC Characteristics

	2709	2809	Units
Open-loop voltage gain	88	90	dB
Input offset voltage	3.0	5.0	mV
Input offset voltage drift	25	35	μV/°C
Input bias current	40	40	pA
Input offset current	15	15	pA
Input resistance	100	100	GΩ
Common-mode voltage range	±10	±12	V pk
Common-mode rejection ratio	−70	−70	dB
Power-supply rejection ratio	−80	−74	dB
Output voltage swing	±12	±11	V pk
Power dissipation (no load)	110	110	mW

The high input impedance and low offset currents of both hybrid operational amplifiers make them ideal basic building blocks for many of the FET-input applications mentioned. A photo of the 2709 is shown in Figure 4.7.

Figures 4.8 and 4.9 illustrate a typical commutator circuit. In this application the input impedance is important, for it affects the transmission ratio of the conducting channel. The transmission ratio is

$$T_r = \frac{R_{in}}{R_{in} + R_s + r_{ds}}. \tag{4.7}$$

Thus, in order to ensure accuracy, R_{in} must be large compared with $R_s + r_{ds}$.

When $E_{in} = 0$, an output offset voltage is generated by the sum of the reverse currents from all of the off switches and the offset current of the amplifier. This voltage is given by

$$E_o = [(N-1)(i_{d(\text{off})}) + i_{os}](R_s + r_{ds}) \tag{4.8}$$

Commutating speed is determined by the "on" resistance of one channel and the total capacitance C_t. Again, a FET input provides a lower capacitance input than a monolithic amplifier, typically 4 pF as compared with 10 pF:

$$T = (R_s + r_{ds})\ (NC_{dg} + C_{in}). \tag{4.9}$$

In Figures 4.10 and 4.11 two conventional digital-to-analog converters are shown. In Figure 4.10 a weighted resistor network is used, whereas in Figure 4.11 a precision ladder network is used.

For biasing an FET in an electrometer application (see Figure 4.12), the source-follower configuration is the best. In this version of the follower-with-gain circuit a ±10-V output with a 2-mA current capability

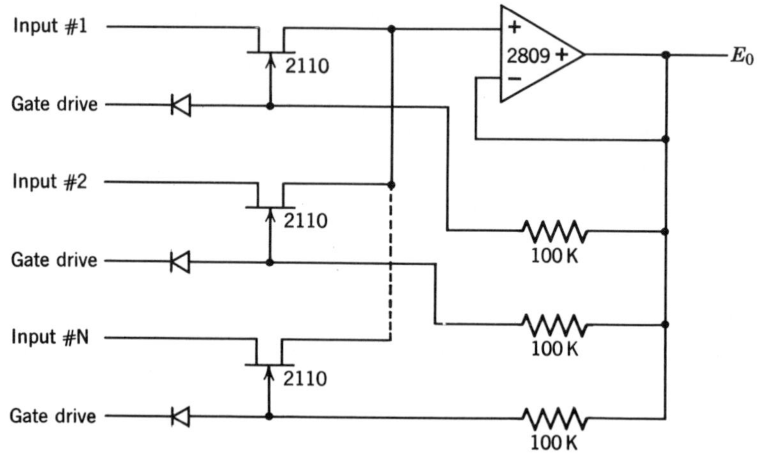

Figure 4.8. Typical commutator circuit.

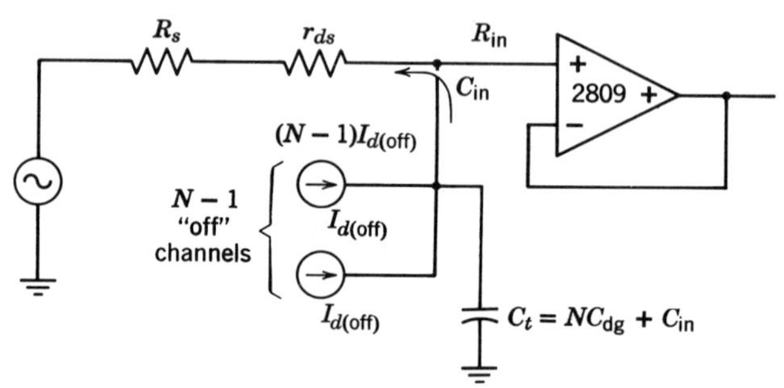

Figure 4.9. Commutator circuit diagram.

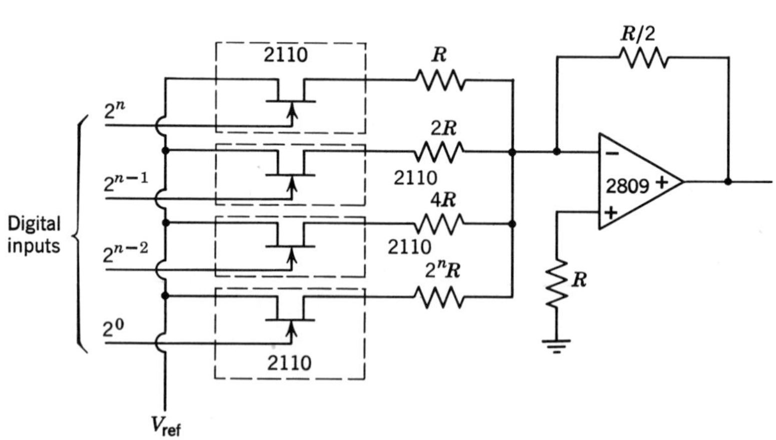

Figure 4.10. Digital-to-analog converter with weighted resistor network.

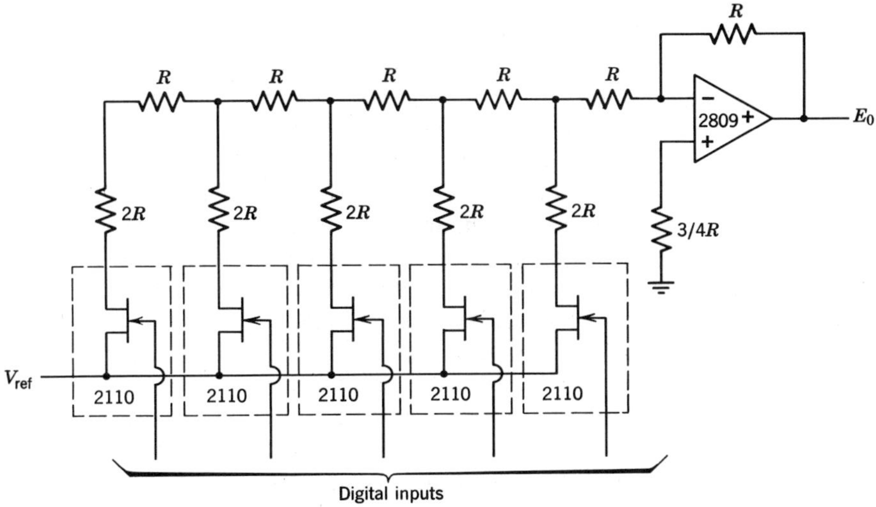

Figure 4.11. Digital-to-analog converter with precision ladder network.

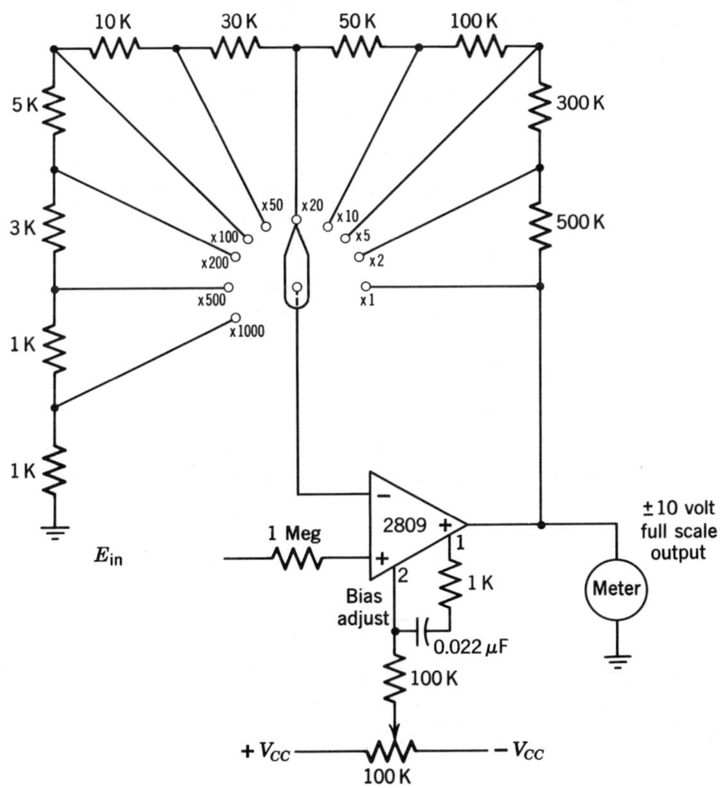

Figure 4.12. Electrometer circuit built with hybrid IC.

is developed from a full-scale input voltage as small as ±10 mV while drawing less than 1 pA from the signal source. This current sensitivity results because of the extremely high input impedance of this circuit.

A high-input-impedance comparator is shown in Figure 4.13. An ac sawtooth reference (can also be dc if the error signal being sampled is ac) is compared with a slowly changing dc error signal. The output is not symmetrical and is proportional to the dc input error signal; that is, if, as in Figure 4.13, the ac reference is ±1 V, then, when the dc signal is equal to zero volts, the output will be a symmetrical square wave. If the dc signal were −0.5 V, the output would be 25% on and 75% off. The output slew rate with a ±1 V, 1000-Hz sawtooth reference in this application is 5 V/μs in both the positive and negative directions. The high input impedance in this application does not load the sawtooth generator; thus the overall system accuracy is increased by having a more linear ramp.

In Figure 4.14, if we add a high-noise-immunity gate to the output of the comparator, we have a simple analog-to-digital converter; for example, with the ±1 V 1000-Hz sawtooth reference previously mentioned and with a 2-MHz clock to the gate (phase-locked to the sawtooth), each pulse on the output of the gate will have a granularity of 1 mV. It is obvious that by using this method any granularity can be obtained.

The output of the comparator can also be used to gate an analog-gate circuit in which the sawtooth function is generated. Thus a self-contained system can be obtained and only the dc error signal need by supplied. Again using the comparator shown in Figure 4.13 and adding a 2114 hybrid single-pole double-throw analog switch (see Figure 4.15), a voltage-controlled oscillator can be made. The 2114 is used to generate the sawtooth reference.

The sawtooth starts at a negative voltage level as determined by the minus voltage reference on the 2114, and a slope determined by the *RC* time constant. The current then charges the capacitor to the level of the dc signal. When the comparison has been accomplished, the switch is

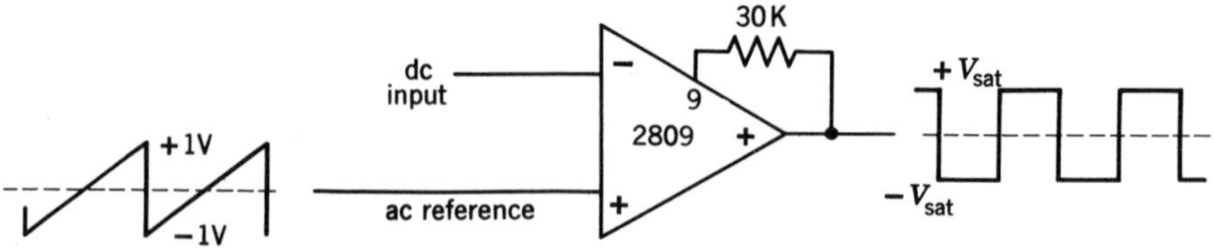

Figure 4.13. High-input impedance comparator includes hybrid IC.

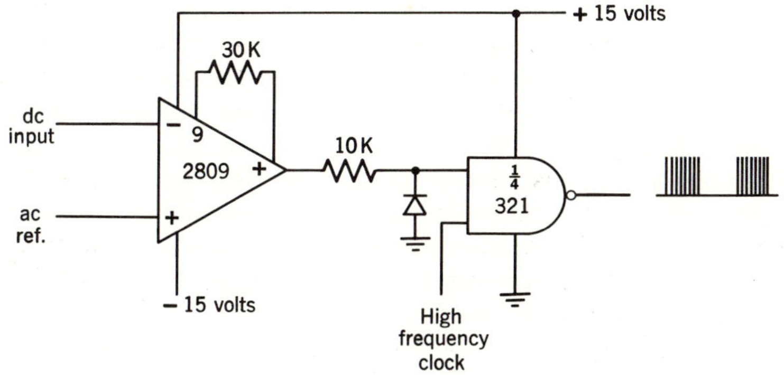

Figure 4.14. Adding gate to comparator converts the circuit into an analog-to-digital converter.

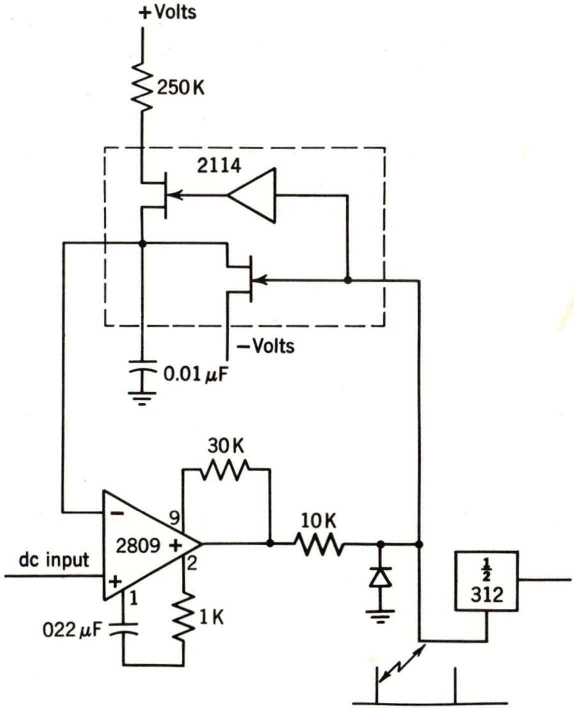

Figure 4.15. Adding switch to comparator converts it to voltage controlled oscillator.

released and the capacitor is discharged to its negative reference, where the process starts over again. The output of the comparator is only a pulse (width dependent upon the propagation delay time). If pulses were used as clock inputs to a J-K flip-flop like the 312, a square wave output would be produced.

The frequency of the oscillation is determined by the RC ($R = 200$ kΩ, $C = 0.1$ μF) time constant (reference frequency should be determined with zero volts applied to the input of the comparator) and the magnitude of the dc signal. The more negative the dc signal, the faster the rate; the more positive the dc signal, the slower the rate.

Figure 4.16 shows another approach to designing a voltage controlled oscillator. Here the dc signal provides a current to a long-term integrator. The integrator's output then goes to a plus-and-minus voltage comparator, the two 805 operational amplifiers. The output of each comparator is gated with the RS flip-flop (½ of a quad 321 2-input gate). The output of the flip-flop is unsymmetrical. The output is used to gate the 2114 switch, which generates a positive and negative reference current for the integrator. As the input current approaches the magnitude of the reference current, either plus or minus, the time between pulses becomes larger, as shown in Figure 4.17.

A long-term integrator circuit with the 2809 is shown in Figure 4.18. By using a high-input impedance amplifier, R can be increased and C can be reduced so that the leakage current of the capacitor can be minimized.

Figures 4.19 and 4.20 show a basic sample-and-hold circuit and the effect of capacitance on slew rate and decay time.

An adder-subtractor application (see Figure 4.21) requires a high input impedance to prevent any loading effects that would decrease the accuracy of the operation. The output voltage for this example is

$$E_o = \frac{R_2}{R_1}(e_4 + e_3 - e_2 - e_1). \tag{4.10}$$

A demodulator-amplifier circuit is shown in Figure 4.22. This circuit provides gain, impedance buffering and effective carrier filtering. The alpha and beta networks are not shown since they depend on specific application requirements.

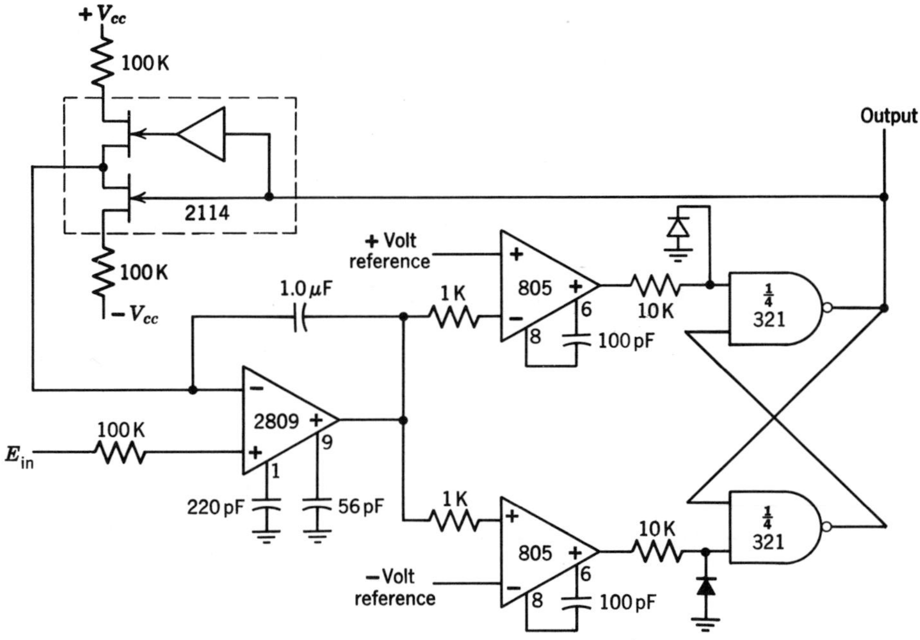

Figure 4.16. Voltage-controlled oscillator built with two conventional operational-amplifier ICs and a hybrid IC.

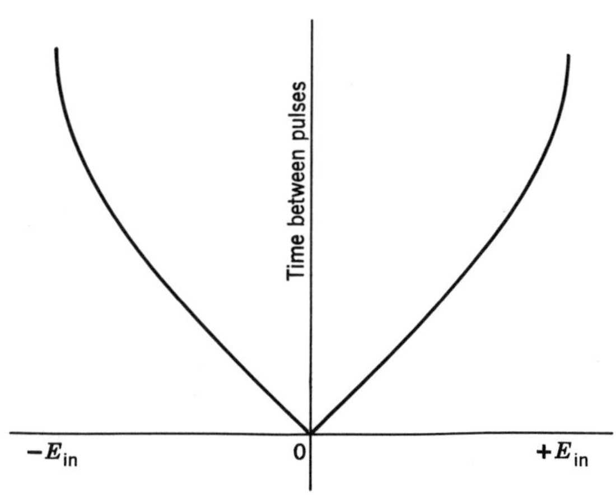

Figure 4.17. Relationship between input signal and time between pulses for oscillator circuit in Figure 4.16.

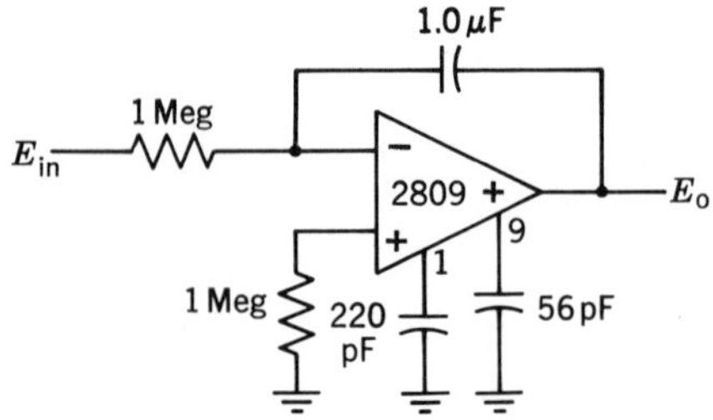

Figure 4.18. Long-term integrator circuit.

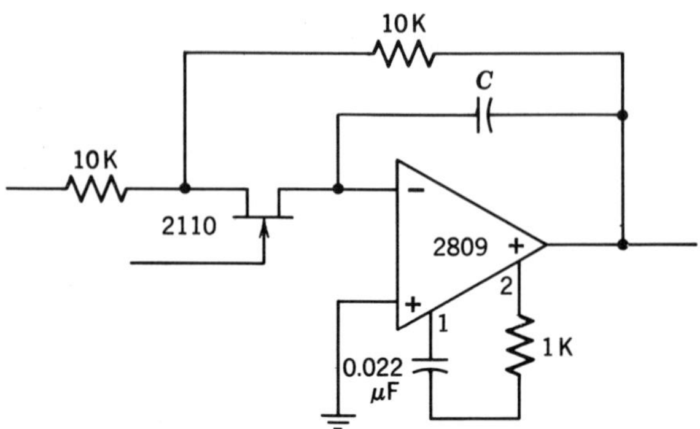

Figure 4.19. Basic sample-and-hold circuit.

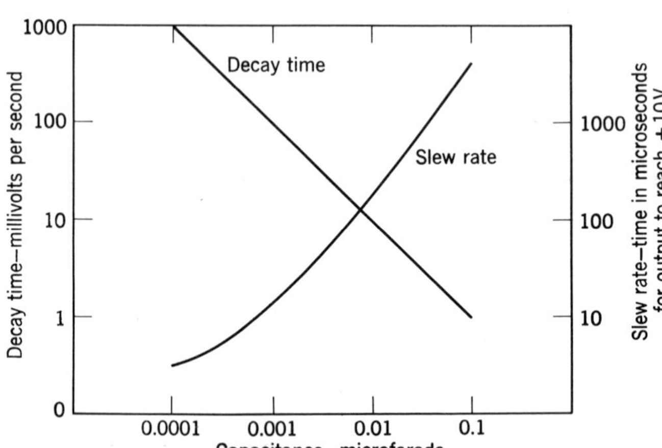

Figure 4.20. Slew rate and decay time relationships with capacitance.

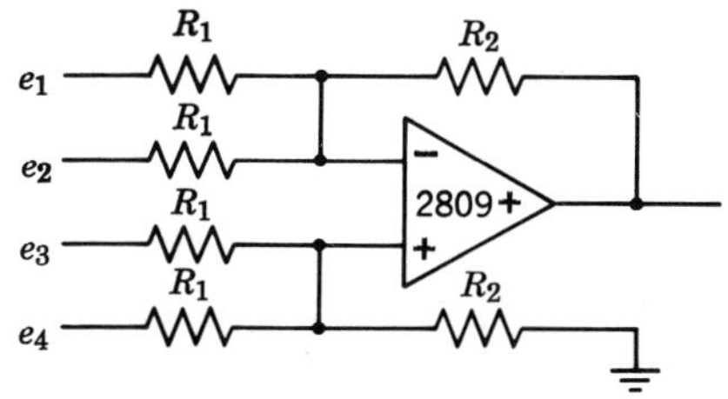

Figure 4.21. Adder-subtractor circuit.

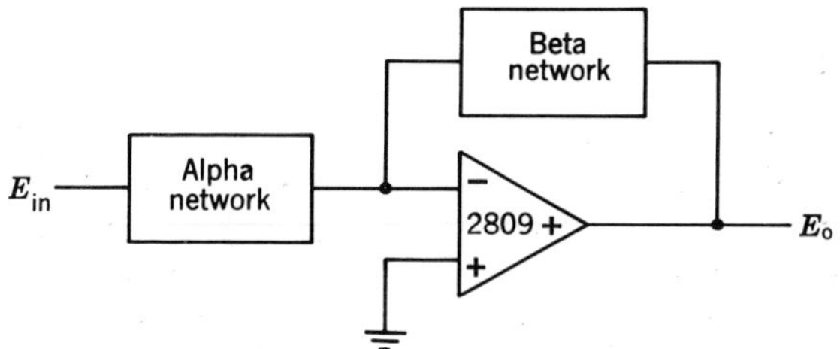

Figure 4.22. Demoldulator-amplifier circuit.

5

Dual High-Gain Differential Amplifiers

by Charles C. Wu and Randy Brandt
Raytheon Company, Semiconductor Operation

5.1 INTRODUCTION

Dual differential-amplifier ICs first made their appearance commercially in late 1967. Typical of this type of monolithic IC is the RM4709 which consists of two separate 709 amplifiers fabricated on the same die. The placing of two amplifiers on the same chip results in a close match of characteristics and has certain advantages over the use of two separate 709 circuits; for example, with the RM4709 there is gain matching within ± 1 dB, phase matching within $\pm 10°$, and excellent dc parameter matching. These characteristics are, of course, extremely difficult to achieve with two discrete 709 amplifiers. Moreover, one RM4709 can perform the functions of two 709s, yet occupies only half the space.

In the first part of this chapter measured characteristic values for the RM4709 are given; emphasis is on channel separation and frequency-response matching. The second part deals with applications of the circuit as a differential-in differential-out (DIDO) amplifier and a noninverting (or potentiometric) amplifier. The input impedance, differential gain, common-mode gain, common-mode rejection ratio, and frequency-stability characteristics are analyzed.

5.2 CHARACTERISTICS OF THE DUAL DIFFERENTIAL AMPLIFIER

A graph of the open-loop voltage gain as a function of frequency for various values of frequency compensation is shown in Figure 5.1 for the RM4709. The typical phase response of the IC vs frequency is given in Figure 5.2*a* and 5.2*b*. As these Figures show, the open-loop voltage gain and phase responses of the dual version are very similar to those of the 709. The gains of the dual 709 are typically matched within ± 1 dB from direct current to 10 MHz. The phase responses of the two amplifiers also track each other closely and are typically within $\pm 10°$ from 1 kHz

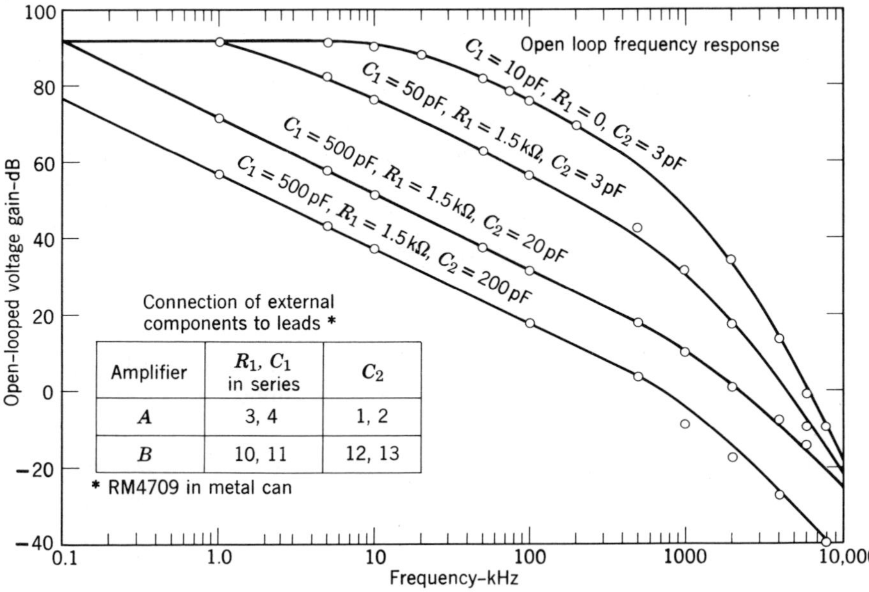

Amplifier	R_1, C_1 in series	C_2
A	3, 4	1, 2
B	10, 11	12, 13

Figure 5.1. Open-loop voltage gain as a function of frequency.

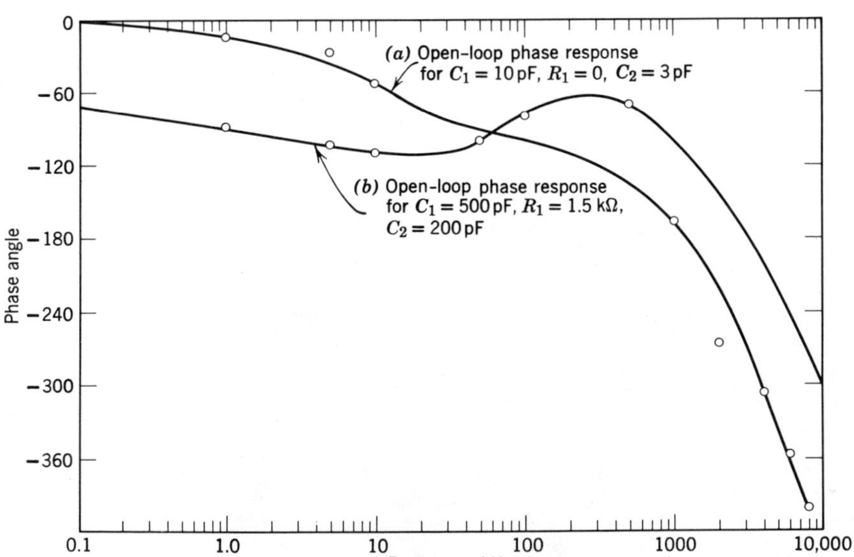

Figure 5.2. Typical phase response.

to 10 MHz. These two features make the dual 709 extremely valuable in applications in which gain and phase tracking are critical.

The offset voltages V_{os} are better matched than randomly selected amplifiers due to the fact that offset contributions of the later stages are matched, even though local V_{BE} mismatch still exists. The offset currents I_{os} are better matched because of the matched β's. These improvements are reflected in the specifications for the RM4709.

5.3 TRANSIENT RESPONSE

The unity-gain transient response of the RM4709 is given in Figure 5.3 for a load consisting of a 100-pF capacitor in parallel with a 2 kΩ resistor. The rise time for a 20-mV step input is found to be about 700 ns. Given the overshoot and rise time for a second-order system, the phase margin, hence stability, and bandwidth are uniquely determined for that system; however, for a higher order system, the relationships between the quantities mentioned above are much more complex and usually can be obtained only empirically. Measurements have shown that an overshoot of less than 30% will ensure the phase margin necessary for stability with the RM4709.

5.4 CROSSTALK

Since the RM4709 is built on a single chip, interaction between amplifiers A and B will occur because of their close proximity and common substrate. This interaction may be defined as crosstalk or separation.

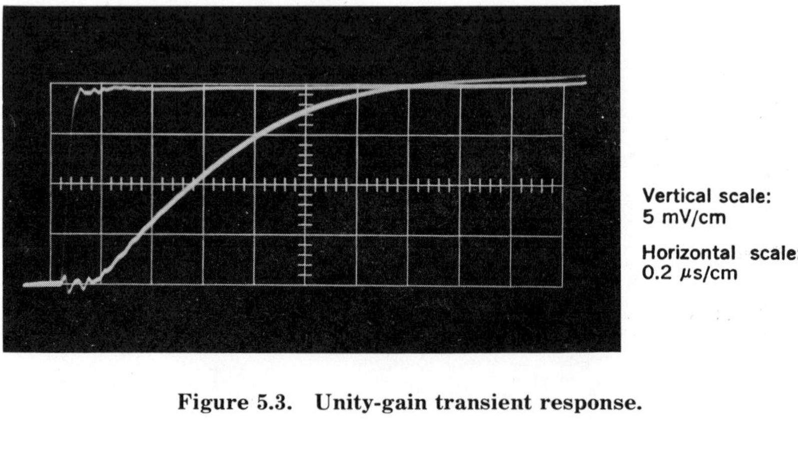

Figure 5.3. Unity-gain transient response.

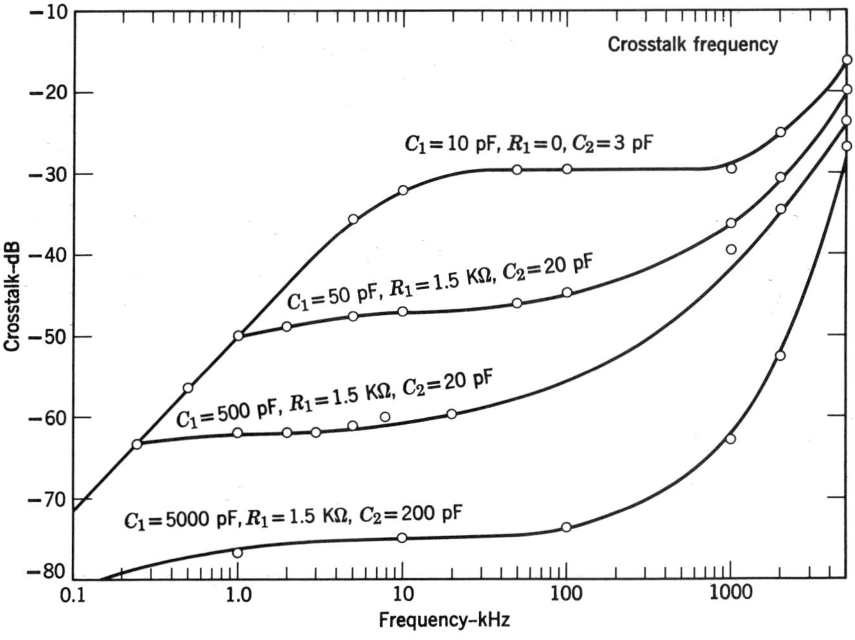

Figure 5.4. Crosstalk versus frequency.

Crosstalk is defined as the ratio of the output voltages of the two amplifiers when a signal is applied to the input of one with the input of the other shorted. Analytically, crosstalk can be written as

$$T_C = 20\ \mathrm{Log}_{10} \left| \frac{V_{02}}{V_{01}} \right| \qquad (5.1)$$

where T_C = crosstalk in dB, V_{02} = output voltage of the amplifier with shorted input, V_{01} = output voltage of the amplifier with signal applied at its input.

In Figure 5.4 an experimentally determined graph of crosstalk vs frequency for four values of frequency compensation is shown. A graph of the ideal asymptote for open-loop voltage gain as a function of frequency is shown in Figure 5.5 along with a probable capacitive-coupling curve that assumes the primary pole to be at the input of the amplifier with the "shorted input." Due to differences in magnitude the scales for gain and crosstalk are located on the left side of the graph. The scale for coupling is shown on the right.

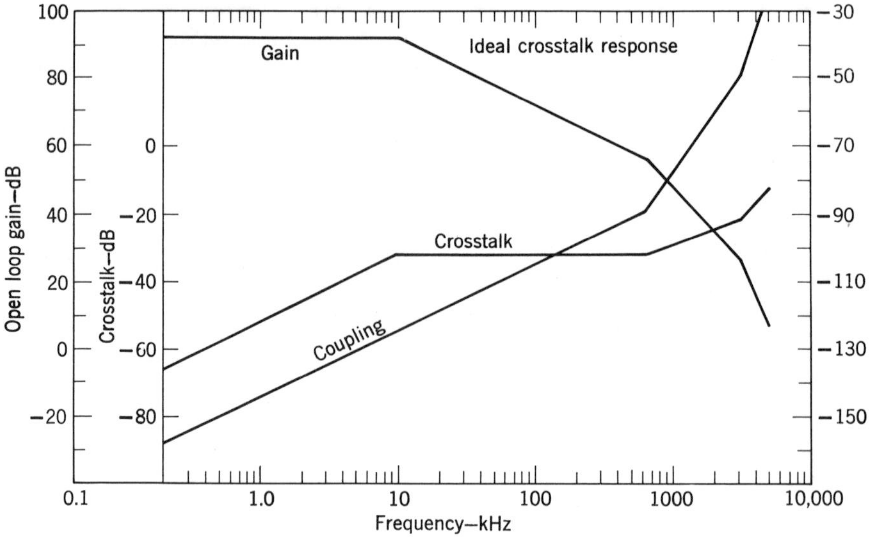

Figure 5.5. Ideal open-loop voltage gain as a function of frequency.

The coupling curve has a slope of +6 dB/octave from 100 Hz to 1 MHz. Above 1 MHz the slope increases to about +18 dB/octave. At 100 Hz the coupling is extremely low and is approximately −164 dB. The gain curve, on the other hand, is +92 dB from 100 Hz to 10 kHz and then starts to roll off at the rate of −6 dB/octave. If the coupling and the gain are added algebraically, the resultant curve approximates the experimentally measured crosstalk characteristic in Figure 5.4. At very low frequencies, coupling between the amplifiers is extremely small but the gain of the amplifier is large. Therefore any signal coupled into the "shorted" input is amplified by the forward gain. Specifically, at 100 Hz, the coupled signal is −164 dB. This small signal is amplified by +92 dB so that the crosstalk is equal to −72 dB at this frequency. As the frequency is increased, the coupling also increases.

Since the gain of the amplifier remains constant up to 10 kHz, the crosstalk curve will increase at the same rate as the coupling curve. From 10 kHz to about 600 kHz the curve flattens in the horizontal direction. This leveling is due to the gain rolling off at −6 dB/octave and the coupling increasing at +6 dB/octave. Hence the crosstalk remains flat over this range of frequencies. Beyond 600 kHz the gain roll off increases to −12 dB/octave, whereas the slope of the coupling curve climbs at a much steeper rate, because additional coupling paths are becoming significant at higher frequencies. At frequencies above 1 MHz coupling rises very rapidly due to increased effects of stray capacitance, which explains the steep slope of the crosstalk curve.

For heavier frequency compensations the rolloff of the gain of the amplifier starts at a lower frequency. Therefore crosstalk is a strong function of frequency compensation and decreases with heavier compensation.

5.5 APPLICATIONS

The unique characteristics of monolithic dual amplifiers such as the RM4709 have been discussed. In the following sections, circuits taking advantage of these characteristics will be presented. The matching of gain, phase, and dc parameters make the dual-amplifier IC extremely versatile in a variety of applications. In this chapter two circuits in which the RM4709 may be used have been selected for detailed analysis. These circuits are the differential-in differential-out (DIDO) amplifier and the noninverting or potentiometric amplifier.

5.6 DIFFERENTIAL-IN, DIFFERENTIAL-OUT AMPLIFIER

A circuit in which the well-matched characteristics of the RM4709 are best used is a differential-in differential-out amplifier. Some of the important features of an amplifier of this type are high-input impedance, high differential-signal gain, and high rejection of common-mode signal. Two configurations for the amplifier are shown in Figure 5.6. In Figure 5.7 the amplifier of Figure 5.6*a* is shown with two types of signals applied to its inputs.

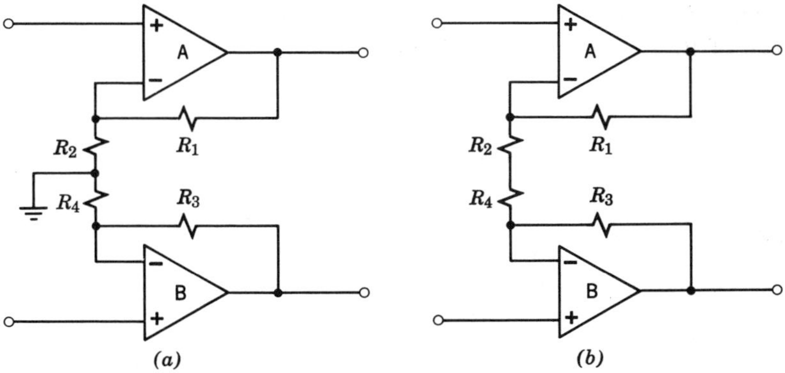

Figure 5.6. Two DIDO amplifier configurations.

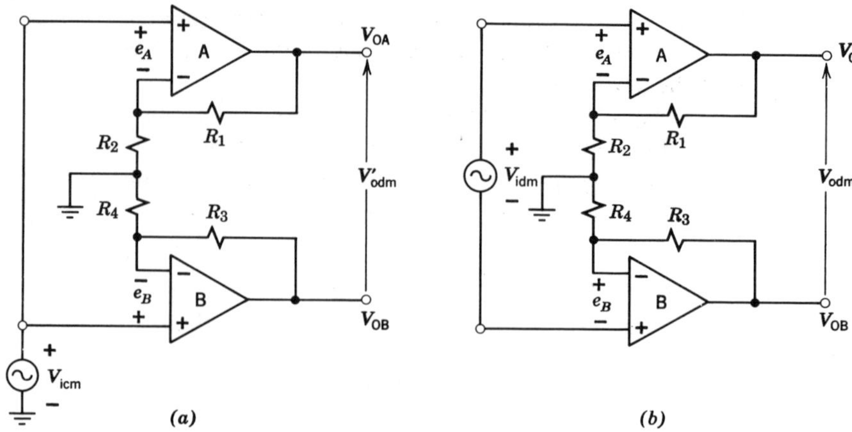

Figure 5.7. Amplifier of Figure 5.6*a* with applied signals.

The common-mode voltage gain G_{cm} for this amplifier is defined as the ratio of the differential output voltage V'_{odm} for a common-mode input signal to the common-mode input voltage V_{icm}. G_{cm} may be expressed as

$$G_{cm} = \frac{V'_{odm}}{V_{icm}}, \tag{5.2}$$

where V'_{odm} and V_{icm} are defined in Figure 5.7. The output voltages V_{OA} and V_{OB} of amplifiers A and B, respectively, are defined as

$$V_{OA} = A_{VA} e_A \text{ and } V_{OB} = A_{VB} e_B, \tag{5.3}$$

where A_{VA} = open-loop voltage gain of amplifier A, A_{VB} = open-loop voltage gain of amplifier B, e_A = input error voltage of A, e_B = input error voltage of B. The common-mode input voltage V_{icm} is

$$V_{icm} = e_A + V_{OA} K_1, \tag{5.4}$$

where

$$K_1 = \frac{R_2}{(R_1 + R_2)}; \tag{5.5}$$

therefore

$$V_{OA} = V_{icm}\left(\frac{A_{VA}}{1 + A_{VA} K_1}\right), \tag{5.6}$$

and similarly

$$V_{OB} = V_{icm}\left(\frac{A_{VB}}{1 + A_{VB} K_2}\right), \tag{5.7}$$

where

$$K_2 = \frac{R_4}{(R_3 + R_4)}. \tag{5.8}$$

The differential output voltage for a common-mode signal is

$$V'_{odm} = V_{OA} - V_{OB}. \tag{5.9}$$

In terms of (5.6) and (5.7),

$$V'_{odm} = V_{icm}\left[\left(\frac{A_{VA}}{1 + A_{VA}K_1}\right) - \left(\frac{A_{VB}}{1 + A_{VB}K_2}\right)\right].$$

The common-mode voltage gain is therefore

$$G_{cm} = \left[\left(\frac{A_{VA}}{1 + A_{VA}K_1}\right) - \left(\frac{A_{VB}}{1 + A_{VB}K_2}\right)\right]. \tag{5.10}$$

For large values of A_{VA} and A_{VB}, (5.10) reduces to

$$G_{cm} \cong \frac{1}{K_1} - \frac{1}{K_2}, \tag{5.11}$$

where K_1 and K_2 are defined by (5.5) and (5.8), respectively.

The differential-mode voltage gain G_{dm} for this amplifier is defined as the ratio of differential output voltage V_{odm} to the differential-mode input voltage. These voltages are shown in Figure 5.7*b*. The differential-mode voltage gain is given by

$$G_{dm} \simeq \frac{V_{odm}}{V_{idm}}. \tag{5.12}$$

From Figure 5.7*b* the differential input voltage is

$$V_{idm} = V_{OA}\left(\frac{1 + A_{VA}K_1}{A_{VA}}\right) + V_{OB}\left(\frac{1 + A_{VB}K_2}{A_{VB}}\right), \tag{5.13}$$

where K_1 and K_2 are again defined by (5.5) and (5.8). For $R_1 = R_3$ and $R_2 = R_4$, and if $A_{VA} = A_{VB}$, (5.13) can be simplified to

$$V_{dm} = V_{odm}\left(\frac{1 + A_{VA}K_1}{A_{VA}}\right). \tag{5.14}$$

Thus the differential mode voltage gain is

$$G_{idm} \simeq \frac{A_{VA}}{1 + A_{VA}K_1}. \tag{5.15}$$

The term common-mode rejection ratio, CMRR, as applied to a differential-in differential-out amplifier is ambiguous because the differential outputs can also be used as two single-ended outputs with respect to ground. A phase splitter is an example of a differential-in amplifier with its differential outputs used as single-ended outputs. Therefore, to prevent confusion the term common-to-differential mode rejection ratio (CDMRR) is defined as the ratio of the differential-mode voltage gain G_{dm} to the common-mode voltage gain G_{cm} of a DIDO amplifier, or

$$\text{CDMRR} = \frac{G_{dm}}{G_{cm}}. \tag{5.16}$$

For the amplifier in this discussion

$$\text{CDMRR} = \frac{A_{VA}/(1 + A_{VA}K_1)}{(1/K_1) - (1/K_2)}. \tag{5.17}$$

Equation (5.17) states that in order to achieve a high CDMRR K_1 and K_2 must be equal.

The input impedance for this amplifier is derived from (5.13) and is found to be

$$Z_i = 2R_{iA}\,(1 + A_{VA}K_1). \tag{5.18}$$

For $R_{iA} = R_{iB}$, $e_A = e_B$, $A_{VA} = A_{VB}$ and $K_1 = K_2$, R_{iA} is the input resistance of amplifier A.

The common-mode voltage gain G_{cm} for the DIDO amplifier of Figure 5.6*b* is also defined by (5.2). The CDMRR is a function of matching to the point where intrinsic impedances of the amplifiers become significant. Further refinement requires inclusion of these impedances and results in equations that predict CDMRR as a function of common and differential mode input-impedance match. From these equations it can be seen that the matching inherent in a monolithic dual is of significant advantage.

The differential output voltage from Figure 5.8 is

$$V_{odm} = V_{OA} + V_{OB}. \tag{5.19}$$

The input voltage is clearly

$$V_{idm} = e_A + V_R + e_B, \tag{5.20}$$

where the voltage V_R is

$$V_R = (V_{OA} + V_{OB})\left(\frac{R_2 + R_4}{R_1 + R_2 + R_3 + R_4}\right) \tag{5.21}$$

Substitution of (5.3) and (5.21) into (5.20) yields

$$V_{idm} = (e_A + e_B)\left[1 + A_{VA}\left(\frac{R_2 + R_4}{R_1 + R_2 + R_3 + R_4}\right)\right]. \tag{5.22}$$

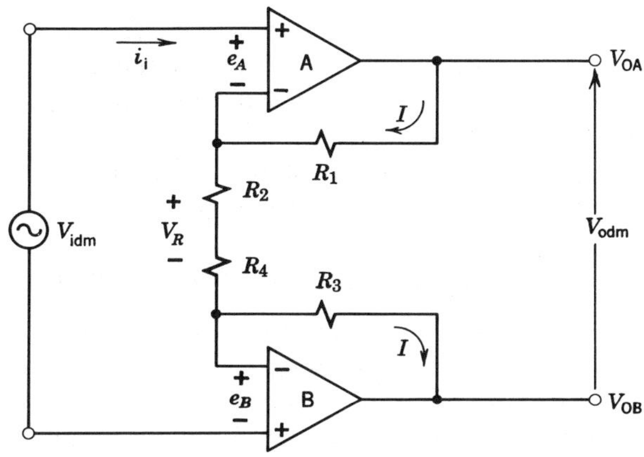

Figure 5.8. Amplifier of Figure 5.6*b* with applied signals.

If (5.22) and (5.19) are substituted into (5.12), the differential-mode gain can be expressed as

$$G_{dm} = \frac{A_{VA}}{1 + A_{VA}\,[(R_2 + R_4)/(R_1 + R_2 + R_3 + R_4)]}. \tag{5.23}$$

The input impedance of this amplifier for differential signals is

$$Z_i = \left(\frac{e_A + e_B}{e_A}\right) R_{iA}\,(1 + A_{VA}K), \tag{5.24}$$

where

$$K = \frac{R_2 + R_4}{R_1 + R_2 + R_3 + R_4}; \tag{5.25}$$

for a well-matched amplifier, that is, $e_A = e_B$,

$$Z_i = 2R_{iA}\,(1 + A_{VA}K). \tag{5.26}$$

For $R_1 = R_3 = 100$ kΩ, $R_2 = R_4 = 1$ kΩ, $A_{VA} = 3.55 \times 10^4$ V/V, and $R_{iA} =$ 500 kΩ the calculated input impedance is 355 MΩ. The measured value was 360 MΩ.

A comparison of the two DIDO amplifiers in Figure 5.6 shows that both have high input impedances. In the case of Figure 5.6*a* the CDMRR can be made extremely large by making K_1 equal to K_2. A high common-mode input signal, however, could severely limit the output-voltage swing. The analysis in Appendix 1 shows that a higher CDMRR can be obtained for the configuration in Figure 5.6*b*, since only the feedback resistors and the inherent amplifier characteristics have to be matched. A high common-mode input signal does not have the same detrimental effect on this amplifier as on the one previously discussed; but heavy compensation must be used to prevent frequency instability. Therefore the bandwidth of this DIDO amplifier is somewhat restricted.

5.7 NONINVERTING OR POTENTIOMETRIC AMPLIFIER

Another amplifier which has an extremely high common-mode rejection ratio is the noninverting or potentiometric amplifier shown in Figure 5.9. By selecting the correct ratios of resistors R_1 to R_2 and R_3 to R_4, the common-mode rejection ratio of this amplifier can be made very large.

In Figure 5.10 the amplifier is shown with a common-mode voltage V_{icm} applied to its inputs. The common-mode voltage gain G_{cm} is defined as the ratio of the common-mode output voltage V'_{odm} to the common-mode input voltage V_{icm}, or

$$G_{cm} = \frac{V'_{odm}}{V_{icm}},$$

where V'_{odm} and V_{icm} are defined in Figure 5.10.

With reference to amplifier B, V_{icm} can be expressed as

$$V_{icm} = e_B + V_{OB}\left(\frac{R_4}{R_3 + R_4}\right); \qquad (5.28)$$

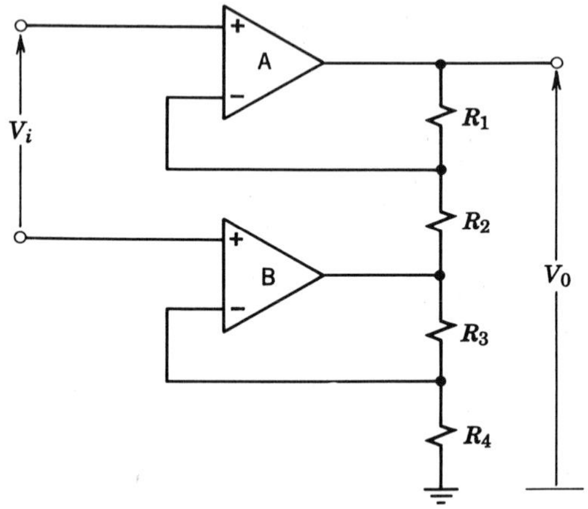

Figure 5.9. Noninverting or potentiometric amplifier.

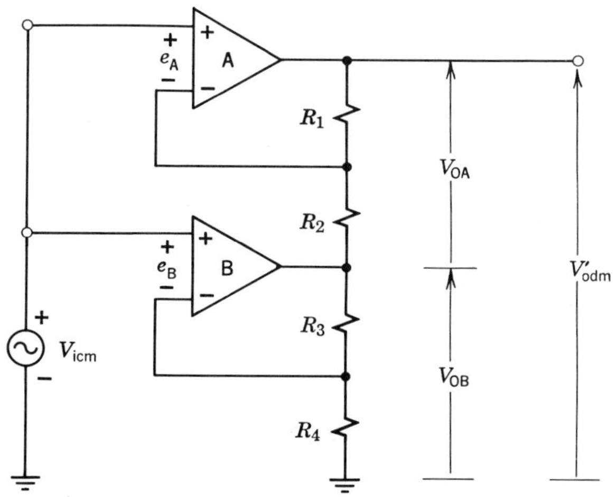

Figure 5.10. The amplifier of Figure 5.9 with a common-mode voltage applied to its inputs.

since $V_{OB}=A_{VB}e_B$, where A_{VB} is the open-loop voltage gain of amplifier B, V_{OB}/A_{VB} can be substituted into (5.28) for e_B so that

$$V_{icm}=V_{OB}\left(\frac{1+A_{VB}K_2}{A_{VB}}\right), \tag{5.29}$$

where

$$K_2 = \frac{R_4}{R_3+R_4}. \tag{5.30}$$

Solving for V_{OB}, the output voltage of B is

$$V_{OB} = V_{icm}\left(\frac{A_{VB}}{1+A_{VB}K_2}\right). \tag{5.31}$$

In Figure 5.11 amplifier B is replaced with an equivalent voltage source V_{OB}. The current I that flows out of A can be defined as

$$I=\frac{V_{OA}}{R_1+R_2}. \tag{5.32}$$

Summing the voltages at the input of A,

$$V_{icm} = e_A + IR_2 + V_{OB}. \tag{5.33}$$

Substituting for e_A with V_{OA}/A_{VA} and for I with (5.32), V_{icm} can be written as

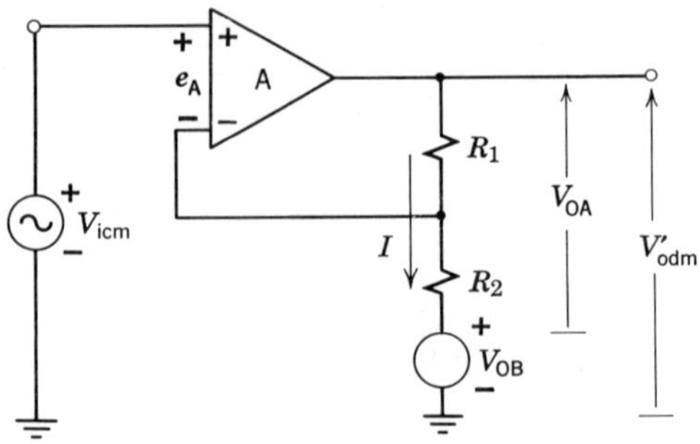

Figure 5.11. Amplifier of Figure 5.10 with amplifier B replaced by an equivalent voltage source.

$$V_{icm} = V_{OA}\left(\frac{1+A_{VA}K_1}{A_{VA}}\right) + V_{OB}, \tag{5.34}$$

where

$$K_1 = \frac{R_2}{R_1+R_2}. \tag{5.35}$$

The common-mode output voltage can be expressed as

$$V'_{odm} = V_{OA} + V_{OB}, \tag{5.36}$$

or solving for V_{OA},

$$V_{OA} = V'_{odm} - V_{OB}. \tag{5.37}$$

Replace V_{OA} in (5.34) with (5.37) and combining like terms,

$$V_{icm} = V'_{odm}\left(\frac{1+A_{VA}K_1}{A_{VA}}\right) + V_{OB}\left(1 - \frac{1+A_{VA}K_1}{A_{VA}}\right). \tag{5.38}$$

Equation (5.31) can now be substituted into (5.38) so that

$$V_{icm} = V'_{odm}\left(\frac{1+A_{VA}K_1}{A_{VA}}\right) + \frac{V_{icm}A_{VB}}{1+A_{VB}K_2}\left(1 - \frac{1+A_{VA}K_1}{A_{VA}}\right),$$

which reduces to the common-mode voltage gain of (5.27). Therefore

$$G_{cm} = \left[1 - \left(\frac{A_{VB}}{1+A_{VB}K_2}\right)\left(1 - \frac{1+A_{VA}K_1}{A_{VA}}\right)\right]\frac{A_{VA}}{1+A_{VA}K_1}. \tag{5.39}$$

The common-mode rejection ratio CMRR is defined as the ratio of differential-mode voltage gain to the common-mode voltage gain, or

$$\mathrm{CMRR} = \frac{G_{dm}}{G_{cm}}. \tag{5.40}$$

To achieve an infinite CMRR G_{cm} must be equal to zero. Therefore (5.39) is set equal to zero and the following expression is obtained:

$$K_1 + K_2 = 1 - \frac{1}{A_{VA}} - \frac{1}{A_{VB}},$$

or

$$K_1 + K_2 = 1 \text{ for } A_{VA}, A_{VB} \gg 1, \tag{5.41}$$

where K_1 and K_2 are defined by (5.35) and (5.30), respectively.

The differential input voltage V_{idm} is given by

$$V_{idm} = V_A + V_B. \tag{5.42}$$

The equations for V_A and V_B are (see Figure 5.12)

$$V_B = e_B + V_{OB} K_2, \tag{5.43}$$

$$V_A = e_A + V_{OA} K_1 - V_{OB}, \tag{5.44}$$

where K_1 and K_2 are once again defined by (5.35) and (5.30), respectively. The differential mode input voltage is the sum of (5.43) and (5.44), or

$$V_{idm} = e_A + V_{OA} K_1 - V_{OB} + e_B + V_{OB} K_2. \tag{5.45}$$

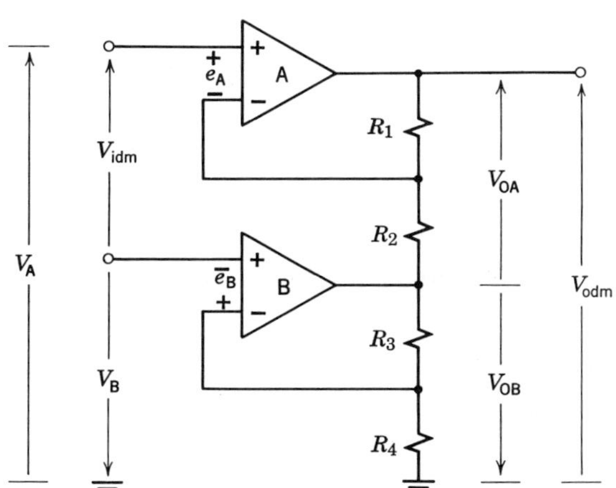

Figure 5.12. Noninverting amplifier with differential signal applied.

The differential-mode output voltage V_{odm} is given by the equation

$$V_{odm}=V_{OA}-V_{OB}; \tag{5.46}$$

solving for V_{OA} in (5.46) and substituting it into (5.45) yields

$$V_{idm}=e_A+e_B+V_{odm}K_1+V_{OB}(K_1+K_2-1). \tag{5.47}$$

Because the input resistances of amplifiers A and B are well matched, $e_A=-e_B$, and since $K_1+K_2-1=0$ is a condition for infinite CMRR (5.47) reduces to

$$V_{idm}=V_{odm}K_1. \tag{5.48}$$

Clearly the differential-mode voltage gain G_{dm} from (5.48) is

$$G_{dm}=\frac{1}{K_1}. \tag{5.49}$$

It can also be shown from (5.41) that

$$\frac{R_1}{R_2}=\frac{R_4}{R_3}. \tag{5.50}$$

The input impedance of this amplifier is given by the following equation:

$$Z_i=\frac{V_{idm}}{i_1}, \tag{5.51}$$

where

$$i_i=\frac{e_A}{R_{iA}}, \tag{5.52}$$

and R_{iA} is the input resistance of amplifier A. From (5.48)

$$V_{idm}=V_{odm}K_1.$$

By substituting (5.46) and $e_B=-e_A$ into (5.48),

$$V_{idm}=e_AK_1(A_{VA}+A_{VB}). \tag{5.53}$$

Now, substituting (5.53) and (5.52) into (5.51), the expression for the input impedance becomes

$$Z_i=R_iK_1(A_{VA}+A_{VB}). \tag{5.54}$$

For $A_{VA}=A_{VB}$ (5.54) reduces to

$$Z_i=2A_{VA}K_1R_i. \tag{5.55}$$

Since the gain of this amplifier is determined by the ratio of R_1 to R_2, amplifier A must be compensated according to the gain desired. Amplifier B requires heavy compensation, however, because it is operating as a unity-gain amplifier. To ensure better matching of the amplifier, A and B may be compensated equally.

ACKNOWLEDGMENTS

The authors would like to acknowledge the contributions of the application configurations and general analytical suggestions by J. A. Livezey and the layout and processing of the devices by O. F. Maxwell.

6

High-Input Impedance Monolithic Amplifiers

by D. R. Sullivan and M. A. Maidique
Transitron Electronic Corporation

6.1 INTRODUCTION

When faced with a requirement for high input impedance, designers have usually turned to a discrete amplifier with an FET front end and its attendant degradation of the input offset voltage and offset drift characteristics. Bipolar monolithic operational amplifiers, which have inherently low input offset voltage and drift, have been restricted in their range of input impedances by the necessity of operating the input transistor at moderate collector current. Even when Darlington (or quasi-Darlington) configurations have been employed, input impedances have been limited to the 1-MΩ range and the difficulty in matching the four input transistors has usually resulted in a degraded input specification.

Recent improvements in low-level device characteristics and careful matching permit the design of all-bipolar operational-amplifier ICs that retain the input offset voltage characteristic of a high quality bipolar IC, but which can achieve input impedances in the 100 mΩ range. An example of an IC is the TOA7709; the circuit diagram for this IC is shown in Figure 6.1. The IC's circuit is similar to the configuration for the TOA1709 except that it has two Darlington pairs in the input stage.

Outside of input impedance, bias current, and offset current, which are considerably improved, the amplifier parameters, including input offset voltage and input offset voltage drift, remain well within the specification of the standard 1709. Typical specifications for the TOA7709 are included in Table 6.1.

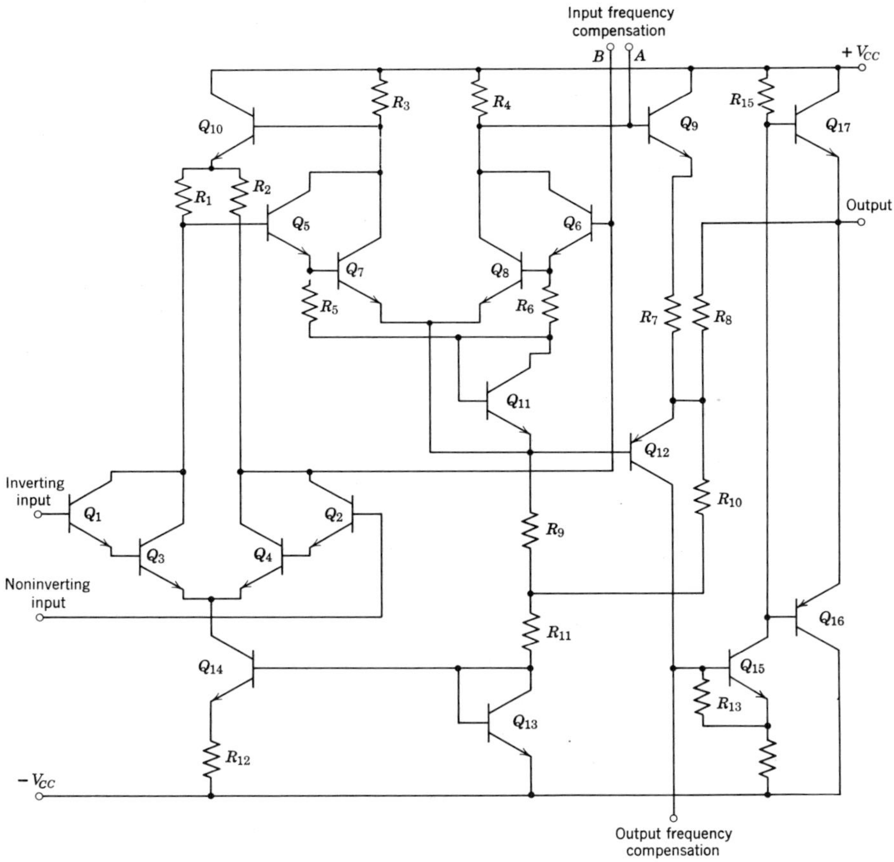

Figure 6.1. High-impedance operational amplifier (TOA7709/TOA8709).

6.2 THE INPUT STAGE

The input circuit consists of a differential amplifier using a Darlington pair in each half. The resistive component of the input impedance can be calculated easily using the simplified model shown in Figure 6.2. By summing voltages around the input loop, the following relationship is established:

$$V_{\text{in}} = (\beta_1 + 1) r_{e1} I_{B1} + (\beta_3 + 1) r_{e3} I_{B3}. \tag{6.1}$$

Since $I_{\text{in}} = I_{B1}$ and $(\beta_1 + 1) I_{B1} = I_{B3}$, the equation can be rewritten as

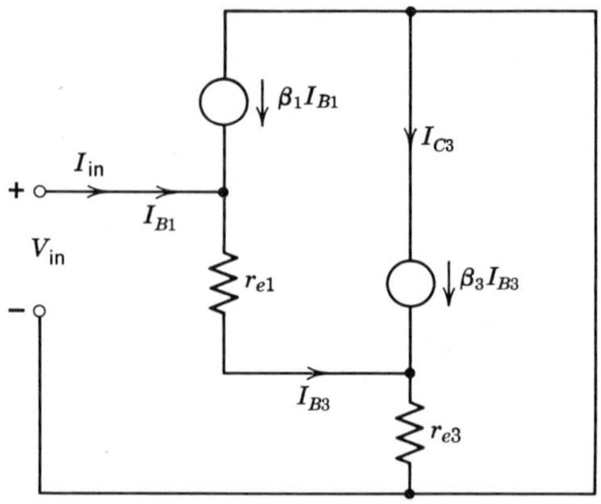

Figure 6.2. Simplified T-model representation of Darlington connection.

$$V_{in} = (\beta_1 + 1) r_{e1} I_{in} + (\beta_3 + 1) r_{e3} (\beta_1 + 1) I_{in}. \tag{6.2}$$

Because only one half of a balanced input circuit has been considered, the total input resistance is

$$Z_{in} = 2\frac{V_{in}}{I_{in}} = 2(\beta_1 + 1)\,[\,r_{e1} + (\beta_3 + 1) r_{e3}\,]. \tag{6.3}$$

Since $r_{e1} = 26\text{ mV}/I_{B3}$ and $r_{e3} = 26\text{ mV}/[\,(\beta_3 + 1) I_{B3}]$, the ratio of r_{e1} to r_{e3} is given by

$$\frac{r_{e1}}{r_{e3}} = (\beta_3 + 1) \text{ or } r_{e1} = r_{e3}\,(\beta_3 + 1). \tag{6.4}$$

Substituting this expression for r_{e1} into (6.3) results in

$$Z_{in} = 2(\beta_1 + 1)\,[\,(\beta_3 + 1) + (\beta_3 + 1)\,]\, r_{e3}. \tag{6.5}$$

For $\beta_3 \gg 1$ and $\beta_1 \gg 1$, (6.5) reduces to

$$Z_{in} = 4(\beta_1 + 1)(\beta_3 + 1) r_{e3} = 4\beta_1\beta_3 r_{e3}. \tag{6.6}$$

Substitution of 26 mV/I_{E3} for r_{e3} in (6.6) yields

$$Z_{in} = 4\beta_1\beta_3 \frac{26}{I_{E3}(\text{mA})} = \frac{104\beta_1\beta_1}{I_{E3}(\text{mA})}. \tag{6.7}$$

If we let $\beta_1\beta_3 = 10{,}000$, and $I_{E3} = 20\ \mu\text{A}$ in (6.7), then

$$Z_{in} = 52\ \text{M}\Omega.$$

The bias current under the above conditions would be

$$I_{\text{in}}=\frac{I_{E3}}{\beta_1\beta_3}=2\text{ nA}. \tag{6.8}$$

and for such an input stage we would expect an offset current of less than 1 nA.

Referring back to (6.8), equation (6.7) can be simplified so that

$$Z_{\text{in}}\,(\text{M}\Omega)=\frac{102}{I_{\text{in}}(\text{nA})}\simeq\frac{100}{I_{\text{in}}(\text{nA})}, \tag{6.9}$$

where Z_{in} is now expressed in megohms and I_{in} in nanoamperes.

If the leakage currents are small enough so that the bias current is mainly produced by transistor action, (6.9) provides a simple method of calculating the dc component of the input impedance from a bias current measurement. Even at ultralow operating currents, it has been verified that the leakage components are only a small percentage of the input current, thus allowing the application of (6.9).

This discussion also neglects the shunting effect of the collector-to-base resistance, which for input impedances of up to 100 MΩ is a reasonable approximation.

Calculation of the reactive component of the input impedance is somewhat more complicated. Because of the extremely low operating current of the input transistors, however, device capacitances are held to a minimum and the dominant capacitance effect is due to stray and header capacitance.

The device can then be modeled by two equal lumped capacitors from each input base terminal which are terminated at a common point, as shown in Figure 6.3. The differential input capacitance C_D is then $C_D=$

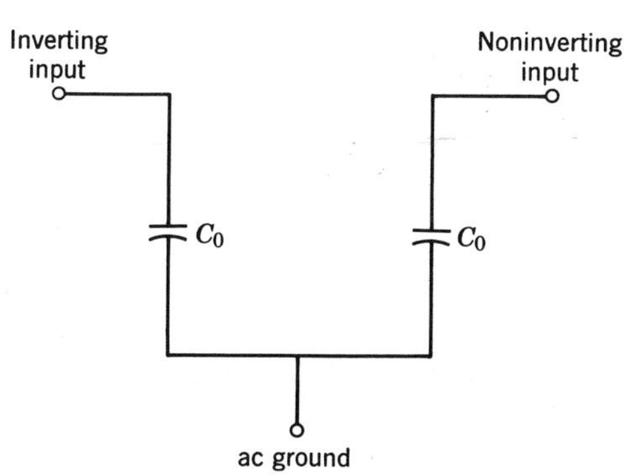

Figure 6.3. Lumped model of input capacitance.

$C_0/2$ and the common-mode capacitance C_c is $C_c = C_0$. Measurements yield a value of $C_0 = 3$ pF; therefore $C_D = 1.5$ pF and $C_c = 3$ pF.

The breakpoint of the differential input-impedance function occurs above 1 kHz even if the resistive component of the impedance is 100 MΩ. Thus a situation exists in which the impedance is flat over a bandwidth which is three to four orders of magnitude broader than that obtained with a matched-FET pair front end. This is, of course, simply a consequence of the ultrahigh FET impedance. But what is more important, the absolute values of input capacitance for the TOA7709 compares quite favorably to typical FET capacitances of 3 to 5 pF.

To assess the performance of the Darlington input stage, bias current, and offset voltage, measurements were made on a randomly selected lot of 200 TOA7709 devices. The results of this test are plotted in Figures 6.4 and 6.5. From Figure 6.4, it can be determined that more than 15%

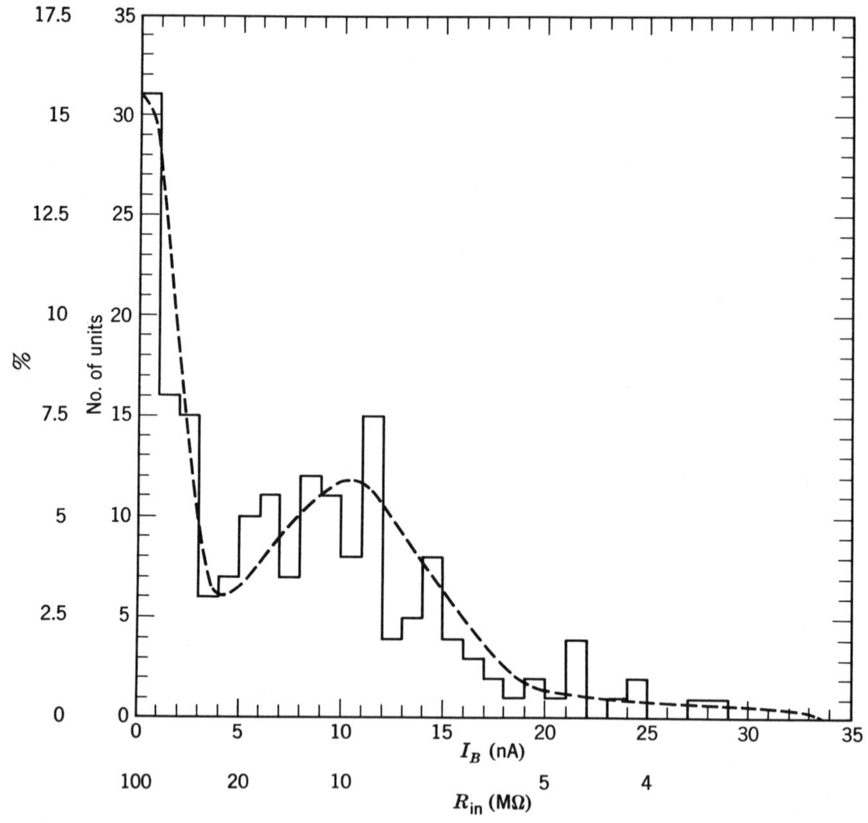

Figure 6.4. Distribution of TOA7709 ICs. Input impedance and bias current at 25°C (random lot).

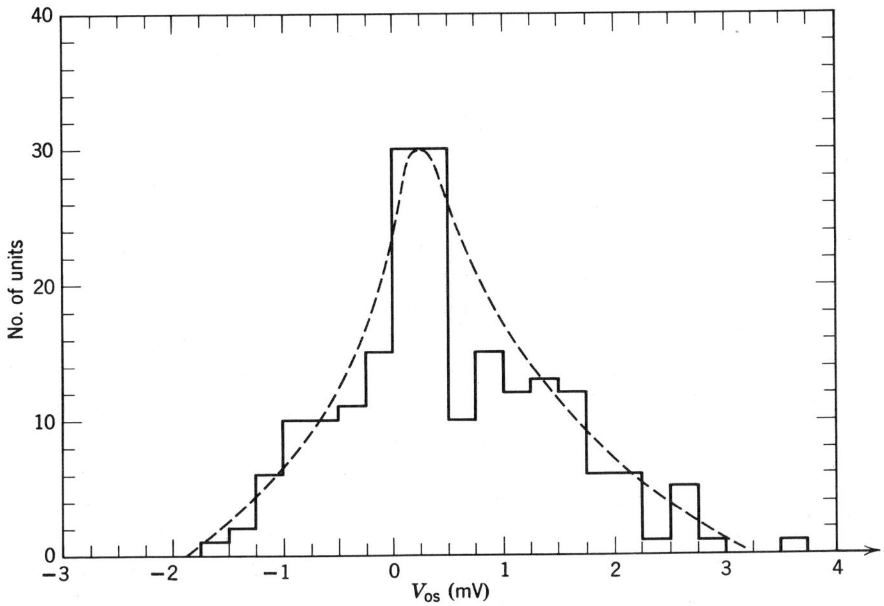

Figure 6.5. Distribution of TOA7709. Input offset voltage at 25°C (random lot).

of the units tested had input bias currents less than 1 nA and input impedances greater than 100 MΩ, while 63% of the devices had bias currents less than 10 nA and input impedances greater than 10 MΩ. The results plotted in Figure 6.5 indicate that 65% of the units had offset voltages less than ±1 mV, whereas 84% had offset voltages less than ±2 mV.

The sensitivity of a high-impedance amplifier circuit, or of any amplification device, is ultimately limited by the circuit's equivalent input noise. Fortunately, the same surface conditions that apply for obtaining good low current betas are of importance in minimizing noise. Consequently, a secondary effect of high beta in low-level devices is low noise; therefore it is not surprising that the Darlington configuration provided only a moderate increase in equivalent noise input voltage (i.e., = 50%) over the typical value for the standard 709 of about 8 μV rms for a 50-kHz noise bandwidth.

6.3 VOLTAGE FOLLOWER

Some interesting insight into the operation of the Darlington front end can be gained by considering the use of the TOA7709 in a voltage-follower configuration (Figure 6.6). For a voltage follower the differential component of the input impedance appears multiplied by the loop gain, which is essentially equal to the open-loop gain since the amplifier has full feedback. The low-frequency input impedance for a voltage gain of 40,000 and a bias current of 2 nA will be $Z_{\text{inf}} = AZ_{\text{in}} = 2 \times 10^{12}\Omega = 2 \times 10^6$ MΩ, where Z_{inf} is the differential input impedance of the voltage follower and Z_{in} is the differential input impedance of the TOA7709. This impedance is shunted by the common-mode input impedance of the TOA7709; this is important in this case since both inputs are always at about the same voltage level. The common-mode impedance component can be found by characterizing the differential amplifier current source by a Thevenin equivalent R_E which has been measured to be in the order of 1 MΩ at the current levels of interest. The common-mode input impedance of the follower Z_{cf}, is then just the common-mode input impedance of a differential amplifier with current gain $\beta_1\beta_3$ and the common emitters returned to ground through a resistance R_{E}. Therefore $Z_{\text{cf}} = 2\beta_1\beta_3R_E = 20{,}000$ MΩ, where $\beta_1\beta_3 = 10{,}000$ as before. The common-mode impedance then limits the achievable dc input impedance of the voltage follower to about 20,000 MΩ.

6.4 LOGARITHMIC AMPLIFIER

A logarithmic amplifier can use the capabilities of the TOA7709 to advantage. Typical circuits for logarithmic amplifiers (see Figure 6.7) use several resistors and potentiometers to null the effects of amplifier unbalance and bias current. The elimination of this nulling circuitry would be highly desirable, particularly in an attempt to integrate the complete

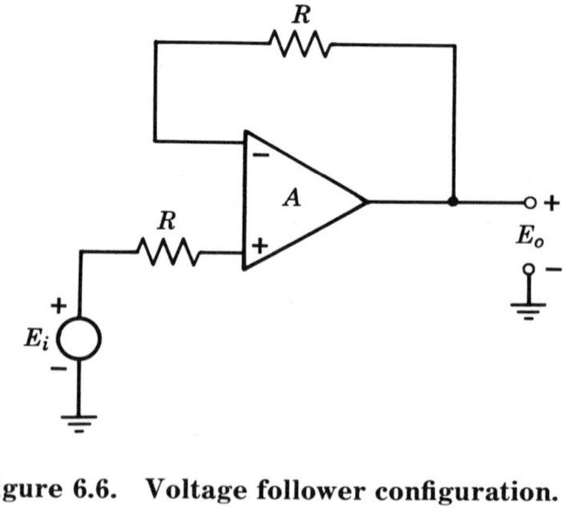

Figure 6.6. Voltage follower configuration.

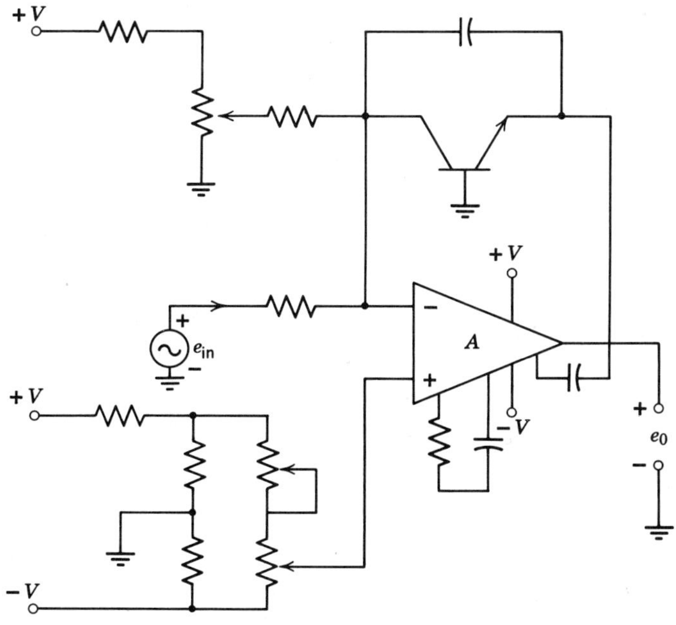

Figure 6.7. Typical compensated logarithmic-amplifier circuit.

logarithmic function. The basic logarithmic amplifier, shown in Figure 6.8, was built to test the performance of the TOA7709 in an uncompensated system.

The TOA7709 used in the test circuit of Figure 6.8 had an input bias current of 2 nA and an input offset voltage of 1.5 mV. Since the principal contributor of error is the bias current, an input of $100 \times 10^3 \times 2 \times 10^{-9}$ or 0.2 mV can be expected for an input bias current of 2 nA and a source resistance of 100 kΩ. This corresponds to an accuracy of 1% at a 20-mV input level.

The results of testing the uncompensated logarithmic amplifier described above are plotted in Figure 6.9. Also included in Figure 6.9 is a plot of an uncompensated logarithmic amplifier using a TOA1709. Three decades of logarithmic operation performance with an accuracy of ±1% are achieved with the TOA7709, whereas the use of the standard TOA1709 in the same circuit restricts the range to about one decade. Note that the upper part of the range has been limited not by current carrying of the logging transistor, but by the desire to use realistic values of input voltage and resistance.

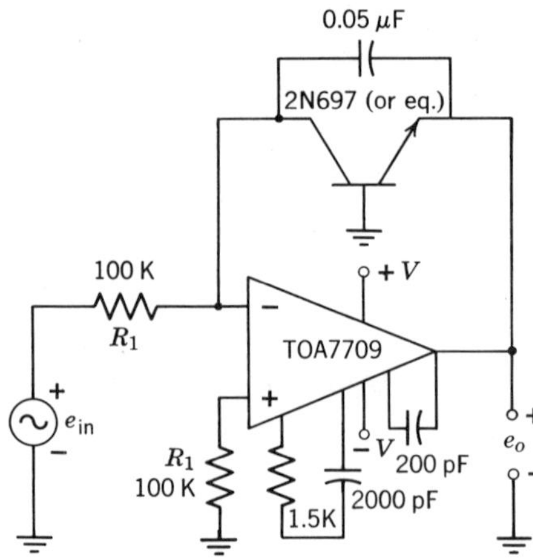

Figure 6.8. Uncompensated logarithmic-amplifier circuit using TOA7709.

6.5 D/A AND A/D CONVERTERS

The input characteristics of the TOA7709 can be used to advantage in precision D/A and A/D converters. Applications include D/A converter output buffers as shown in Figures 6.10 and 6.11, and A/D converter comparators as shown in Figure 6.12; a ladder network is used for the basic D/A converter in each case. The accuracy of each of the functions depends directly on the magnitude and stability of the amplifier characteristics, as discussed in the following paragraphs.

6.6 D/A CONVERTER OUTPUT BUFFER

A noninverting voltage-follower configuration is used as the output buffer amplifier for the D/A converter shown in Figure 6.10. The input impedance of the voltage follower, as discussed previously, is approximately 20,000 MΩ; as a result the error due to impedance loading of the ladder can be neglected for all practical purposes. The only significant sources of error due to the amplifier characteristics are amplifier gain, input offset voltage, input offset current, input bias current and common-mode rejection ratio.

Gain error is proportional to the loop gain of the amplifier. The voltage follower has a loop gain essentially equal to the open-loop gain of the TOA7709 and a differential input impedance high enough to make the feedback factor practically one. Bias current errors are eliminated by connecting a resistor in the feedback path which is equal to the output impedance of the ladder network. The voltage error at the amplifier output due to the input offset current is equivalent to the input offset current multiplied by the output impedance of the ladder, while the input offset voltage and offset voltage drift are referred to the amplifier's output by gain of one.

The finite common-mode rejection ratio of the amplifier produces an output error proportional to the CMRR for the case of a voltage follower, and the amplifier input signal is restricted to the common-mode range of the amplifier. For an ±8 V swing at the output of the ladder and the typical TOA7709 specifications shown in Table 6.1 the following is a

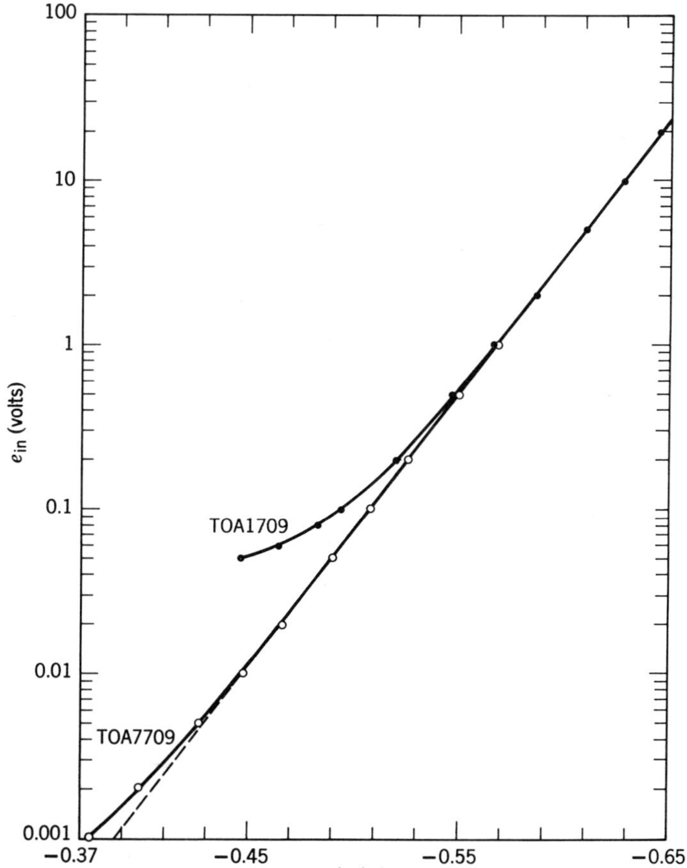

Figure 6.9. Uncompensated logarithmic-amplifier performance.

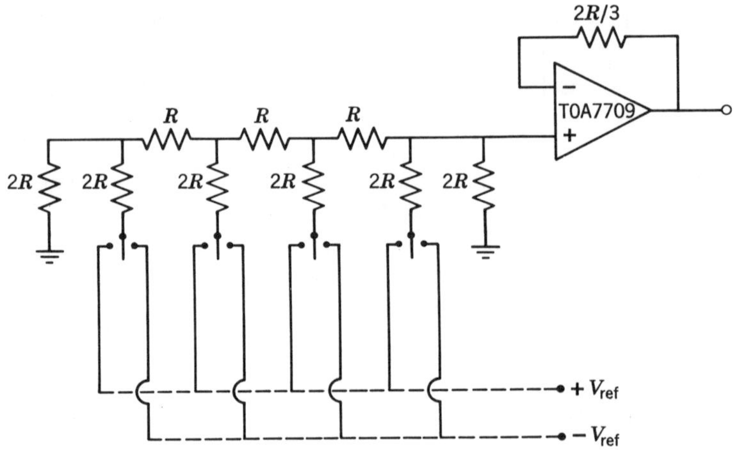

Figure 6.10. Noninverting D/A buffer.

summary of the errors associated with the noninverting buffer amplifier: gain error, ±20 ppm; bias current error, 0; offset-current error per kΩ of ladder-output impedance, ±0.12 ppm; offset voltage error, ±6 ppm; offset-voltage drift error/°C, ±0.4 ppm; common-mode rejection-ratio error, ±25 ppm.

TABLE 6.1 Typical Characteristics of the TOA7709

Voltage gain	45,000
Input bias current	10 nA
Input offset current	±2 nA
Input offset voltage	±1 mV
Input offset voltage drift	±6 μV/°C
Input impedance	10 MΩ
Common-mode rejection ratio	90 dB
Common-mode input range	±10 V
Differential-mode input range	±10 V

An inverting-amplifier configuration rather than a voltage follower can also be used as a ladder output buffer, as shown in Figure 6.11. The ladder output impedance $2R/3$ serves as the input resistor to the amplifier. For a gain of one the feedback resistor is also $2R/3$, and the noninverting input is returned to ground through a resistor whose value is $R/3$; this eliminates bias current errors.

The biggest advantage of the inverting configuration is that the error due to the amplifier's common-mode rejection ratio is essentially eliminated. The restrictions on input swing are determined by the available output swing of the amplifier rather than the input common-mode range. The disadvantage of this connection is that all other sources of error are doubled, since the feedback factor is one-half, rather than one, as in the case of the noninverting configuration.

6.7 A/D CONVERTER COMPARATOR

The TOA7709 is connected as a high-gain amplifier when used as a comparator in the successive approximation A/D converter shown in Figure 6.12. The dc signal to be digitalized is fed into one input of the amplifier while the output of the ladder is fed into the other input. The output of the amplifier is connected to the logic which in turn determines the state of the ladder switches necessary to reproduce the dc input signal at the output of the ladder. The binary state of the ladder switches represents the digitized analog input.

Gain error for the comparator is of no consequence, since all that is required is a gain high enough to give the proper resolution. All other A/D errors associated with the comparator are the same as those described for the noninverting buffer for the D/A converter. This is true because, for the comparator, the errors of interest are referred to the amplifier input. There is one other very important parameter, however, that must be considered for the comparator case; this is the overdrive recovery time. Tests made on the TOA7709 indicate that it will come out of hard saturation with basically the same speed as a standard 709 opera-

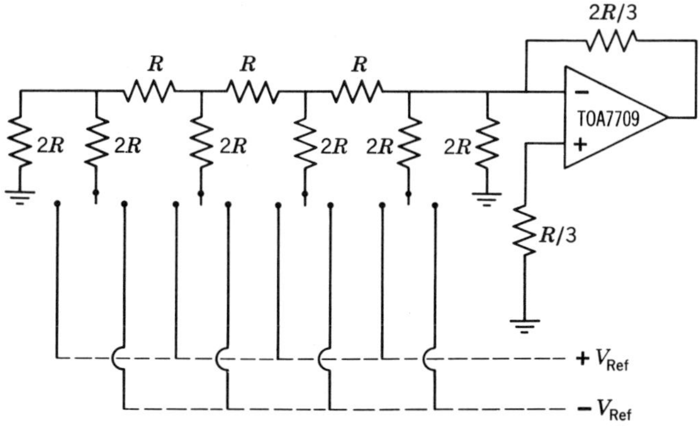

Figure 6.11. Inverting D/A buffer.

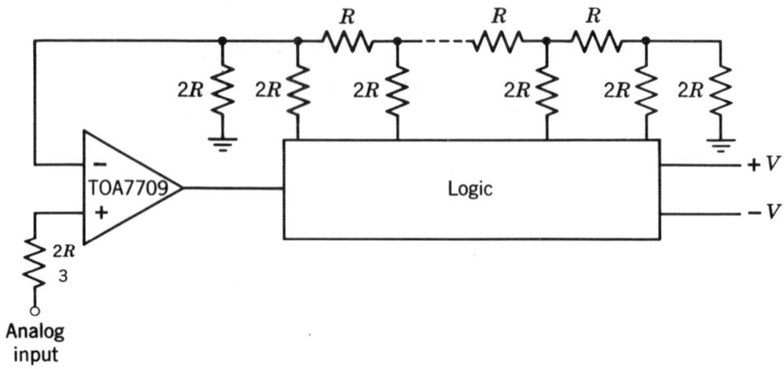

Figure 6.12. A/D comparator.

tional amplifier. Recovery from a 5-V input to 1-mV input of the opposite polarity takes less than 5 μs; 90% of this time is storage and 10% is rise time. Here the total recovery time considered is that time necessary for the output to cross through zero. Offset voltage and current effects were eliminated by balancing them out at the input stage. Thus the comparator is suitable for systems with bit rates as high as 100 kHz.

6.8 CONCLUSIONS

The TOA7709 high-impedance Darlington amplifier described in this chapter extends the application realm of monolithic operational amplifiers. Typical applications, including voltage follower, logarithmic amplifier, D/A output buffer, and A/D comparator have been discussed.

The combination of low bias current, low offset voltage, good temperature stability and high input impedance also make this type of IC useful in other applications where these error sources must be minimized.

7

Integrated Voltage Follower with Low Input Current

by Robert J. Widlar
National Semiconductor

7.1 INTRODUCTION

Most integrated operational amplifiers on the market today have serious limitations that impede their performance in many voltage-follower applications. Their speeds are often too slow because a voltage follower requires maximum frequency compensation, reducing slew rate to somewhere between 0.1 and 1 V/μs [1,2]. Voltage followers are most frequently used as buffer amplifiers from high impedance sources; as a result, the input current of popular amplifiers gives excessive dc offset when operated with source resistances much above 10 kΩ.

The use of ICs specifically designed for voltage-follower service can help overcome the limitations. This is demonstrated in the following paragraphs using as an example a monolithic voltage follower that combines low offset voltage with an input current of 2 nA and a 10 V/μs slew rate. The performance described for this circuit is obtained by using improved bipolar transistors along with an operational-amplifier circuit design which is optimized for the voltage-follower configuration. The device is designed to operate from supply voltages between ± 12 and ± 15 V; it features a 10-MHz bandwidth along with a 3-pF input capacitance and a minimum input resistance of 10,000 MΩ. In addition it requires no external components for frequency compensation and incorporates continuous short-circuit protection.

7.2 CIRCUIT DESCRIPTION

Fewer problems are encountered in designing a high-performance voltage follower than in designing a similar general-purpose amplifier. One reason for this is that level shifting is not required, so that comple-

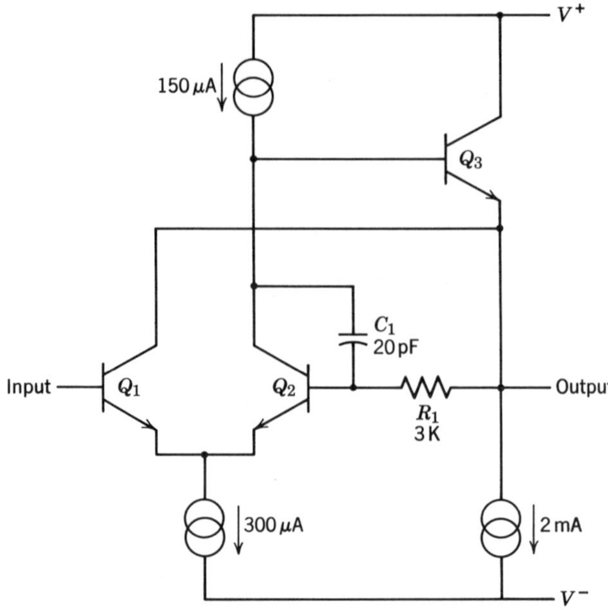

Figure 7.1. Basic configuration for the voltage follower.

mentary transistors are unnecessary as gain stages. Hence it is possible to get better high-frequency performance. Good performance at high frequencies has been limited in the past by the quality of the *pnp* [3] transistors that can be made in monolithic circuits. Also, because 100% feedback is used in a voltage regulator, the open-loop gain does not have to be so high as that of a general-purpose amplifier. This permits the use of a simpler circuit which is easier to frequency compensate. In addition, with a fixed configuration such as a voltage follower, the input stage can be included within the compensation network. This makes it easier to get fast slewing without having to provide unreasonably large small-signal bandwidths that would make the amplifier more prone to instabilities.

Figure 7.1 demonstrates how simple a voltage-follower circuit can be. This circuit uses a single-stage differential amplifier with an emitter-follower output. Since current sources are used at the emitter of the differential pair and as a collector load, it is practical to get an open-loop gain of 3000 from a single stage. The collector of the input transistor Q_1 is bootstrapped to the output to increase gain and raise the input resistance. It also eliminates leakage currents by operating the input at zero collector-to-base voltage. A class-A output stage is used because it behaves better at high frequencies with capacitive loads. Although frequency compensation is not always required with this configuration, R_1 and C_1 have been included to improve stability with capacitive loading. The compensation network is placed so that the circuit has good transient rejection on both the positive and the negative supplies.

7.3 INPUT STAGE

In order to get fast slewing, it is necessary to operate the differential amplifier at a fairly high current for an input stage. Therefore a Darlington connection is used on the input transistors to get low input current. As can be seen from Figure 7.2, however, bleed resistors R_1 and R_2 operate the input transistors at a current which is large by comparison to the base current of Q_3 and Q_4. This keeps Q_1 and Q_2 from seeing mismatches in the base currents of Q_3 and Q_4, which is the largest source of offset

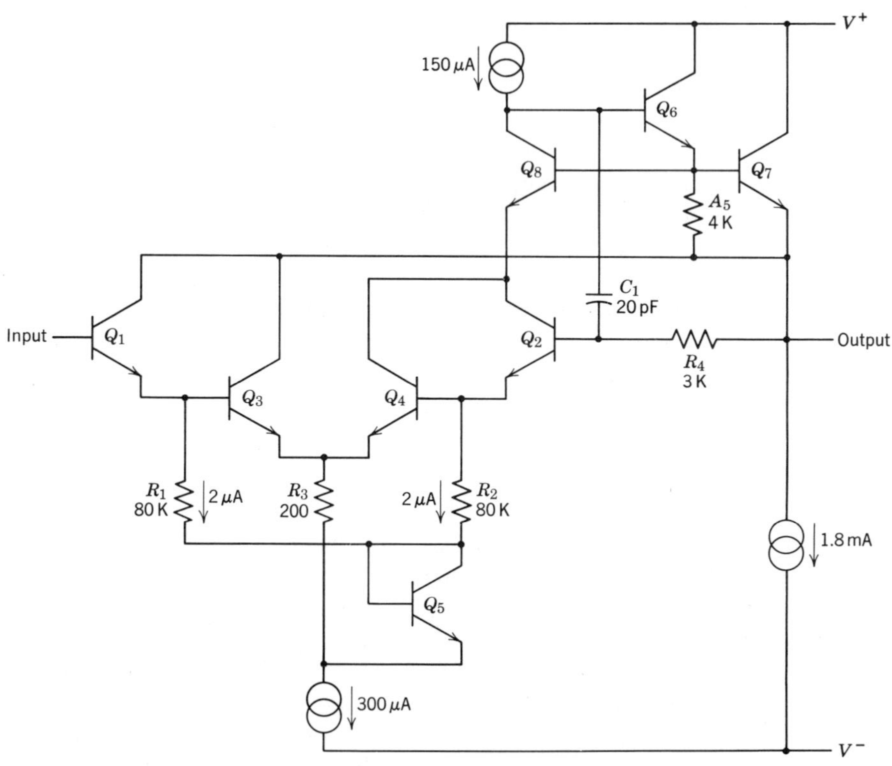

Figure 7.2. Partial schematic of the LM102 voltage follower.

voltage in an ordinary Darlington differential stage. This bleed current also doubles the gain of the stage and improves the high frequency performance.

Use of a Darlington stage is not the entire secret to getting low input currents [4]. With the integrated-circuit transistors that have been available in the past, reducing the collector current by a factor of 10 would only reduce the base current by a factor of 3, since the current gain falls off rapidly at low collector currents. In order to get any real improvement from operating at low currents, it has been necessary to make better transistors. The devices used here have a typical current gain of 1000 at 2-μA collector current.

Operating at 2-μA currents requires large resistance values which are not easily fabricated in integrated circuits. Therefore the bleed circuit on the input stage had to be designed to minimize this resistance. R_1 and R_2 are operated with a 160-mV drop across them, which is determined by the drop across R_3 plus the emitter-base voltage difference between Q_5 and the differential transistors Q_3 and Q_4. This difference is 100 mV since the differential transistors are operated at roughly 35 times the current through Q_5 [5].

Pinch resistors had to be used for R_1 and R_2 to get 80 kΩ within a reasonable surface area. The pinch resistors were also necessary to keep the parasitic capacitance of the resistors small, for this capacitance could severely degrade the large-signal pulse response. Pinch resistors, however, do have a large positive temperature coefficient that causes the operating current of Q_1 and Q_2 to increase to 3.5 μA at -55°C and decrease to 1.4 μA at 125°C.

Figure 7.2 shows that an extra transistor Q_8 has been added on the collectors of Q_2 and Q_4. This forms a cascade stage which operates Q_2, like Q_1, at near zero collector-to-base voltage. An additional emitter follower is included on the output to further reduce output resistance.

7.4 BIASING CIRCUITRY

Figure 7.3 is a simplified schematic of the biasing circuitry that is represented by current sources in Figures 7.1 and 7.2. In order to realize low offset voltage, the current source on the collector of Q_2 must supply a current that is exactly one-half of the input-pair emitter current.

To do this, diode-connected transistors Q_{14} and Q_{15} provide a bias voltage that is regulated against supply-voltage variations for the current-source transistors Q_{10}, Q_{12} and Q_{13}. Q_{12} is the current source for the input pair, whereas Q_{13} generates a current which is one-half the output current

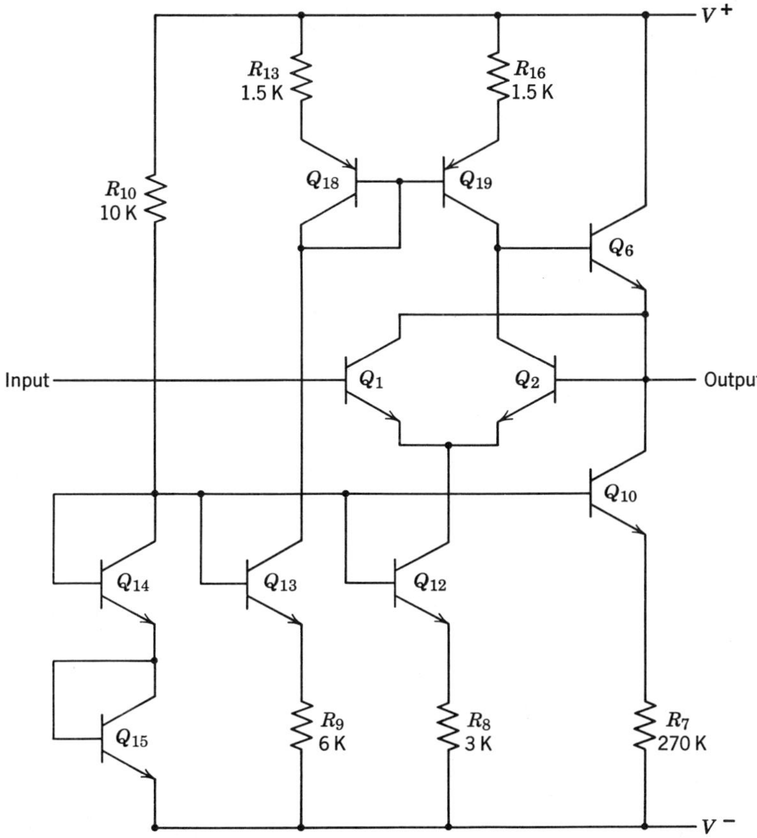

Figure 7.3. Simplified schematic of the biasing circuitry.

of Q_{12}. This is accomplished by making R_9 twice as large as R_8 and Q_{13} one-half the size of Q_{12}. The output current of Q_{13} is fed to Q_{18}, which biases Q_{19}. If it is assumed that Q_{18} and Q_{19} are well matched and have large current gains, the output current of Q_{19} will be equal to the collector current of Q_{13}—or one-half the emitter current of the input pair, as required.

7.5 ADDITIONAL DETAILS

In practice it cannot be assumed that the current gain of the *pnp* transistors, Q_{18} and Q_{19}, is large [6]. In fact the current gain could be as low

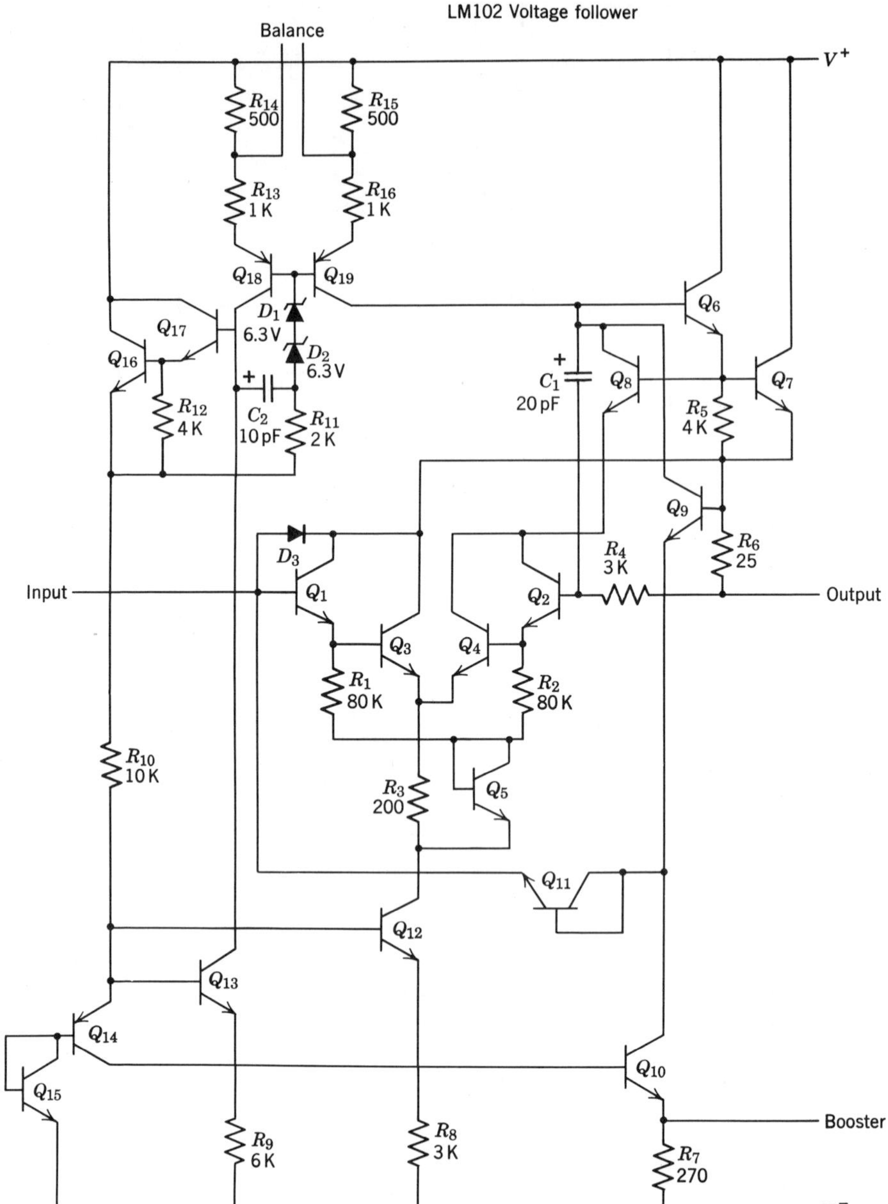

Figure 7.4. Complete schematic diagram.

as unity. As a result additional circuitry is required to get proper operation. Figure 7.4 shows how this is done.

Instead of connecting the base directly back to the collector, emitter-follower buffers Q_{16} and Q_{17} are used to isolate the base current from the collector of Q_{18}. Level shifting diodes D_1 and D_2 are included so that Q_{18} is operated at approximately the same collector base voltage as Q_{19} when the output of the amplifier is at zero, further improving the match.

The RC network, R_{11} and C_2, is included to suppress oscillations in this feedback loop. The voltage drop across C_2 is less than a couple of volts so that a junction capacitor can be fabricated from the emitter and base diffusions of the *npn* transistors. With this, the required capacitance can be obtained in a reasonable area of the chip with no additional process steps, as would be required if an MOS capacitor were used. The same is true, incidentally, for C_1.

A class-A output stage is used primarily for simplicity, although the higher quiescent current in the output stage improves stability with capacitive loads. The emitter of the current sink Q_{10} is brought out so that an external resistor can be connected between it and the negative supply for increased output current for applications in which the higher dissipation can be tolerated. The current source is biased from the collector of a low-gain lateral *pnp* transistor Q_{14}, so that the bias voltage for the input-stage current sources will not be greatly affected when Q_{10} saturates on negative signals.

Resistors are included in series with the emitters of the pnp current-source transistors Q_{18} and Q_{19} to reduce their output conductance, thereby increasing gain. Taps on these resistors are brought out to provide for offset balancing. The tap point is selected to give a smooth ± 20 mV adjustment range when a 1-kΩ potentiometer is connected between the balance terminals and the positive supply.

The output is inherently short-circuit proof in the negative direction. Current limiting for positive outputs is provided by Q_9 and R_6. When operating from low-source resistances, however, a 2- to 10-kΩ resistor must be added in series with the input, since the input is clamped directly to the output through D_3 and Q_{11} which protect the input transistors from overvoltage. This resistor was not included on the chip because it is difficult to locate a diffused resistor in an isolation region where it would be effective but not contribute to input leakage current at high temperatures.

A photomicrograph of the LM102 is shown in Figure 7.5. Although the schematic diagram of the circuit appears complicated, it fits neatly on a 49-by-49 mil-square die.

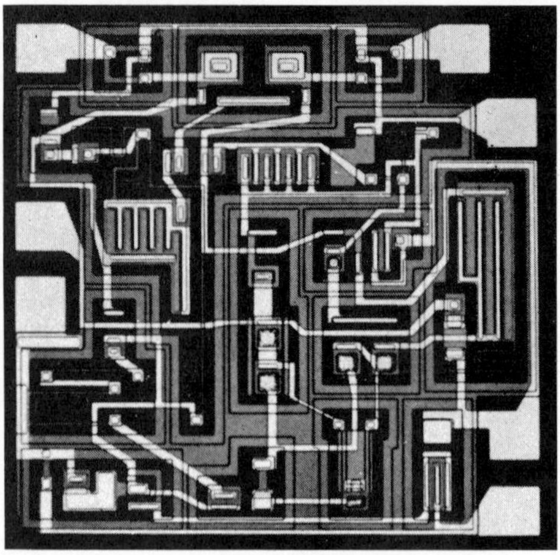

Figure 7.5. Photomicrograph of the LM102.

7.6 PERFORMANCE

The electrical characteristics of the LM102 are summarized in Table 7.1. It is evident from this that the primary design objectives, high speed and low input current, have indeed been achieved.

The low input bias current of the voltage follower is illustrated in Figure 7.6. It can be seen that the input current reaches a minimum at

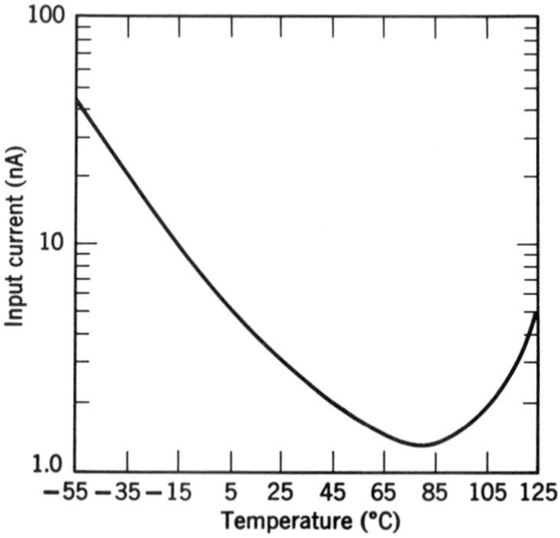

Figure 7.6. Input bias current.

85°C but remains low up to 125°C. This suggests operating the LM102 in a temperature stabilized component oven for wide temperature range applications. If this is done, the LM102 will give input currents which are considerably better than can be realized with FET amplifiers over a −55 to 125°C temperature range. In addition the temperature stabilization will greatly reduce the offset voltage drift.

A plot of the frequency response of the LM102 is depicted in Figure 7.7. The low-frequency gain figure corresponds to an open loop gain of about 2000. Although this sounds low for an operational amplifier, it should be remembered that a voltage follower has 100% feedback so the gain error is only 0.05%. Furthermore, because of its better high-frequency response, the LM102 actually has 10 times more gain than either the LM101 or the LM709 at frequencies greater than 10 kHz. The gain of all these amplifiers is equal at 500 Hz.

It is difficult to measure the low-frequency gain of a voltage follower directly because the gain error is so small. It can be accomplished, however, by grounding the input of the amplifier and driving both power supplies simultaneously with the desired input signal. The amplifier error can then be observed directly on the output.

The response of the amplifier at frequencies up to 10 MHz is given in Figure 7.8. With a 10-kΩ source resistance, the bandwidth is nearly 10 MHz. Some peaking is evident, although it is not serious. At higher source resistances, the bandwidth is reduced by the 3-pF input capacitance as shown in the figure.

Feedback amplifiers generally have a full-signal bandwidth which is considerably less than the small-signal bandwidth. The LM102 is no exception. It can only deliver its rated output swing at frequencies less than 60 kHz, as shown in Figure 7.9. There is no standard way of measuring the frequency limited output swing [7], but the criterion used here was that the total harmonic distortion be less than 5%.

The output resistance of the follower is about 1 Ω, as shown in Figure 7.10. This gives a gain error less than 0.01% with load resistances above 10 kΩ. At high frequencies as well as high temperatures, the output resistance increases because the open-loop gain of the amplifier falls off.

7.7 INCREASED OUTPUT SWING

The function of the booster terminal on the output-stage current sink is illustrated in Figure 7.11. By itself the amplifier can only deliver its rated ±10-V output swing into load resistances greater than 5.7 kΩ at 25°C. With heavier loads it will clip in the negative direction. A 300-Ω resistor between pins 5 and 4 extends the drive capability to 2.5 kΩ,

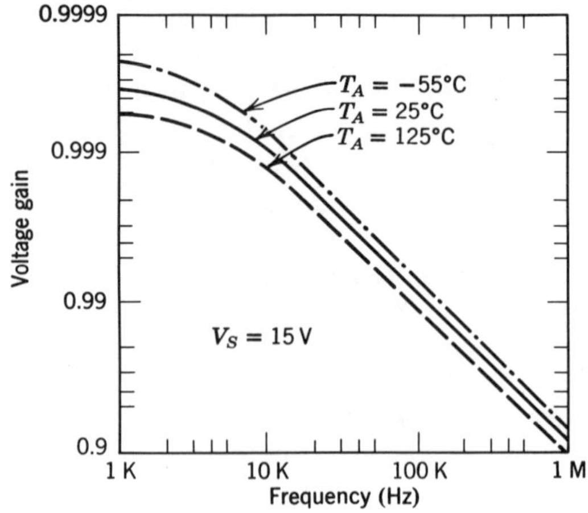

Figure 7.7. Voltage gain at moderate frequencies.

whereas a 100-Ω resistor will enable the amplifier to give a ±10-V swing with 1.4-kΩ loads. The figure also shows the effect of temperature on the drive capability.

It should be remembered that increasing the drive current will increase dissipation in the IC; for example, when the amplifier is set up to drive ±10 V into a 2-kΩ load at 125°C, the worst-case dissipation increase will be 150 mW (for a steady +10-V output with load).

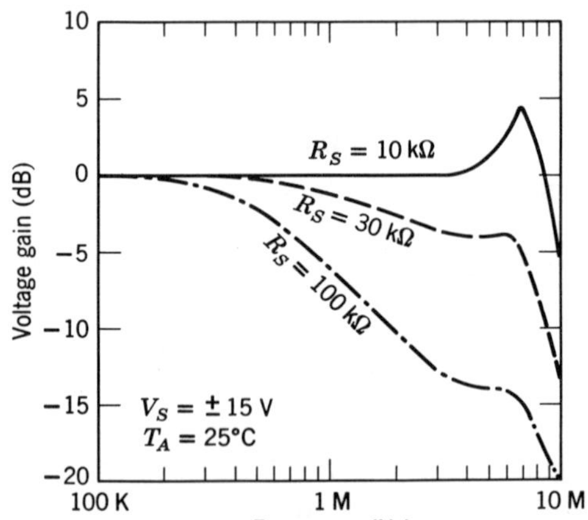

Figure 7.8. High frequency response.

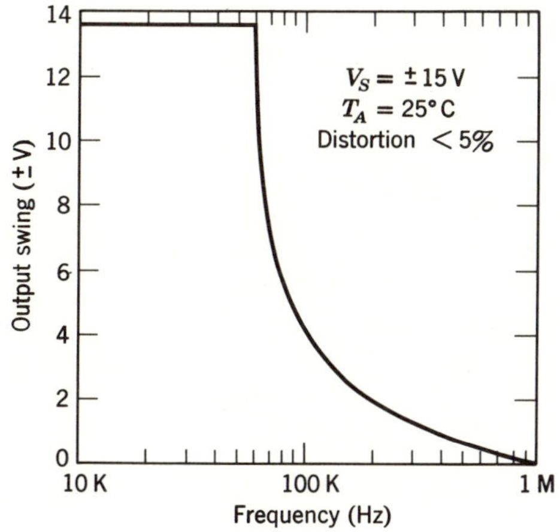

Figure 7.9. Frequency limited output swing.

The current limiting characteristics of the LM102 are shown in Figures 7.12 and 7.13. Figure 7.12, which gives the positive output level as a function of load current, demonstrates the sharpness of the current limiting. The short-circuit current also drops as the chip heats up, reducing power dissipation. The limiting characteristics in the negative direction are illustrated in Figure 7.13. The circuit begins to limit at lower currents because the available current is determined by a fixed-

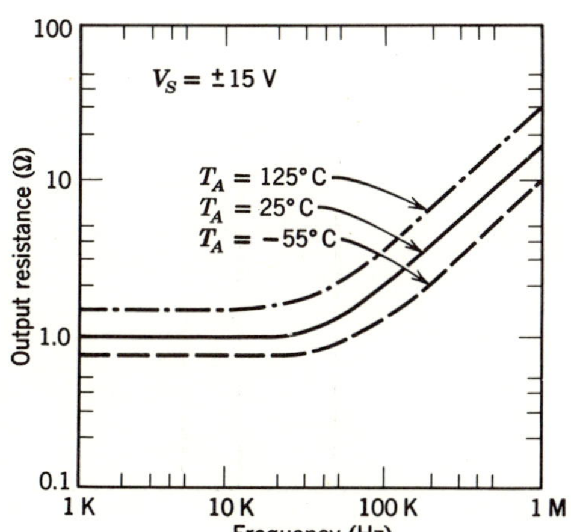

Figure 7.10. Output resistance.

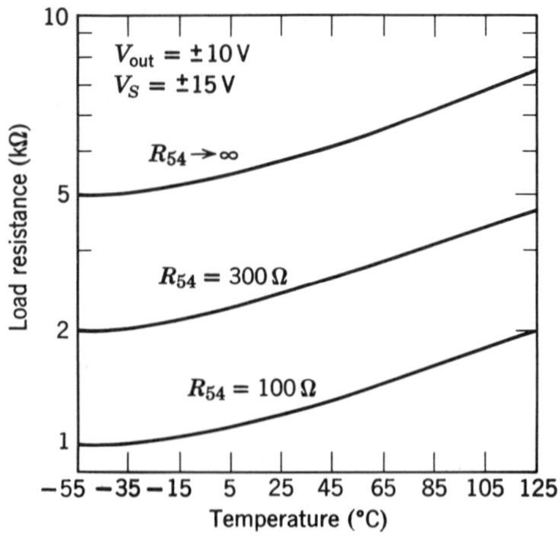

Figure 7.11. Minimum load resistance for rated output swing.

current source. It should be noted that after the output swing first starts to fall off, further increases in load current are supplied by the input through the protective clamp diodes D_3 and Q_{11}.

A plot of the current drain over a −55 to 125°C temperature range is shown in Figure 7.14. The supply current does not increase appreciably over the entire output voltage range, including saturation. It is evident here that fast operation is obtained in the follower without excessive power dissipation.

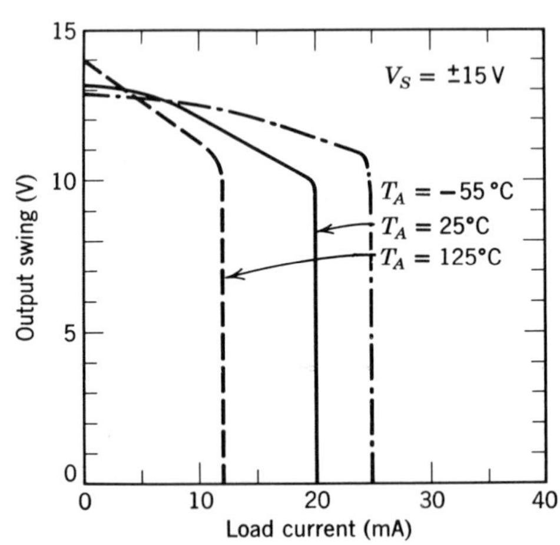

Figure 7.12. Positive current limiting.

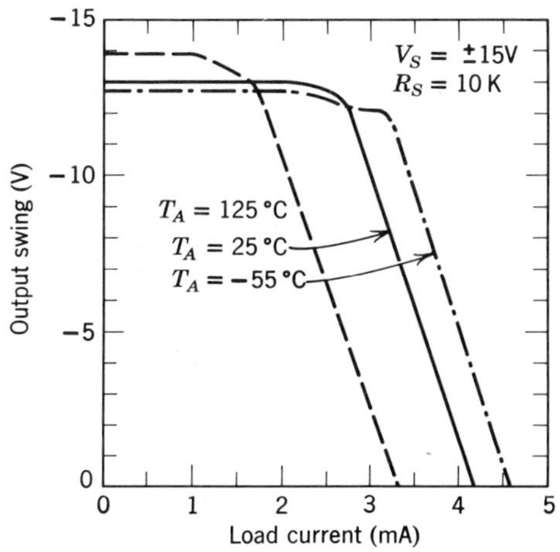

Figure 7.13. Negative current limiting.

7.8 SLEWING

The fast slewing of the follower is demonstrated in Figure 7.15. A fairly large overshoot is evident for positive-going input signals above about 4 V. As shown in the figure, this can be eliminated by using a high-speed clamp diode between the input and the output (with the anode on the input). Although there is an internal clamp diode in this position (D_3 in Figure 7.4), it is of necessity a collector-base diode which stores

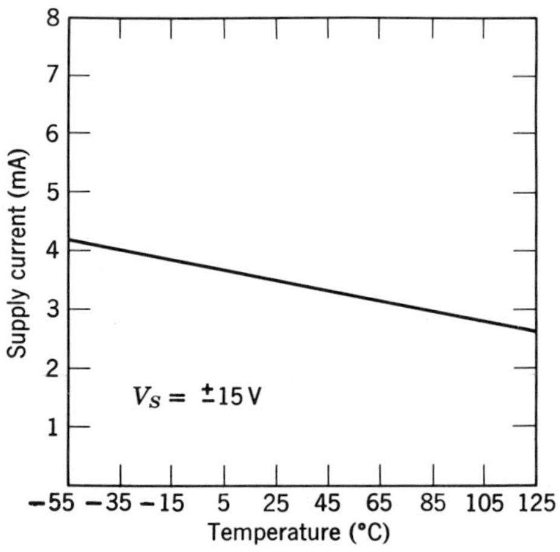

Figure 7.14. Supply current.

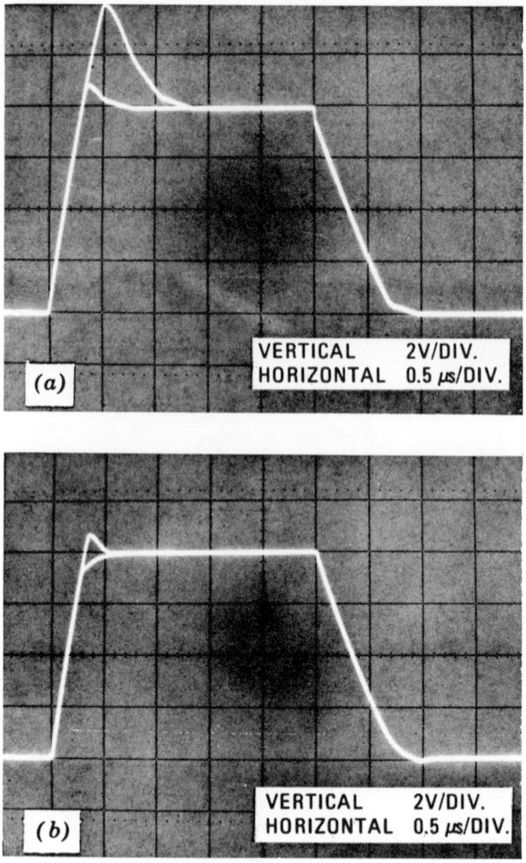

Figure 7.15. Large-signal pulse response with and without a clamp diode for (*a*) 3-kΩ source resistance and (*b*) 30-kΩ source resistance.

excess charge when turned on with input signals that rise faster than the output can follow. This stored charge causes the overshoot.

If the LM102 is driven from source resistances higher than 30 kΩ, the leading edge of the input pulse will always be slowed down enough by the input capacitance that the output can follow the input and the clamp diode is not needed. This is shown in Figure 7.15*b*.

Figure 7.15 demonstrates that the slew rate is about 10 V/μs in the slowest direction, even including the effects of overshoot. But because of its restricted output current swing in the negative direction, the device will not give this slew rate with capacitive loads greater than 100 pF

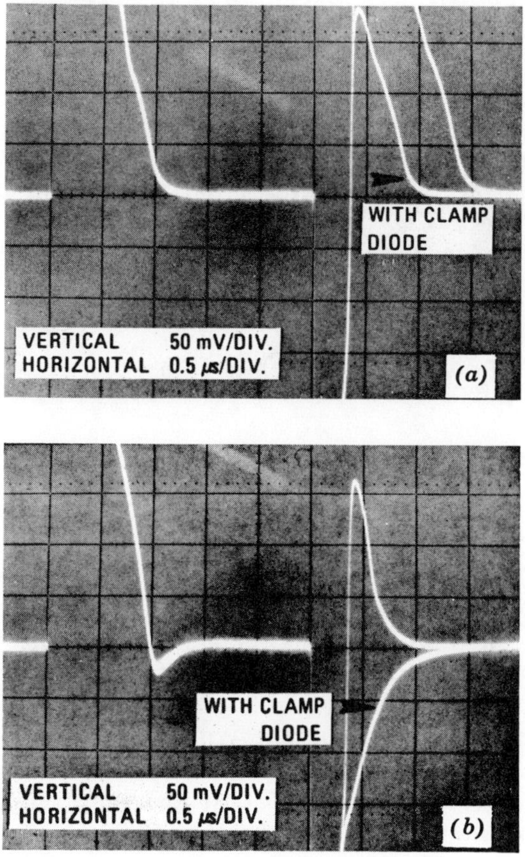

Figure 7.16. Error signal for 8 V input pulse with and without a clamp diode for (*a*) 3-kΩ source resistance and (*b*) 30-kΩ source resistance.

unless the output sink current is increased with an external resistor on the booster terminal.

Figure 7.16 illustrates that the settling time of the LM102, to within 5 mV of its final value, is less than 1.5 μs for an 8-V input pulse. These photographs show the error signal, which is the difference between the input and the output, with a $\pm$4-V rectangular pulse applied.

7.9 STABILITY

The stability of the amplifier under varying conditions of capacitive loading, temperature, and supply bypassing [8] is revealed in Figures

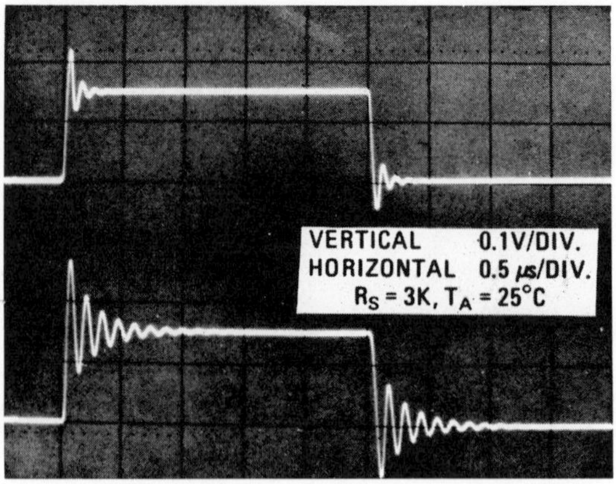

Figure 7.17. Small-signal transient response for $C_L = 10$ pF (top) and $C_L = 200$ pF (bottom).

7.17 through 7.19. Figure 7.17 gives the small-signal transient response with capacitive loading. These pictures were taken with both supplies bypassed to ground with 0.01-μF ceramic capacitors. With loads approaching 200 pF, the circuit tends toward instability. With capacitive loads much above this, it will oscillate, although it will be stable again with more than 0.1 μF on the output. With larger capacitances, however, both the small-signal rise time and the slew rate will be reduced.

The stability over a −55 to 125°C temperature range is depicted in Figure 7.18. Again, the conditions here are 200-pF capacitive load with bypassed supplies.

The effect of unbypassed supplies is demonstrated in Figure 7.19. The response was measured under the same conditions as Figure 7.17, except that there is 16 inches of wire between the device and the bypass capacitors on the power supply. It is evident that the circuit is on the verge of becoming unstable with capacitive loading. This clearly proves the advisability of properly bypassing the supplies on any high-frequency amplifier.

7.10 OPERATING HINTS

A number of precautions concerning the proper use of the LM102 have already been given along with hints on optimizing the performance in certain applications. These are worth repeating here.

1. The output is short circuit protected; however, the input is clamped to the output to prevent excessive voltage from being developed across the input transistors. If the amplifier is driven from low-source impedances, excessive current can flow through these clamp diodes when the output is shorted. This can be prevented by inserting a resistor larger than 3 kΩ in series with the input.
2. The circuit cannot deliver its full slew rate into capacitive loads greater than 100 pF unless more sink current is provided on the output with a resistor between pins 4 and 5.
3. The amplifier may oscillate when operated with capacitive loads between 200 pF and 0.01 μF.
4. As is the case with any high-frequency amplifier, the power-supply leads of the LM102 should be bypassed with capacitors greater than 0.01 μF located as close as possible to the device. This is particularly true if the IC is driving capacitive loads.

An arrangement for getting full output swing into loads less than 8 kΩ is shown in Figure 7.20*a*. The external resistor R_1 should not be less than 100 Ω as this could cause limiting on positive peaks. Figure 7.20*b* illustrates how to connect a potentiometer to balance out the offset voltage. Figure 7.20*c* shows the use of a clamp diode to reduce the overshoot that occurs when the follower is driven with large input pulses having a leading-edge slope greater than 10 V/μs. The diode is only needed, however, when the source resistance is less than 30 kΩ, since the slope

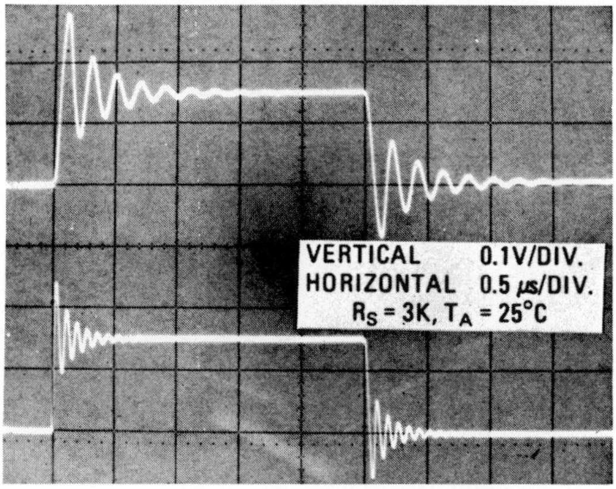

Figure 7.18. Transient response for $C_L = 200$ pF at 125°C (top) and −55°C (bottom).

seen by the amplifier will be reduced by the input capacitance with the higher source resistances.

7.11 APPLICATIONS*

The use of the LM102 in a switch circuit for driving the ladder network in an analog-to-digital converter is shown in Figure 7.21. Simple transistor switches, connected in the reverse mode for low saturation voltage, generate the 0 and 5 V levels for the ladder network. The switch output is buffered by A_2 and A_3 to give a low driving impedance in both the high and low states. The switch transistors can be driven directly from integrated logic circuits. Resistors R_7 and R_8 limit the base drive; the values indicated are for operation with standard TTL or DTL circuits. If necessary, the switching speed can be increased somewhat by bypassing the resistors with 100-pF capacitors.

Even with operation at maximum speed, clamp diodes are not needed on the voltage followers to reduce overshoot. The pullup resistors on the switches R_5 and R_6 can be made large enough so that the LM102 does not see a positive-going input pulse that is much faster than the output slew rate.

The main advantage of this circuit is that it gives much lower output resistance than push-pull switches. Furthermore the drive circuitry for these switches is considerably simpler. The LM102 can also be used as a

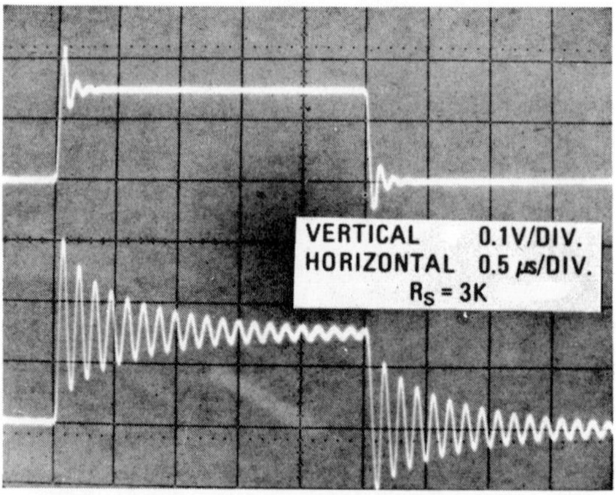

Figure 7.19. Transient response with unbypassed supplies, $C_L = 10$ pF (top) and $C_L = 200$ pF (bottom).

* Other applications are given in Chapter 3.

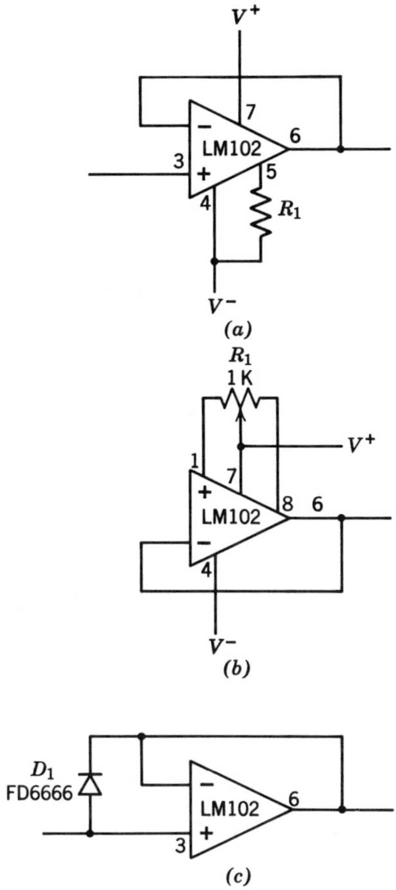

Figure 7.20. Auxiliary circuits for the LM102: (*a*) for increased swing; (*b*) for offset balancing; (*c*) for eliminating overshoot.

buffer for the temperature-compensated voltage reference, as shown in Figure 7.21. The output of the reference diode is divided by a resistive divider, and it can be set to the desired value with R_3.

7.12 ANALOG COMMUTATOR

The low input current and fast slewing of the LM102 make it well suited as a buffer amplifier in high-speed analog commutators. The low input current permits operation with switch resistances even higher than 10 kΩ without affecting the dc stability.

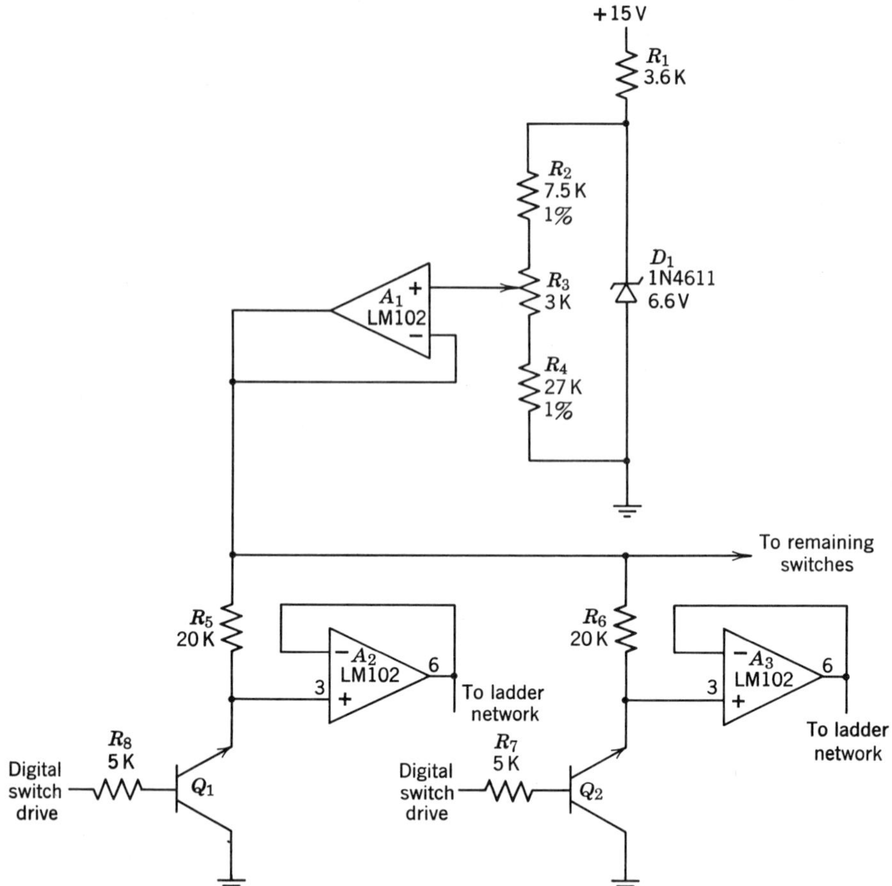

Figure 7.21. Using an LM102 to drive the ladder network in an analog-to-digital converter.

An expandable four-channel analog commutator is shown in Figure 7.22. Two DM7501 dual flip-flops form a four-bit static shift register. The parallel outputs drive DM7800 level translators that convert the TTL logic levels to voltages suitable for driving MOS devices; these voltages are coupled into an MM451 four-channel analog switch. An extra gate on the input of the translator can be used, as shown, to shut off all the analog switches.

In operation, a bit enters the register and cycles through at the clock frequency, turning on each analog switch in sequence. The "clear" input is used to reset the register such that all analog switches are off. The channel capacity can be expanded by connecting registers in series and hooking the output of additional analog switches to the input of the buffer amplifiers.

When the output of a large number of MOS switches are connected together, the capacitance on the output node can become high enough to reduce accuracy at a given operating speed. This problem can be avoided, however, by breaking up the total number of channels, buffering these segments with voltage followers, and then subcommutating them into the A/D converter.

7.13 SAMPLE AND HOLD

Although there are many ways to make a sample-and-hold device, the circuit shown in Figure 7.23 is undoubtedly one of the simplest. When a negative-going sample pulse is applied to the MOS switch, it will turn

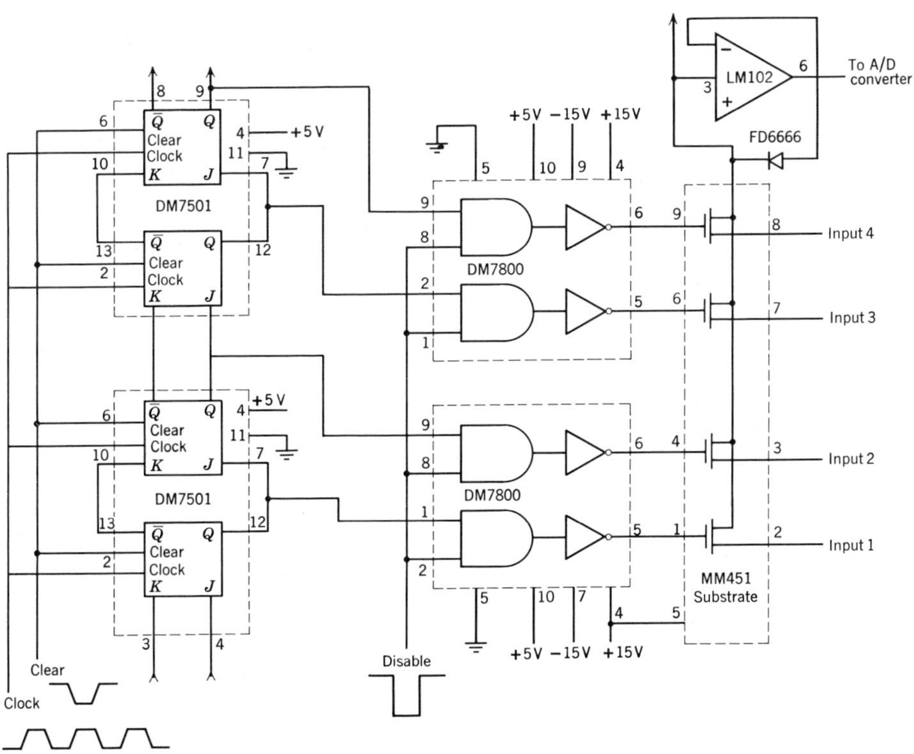

Figure 7.22. Analog commutator with buffered output.

on hard and charge the holding capacitor to the instantaneous value of the input voltage. After the switch is turned off, the capacitor is isolated from any loading by the LM102; and it will hold the voltage impressed upon it.

The maximum input current of the LM102 is 10 nA, so with a 10-μF holding capacitor the drift rate in hold will be less than 1 mV/sec. If accuracies of about 1% or better are required, it is necessary to use a capacitor with polycarbonate, polyethylene, or teflon dielectric. Most other capacitors exhibit a polarization phenomenon [8] which causes the stored voltage to fall off after the sample interval with a time constant of several seconds; for example, if the capacitor is charged from 0 to 5 V during the sample interval, the magnitude of the falloff is about 50 to 100 mV.

7.14 AC AMPLIFIER

The LM102 has a minimum input resistance of 10,000 MΩ, so that for dc amplifier applications this can be completely neglected. With an ac coupled amplifier, however, a biasing resistor must be used to supply the input current. This drastically reduces the input resistance.

Figure 7.24 illustrates a method of bootstrapping the bias resistor to get higher input resistance. Even though a 200-kΩ bias resistor is used for good dc stability, the input resistance is about 12 MΩ at 100 Hz, increasing to 100 MΩ at 1 kHz.

7.15 ACTIVE FILTERS

Active RC filters have been replacing passive LC filters at an ever-increasing rate because of the declining price and smaller size of active components. Figure 7.25 is a low-pass filter which is one of the simplest forms of active filters. The circuit has the filter characteristics of two isolated RC filter sections and also has a buffered, low-impedance output.

The attenuation is roughly 12 dB at twice the cutoff frequency and the ultimate attenuation is 40 dB/decade. A third low-pass RC section can be added on the output of the amplifier for an ultimate attenuation of 60 dB/decade [9], although this means that the output is no longer buffered.

There are two basic designs for this type of filter. One is the Butterworth filter with maximally flat frequency response. For this characteristic the component values are determined from [10]

$$C_1 = \frac{R_1 + R_2}{\sqrt{2} R_1 R_2 \omega_C}$$

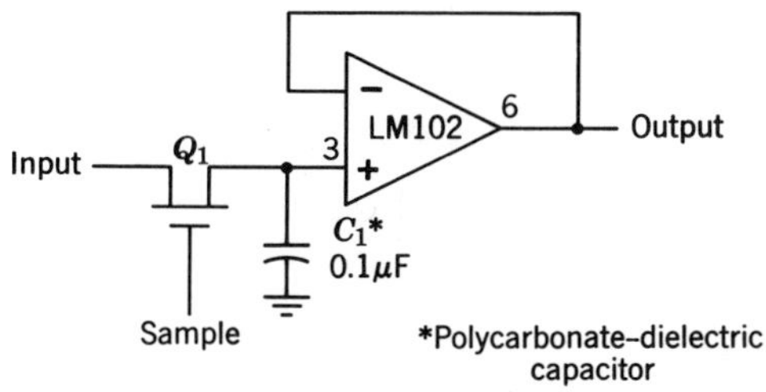

Figure 7.23. Sample-and-hold circuit.

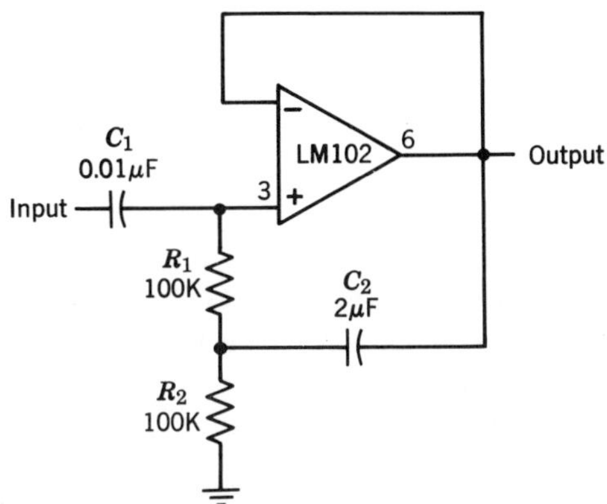

Figure 7.24. High input impedance ac amplifier.

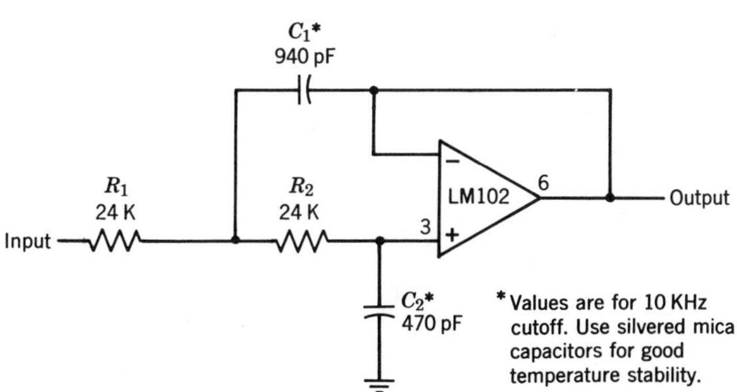

Figure 7.25. Low-pass active filter.

and

$$C_2 = \frac{\sqrt{2}}{(R_1 + R_2)\omega_C}.$$

The second kind is the linear phase filter with minimum settling time for a pulse input. The design equations for this are

$$C_1 = \frac{R_1 + R_2}{\sqrt{3}\, R_1 R_2 \omega_C}$$

and

$$C_2 = \frac{\sqrt{3}}{(R_1 + R_2)\,\omega_C}.$$

Substituting capacitors for resistors and resistors for capacitors in the circuit of Figure 7.25 produces a similar high-pass filter. This circuit is shown in Figure 7.26.

7.16 CONCLUSIONS

The LM102 incorporates transistors which have higher current gain than is available with discrete components. Further, a factor of three-to-five improvement can be expected in the near future. The performance realized challenges that of field-effect transistors, if operation over the military temperature range is considered. This is especially true if the components are included in a temperature-stabilized oven.

Although the circuit described in this chapter is restricted to voltage-follower applications, many of the techniques used here can be applied to general-purpose amplifiers. Even though it is only a voltage follower, the

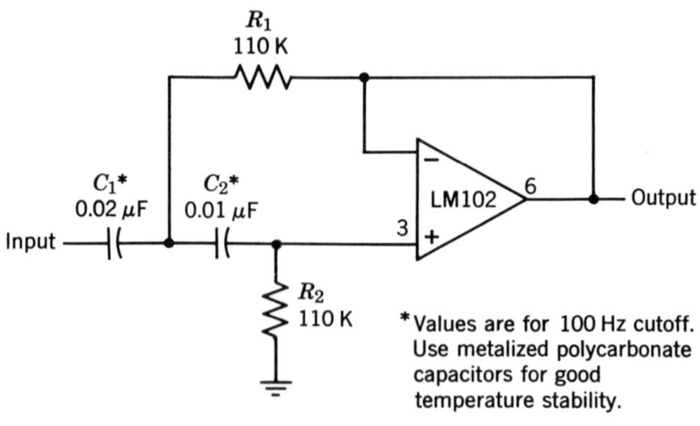

Figure 7.26. High-pass active filter.

LM102 can be used in a wide variety of applications ranging from low drift sample-and-hold circuits to a buffer amplifier for high-speed analog commutators. Its usefulness is enhanced by the fact that it is a plug-in replacement for both the LM101 and the LM709 in voltage-follower applications. The circuit will work in the same socket, unaffected if the compensation components for the other amplifiers are installed or not.

REFERENCES

[1] Widlar, R. J., "A Unique Circuit Design for a High Performance Operational Amplifier Especially Suited to Monolithic Construction," *Proc. NEC,* **XXI**, 85-89 (October 1965).

[2] Widlar, R. J., "Monolithic Op Amp with Simplified Frequency Compensation," *EEE,* **15**, No. 7, 58-63 (July 1967).

[3] Lindmayer, J. and W. Schneider, "Theory of Lateral Transistors," *Solid State Electronics,* **10**, 225-234 (1967).

[4] Widlar, R. J., "Future Trends in Dc Amplifiers," *National Semiconductor Corporation TP-4.*

[5] Widlar, R. J., "Some Circuit Design Techniques for Linear Integrated Circuits," *IEEE Trans. on Circuit Theory,* **CT-12**, No. 4, 586-590 (December 1965).

[6] Lin, H. C., T. B. Tan, G. Y. Chang, B. Van Der Leest, and N. Formigoni, "Lateral Complementary Transistor Structure for the Simultaneous Fabrication of Functional Blocks," *Proc. of the IEEE,* 1491-1495 (December 1964).

[7] Stata, Ray, "User's Guide to Applying and Measuring Operational Amplifier Specifications," *Analog Dialogue,* **1**, No. 3, 1-8 (September 1967).

[8] Dow, Paul C., Jr., "An Analysis of Certain Errors in Electronic Differential Analyzers, II—Capacitor Dielectric Absorption," *IRE Trans. on Electronic Computers,* 17-22 (March 1958).

[9] Scott, L., "Criteria for the Design of Active Filters Using Resistance and Capacitance Elements in Feedback Circuits," *Solid State Electronics,* **9**, 641-651 (1966).

[10] Melsheimer, R. S., "If You Need Active Filters," *Electronic Design,* 78-82 (April 12, 1967).

ACKNOWLEDGMENTS

The author acknowledges the contributions of Dave Talbert in developing and establishing the manufacturing processes for this integrated circuit. In addition, invaluable contributions were made by Mineo Yamatake in the design, development and evaluation of the device.

8

Medium-Power Operational-Amplifier ICs

by H. A. Wittlinger
RCA Electronic Components

8.1 INTRODUCTION

Power outputs in excess of 250 mW into a 500-Ω load resistance with harmonic distortion of less than 0.2% are provided by the CA3033 and CA3033A. The typical input impedance is 1 MΩ and the voltage gain is at least 90 dB. Offset voltage is less than 5 mV and offset current is typically 9 nA. Input bias current is typically 103 nA. These features make these amplifiers especially suitable for systems in which an operational amplifier and power amplifier or driver were formerly required.

8.2 CIRCUIT DESCRIPTION

The RCA CA3033 and CA3033A consist of two differential-amplifier stages followed by a class-B output stage, as shown in the schematic diagram of Figure 8.1. (The two circuits are designed to permit operation from dual 12- and 18-V dc power supplies, respectively.) Emitter-follower inputs provide the exceptionally high input impedance and low bias current. An additional advantage of the emitter-follower inputs is that the Miller capacitance of the differential amplifier is substantially reduced and the input capacitance of the amplifier is lower than if a similar single-transistor configuration were used.

The output of the first differential-amplifier stage (Q_3 and Q_4) is buffered from the input of the next differential-amplifier stage (Q_6 and Q_7) by emitter-followers Q_{19} and Q_{20}. This arrangement reduces the input-loading effects on the first stage and therefore maintains the first-stage gain.

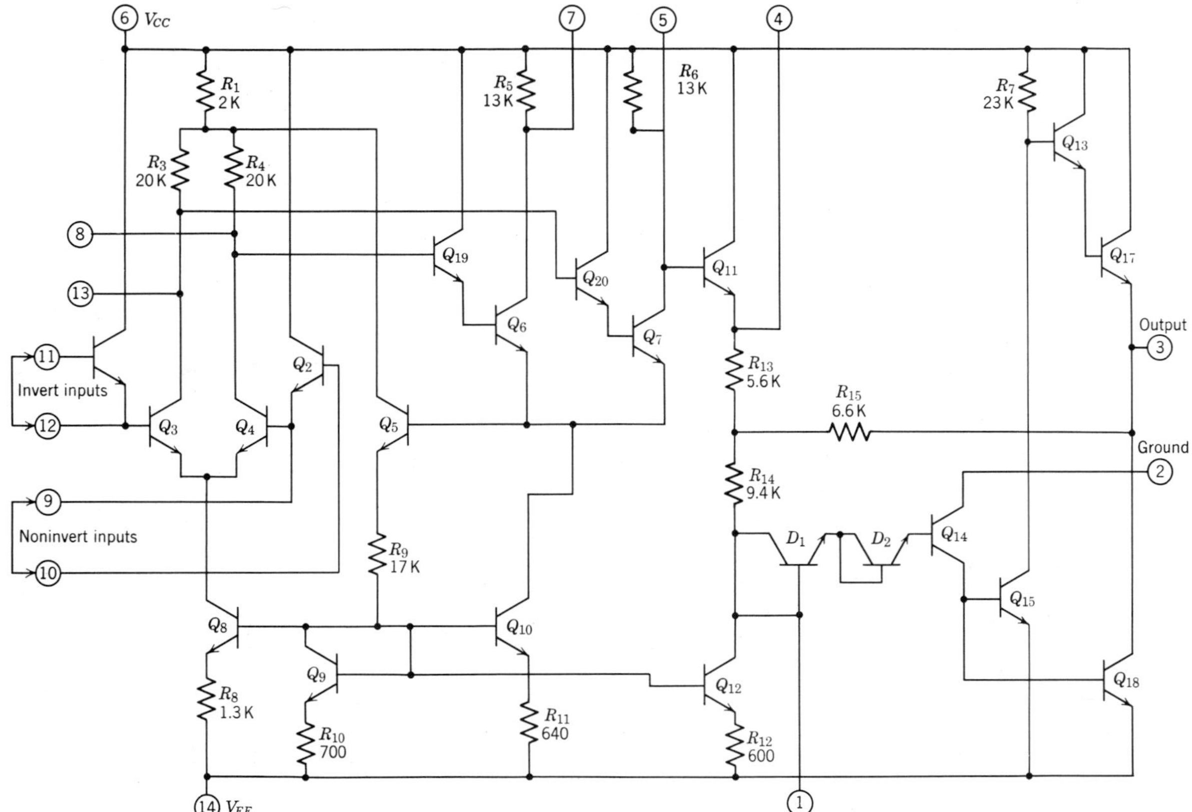

Figure 8.1. Schematic diagram of a CA3033 or CA3033A operational amplifier.

A circuit is incorporated into this design to sense any change in the operating point of the first differential amplifier caused by variations in either the positive or the negative supply voltage. Any changes in the supply voltages are reflected to the base of transistor Q_5, which detects changes in the collector voltage of the first differential amplifier and compensates for them; for example, a rise in the voltage at the emitters of Q_6 and Q_7 increases the bias voltage to Q_5 and thus increases the collector current, countering the apparent rise in the collector voltage of either Q_3 or Q_4. At the same time the emitter current of Q_5 also increases, and increases the voltage drop across the diode-connected transistor Q_9 and resistor R_{10} to increase the collector current of Q_8. Thus any apparent increase in the collector voltage of the first differential stage causes a correction both at the constant-current source Q_8 and at the collector supply voltage through R_1, the common load resistor for Q_3, Q_4, and Q_5.

An emitter-follower Q_{11} buffers the output of the second differential-amplifier stage and drives the divider and summing network to the output stage. Resistor R_{13} may be considered the input resistance of an amplifier to the summing point, the junction of R_{13}, R_{14}, and R_{15}. Resistor R_{14} shifts the operating point of the output stage with little attenuation of the signal as a result of the high collector impedance of the constant-current source Q_{12}.

Diodes D_1 and D_2 provide further dc shifting of the signal to the base of the emitter-follower Q_{14}. This emitter-follower provides further level shifting and a low driving impedance to transistors Q_{15} and Q_{18}.

The excellent matching of the base-to-emitter voltage of the integrated-circuit transistors makes it possible to establish the idling current of the output stage accurately. Because the collector-current characteristics of Q_{15} and Q_{18} as a function of base-to-emitter voltage are matched, the collector current in Q_{15} determines the idling current in Q_{18}. For example, if the operating current of Q_{15} is set at 1 mA for a given base-to-emitter voltage, the operating current of Q_{18} is also 1 mA because the base-to-emitter voltages of both transistors are the same. This type of design results from the excellent transistor matching that is possible with monolithic processing.

8.3 NOISE FIGURE

Figure 8.2 shows noise figure as a function of source resistance for the CA3033A. The curve shows that the optimum noise match occurs for input impedances between 100,000 Ω and 1 MΩ. An improvement in

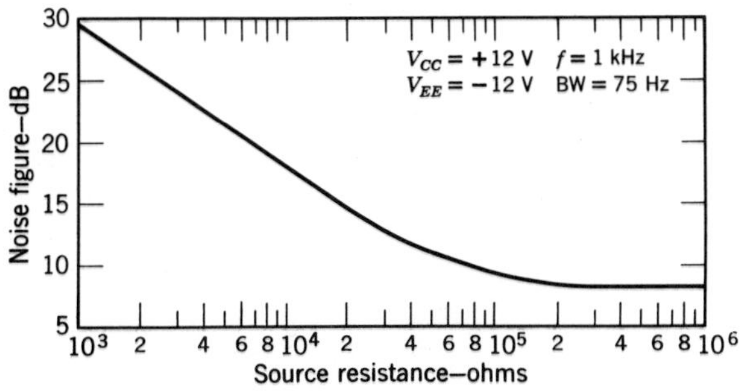

Figure 8.2. Noise-figure curve for CA3033A.

noise figure at a frequency of 1 kHz and an input impedance of 1000 Ω may be obtained by addition of two 0.3-MΩ resistors from the emitters to the V_{EE} supply to increase the operating current of the input transistors.

8.4 COMPENSATION

Basic phase compensation of the CA3033 and CA3033A amplifiers is quite simple. Figure 8.3 shows typical phase-compensation characteristics for the CA3033; data for the CA3033A are similar. The two compensating capacitors, C_x and C_y, are connected from the collectors of the first differential amplifier (terminals 8 and 13) to ground. When capacitance values greater than 0.1 μF are used, however, a lower-voltage capacitor that has a value equal to half that given on the curves may be connected between terminals 8 and 13, and a 0.001-μF capacitor connected from either terminal 8 or 13 to ground. This arrangement provides the same gain-phase rolloff shown on the curves and permits the use of lower-voltage ceramic disc capacitors now available. For linear operation the maximum expected difference voltage between the two collectors is less than 1 V.

The dashed lines in Figure 8.3 illustrate the use of the curves for design of a 60-dB amplifier. First, the intersection of the various gain-frequency curves is followed out to the curve for a capacitor value of 0.001 μF. At this point the expected 3-dB amplifier response is approximately 230 kHz and the phase angle ϕ is 118°; the phase margin (180° − ϕ) is thus 62°. For a capacitance value of 300 pF, the expected 3-dB response is 580 kHz, but the phase angle is 175°; the resulting phase margin of only 5° is a most undesirable situation. In the other direction,

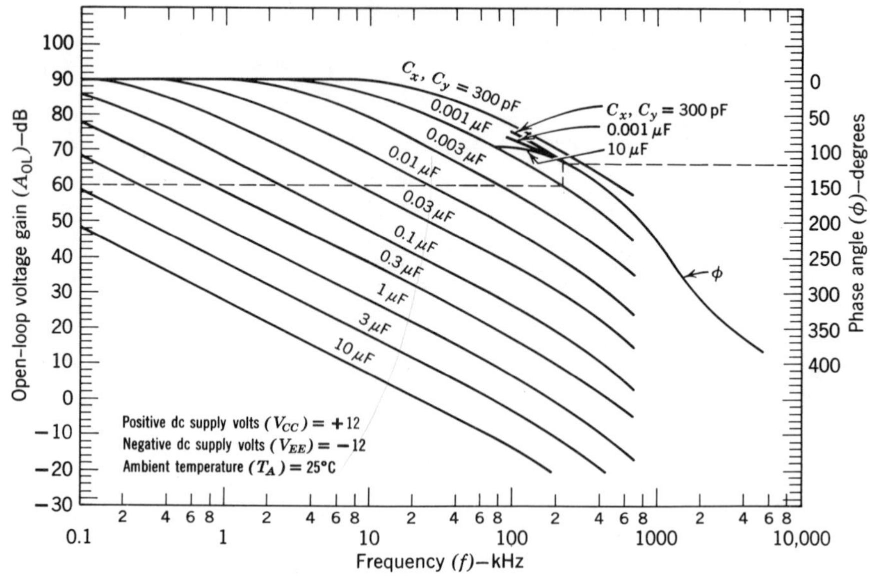

Figure 8.3. Phase-compensation characteristics for CA3033 or CA3033A.

the use of 0.003-μF compensating capacitors provides a 3-dB response of 90 kHz and a phase angle of approximately 90°, with a phase margin of 90°.

8.5 SLEWING RATE

One of the most important considerations in using a high-power operational amplifier such as the CA3033 or CA3033A, is the maximum full power output swing in terms of frequency that can be expected from the unit. This characteristic is important both in low-frequency applications and in pulse applications in which the rise times of a given design must be known. Figure 8.4 shows the maximum rate of change, or slewing rate, of a sine wave in V/μs for various peak-to-peak output voltages [1]. For example, a 30-V peak-to-peak sine wave at a frequency of 10 kHz has a maximum rate of change of 1 V/μs. Therefore, besides being able to swing the value of 30 V peak-to-peak at direct current, an amplifier must also be able to swing this amplitude at a minimum rate of 1 V/μs.

Slewing rate is a function of the phase-compensation circuit, the operational-amplifier design in terms of the gain after phase compensation, and the output-stage design. The phase-compensation circuit is usually placed around the input stages of an operational amplifier. This approach offers two advantages. First, it reduces the amplitude of any higher-frequency components that may overload the following stages (e.g., the residual rf carrier remaining on the output of a video detector with a single RC rolloff could cause serious overload and a distorted output if it were allowed to continue through the amplifier). Second, because of the relatively high gain that follows the compensating circuit in the input stage, the compensating capacitors charge and discharge on a smaller and faster portion of the RC time constant associated with the collector load resistors and compensation capacitors.

Another consideration that influences the slewing rate is the signal-handling capability of the output stage. It is evident that, regardless of how fast the first portion of an amplifier responds to a step input, the output stage can limit the rate of rise and fall time.

A curve of the maximum full-power-output frequency of the CA3033 and CA3033A as a function of the phase-compensation capacitance is given in Figure 8.5. This curve may be used with the curve of Figure 8.4 to determine the amplifier slewing rate. In the case of the 60-dB amplifier shown in Figure 8.6, for example, two 0.002-μF capacitors are used for phase compensation. The curves of Figure 8.3 indicate that the 3-dB bandwidth is 120 kHz. Figure 8.5 shows that full output may be expected up to a frequency of 280 kHz. Therefore the design is capable of full-power output up to the 3-dB point.

A similar approach may be used for a pulse amplifier. Two 0.01-μF phase-compensating capacitors are used to yield a 3-dB response of 30 kHz. The expected rise time is 11.7 μs [2]. The power-output curve of

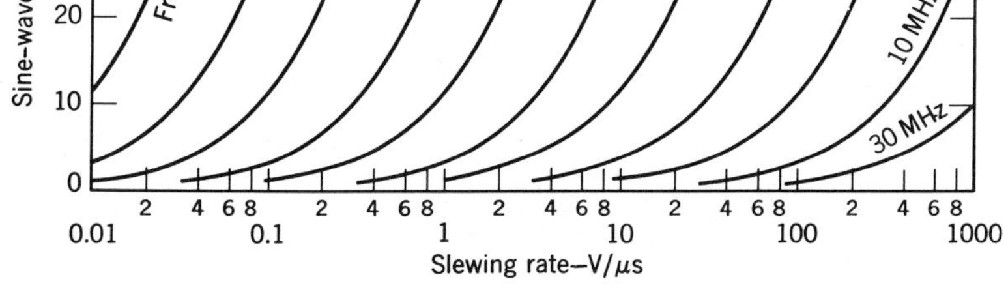

Figure 8.4. Slewing-rate curve.

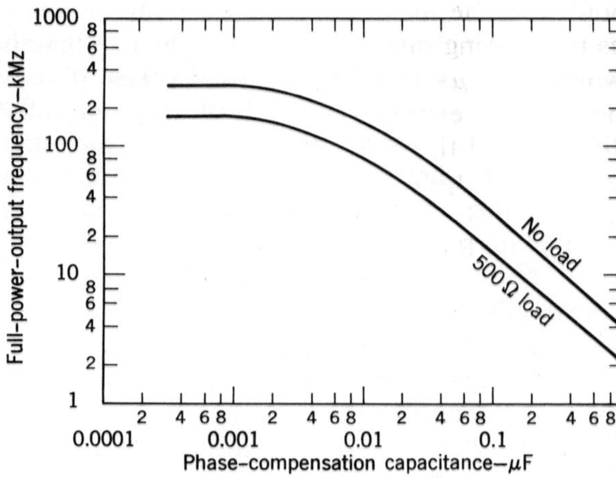

Figure 8.5. Frequency for full power output as a function of phase-compensating capacitance.

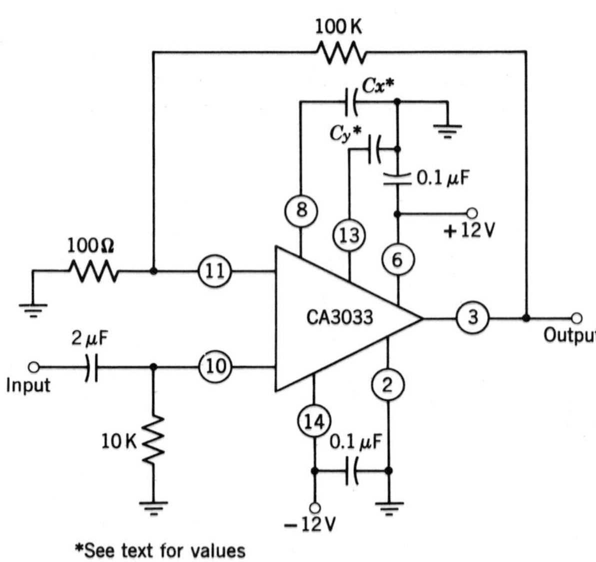

Figure 8.6. 60-dB test amplifier for CA3033.

Figure 8.5 shows that the maximum frequency for full power output is 155 kHz; thus the slewing rate is 10.6 V/μs. The rise time based on this slewing rate would be (1 μs/10.6 V) $\times$ 20 V, or 1.9 μs. Because this value is less than the rise time estimated from the 3-dB point, the design is not slewing-rate limited, and the 3-dB rise time will be met. Thus for maximum high-frequency output the lowest value of phase-compensation capacitors must be used; therefore high closed-loop gains are implicit. Waveforms for the 60-dB amplifier are shown in Figure 8.7.

8.6 APPLICATIONS

Figure 8.8 illustrates the use of the CA3033A in a 20-dB, 255-mW power amplifier operating from a single 36-V supply. Figure 8.9 shows the pulse response of this amplifier under no-load conditions and with a resistive load of 500 Ω; also shown are curves of distortion as a function of frequency. The waveforms show that the pulse response of the amplifier is limited by slewing rate rather than frequency response. Some crossover distortion is evident in the response for the 500-Ω load.

The impedance of the feedback network can have a significant effect on the pulse response of a given amplifier design, particularly when higher-frequency performance is required. The response is influenced by the stray capacitance of the associated wiring, the shunt capacitance of the feedback resistors and the input impedance of the operational amplifier. Because the CA3033 and CA3033A have higher input impedance as a result of the emitter-follower inputs, the input capacitive loading is reduced considerably and higher-impedance feedback networks can be used.

A simple system was fabricated to demonstrate the ease with which the CA3033 and CA3033A operational amplifiers may be applied. The basis of the system was the rudiments of a digital voltmeter, using the linear staircase approach without the associated totalizing circuitry.

A block diagram of the system, together with the waveforms at all interfaces, is shown in Figure 8.10. A squelchable integrated-circuit multivibrator is used as the clock. As described later, the clock frequency is independent of supply voltage. Although this independence is not a necessary condition for circuit operation, it is inherent in this type of operational-amplifier multivibrator circuit and may be considered an integrated-circuit bonus.

The output from the clock drives a linear staircase generator. Input voltage to this circuit is applied directly from the multivibrator to minimize the effects of diode temperature coefficients. The output from the staircase generator is applied to a comparator that compares the staircase with the voltage to be measured.

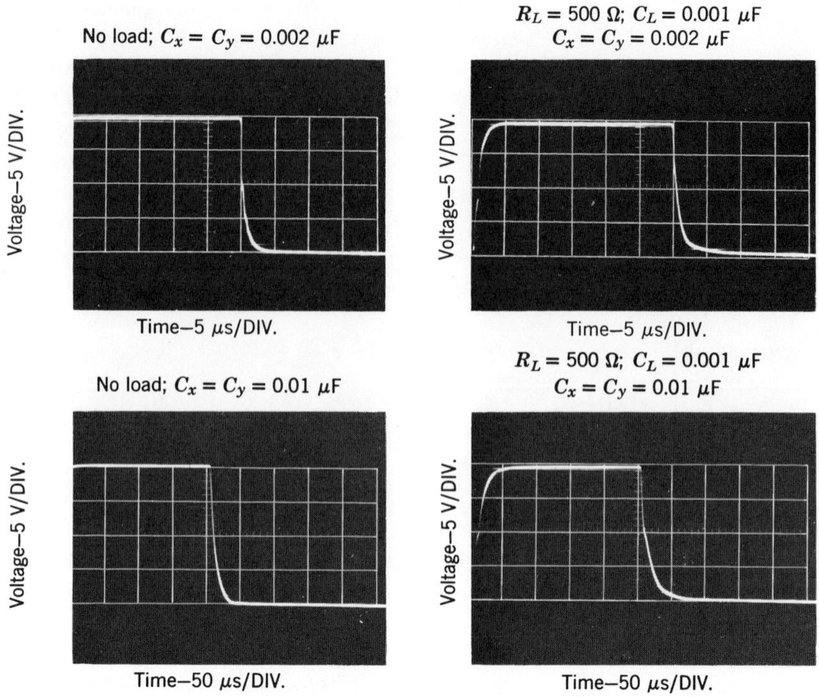

Figure 8.7. Waveforms for the 60-dB amplifier shown in Figure 8.6.

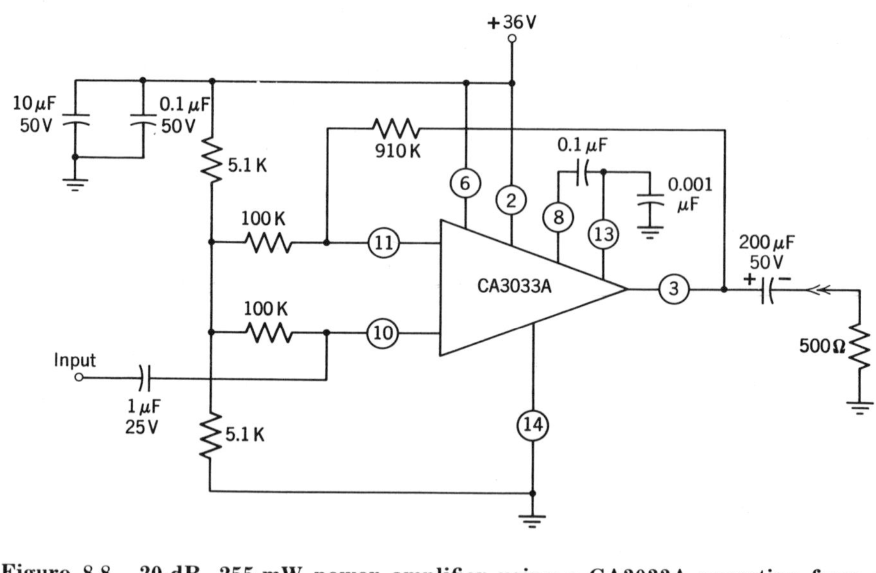

Figure 8.8. 20-dB, 255-mW power amplifier using a CA3033A operating from a 36-V supply.

The comparator output fires a monostable multivibrator that controls the display time, which is variable from about 100 ms to 1 s. An inverter following the multivibrator supplies drive of the correct polarity to the integrator capacitor-discharge switch and the multivibrator squelch circuit.

Circuit operation begins with the staircase generator ramp running down until its voltage is equal to the unknown voltage on the input of the comparator. When the two voltages are equal, the monostable multivibrator is triggered by the output from the comparator. The output of the monostable multivibrator squelches the astable multivibrator and discharges the integrating capacitor through the bistable multivibrator. The wait, or display time, set by the monostable multivibrator allows sufficient time for full discharge of the integration capacitor and appears as a steady reading on the display device.

8.7 ASTABLE MULTIVIBRATOR

A schematic diagram of a squelchable multivibrator is shown in Figure 8.11. The only requirement that must be met by the squelch circuit is that the amplitude of the output pulse not change as a result of the squelch operation; otherwise, the amplitude of the final steps would be different from that of the initial steps and staircase linearity would suffer.

Freedom of the circuit from supply-voltage variations results from the excellent saturation characteristics of the integrated circuit at both positive and negative output swings. The positive and negative thresholds of the circuit are given by the following:

$$\text{positive threshold} = (V_{CC} - V_{CE+\text{sat}})\frac{R_1}{R_1+R_2},$$

$$\text{negative threshold} = (V_{EE} + V_{CE-\text{sat}})\frac{R_1}{R_1+R_2},$$

where V_{CC} and V_{EE} are the positive and negative supply voltages and $V_{CE+\text{sat}}$ and $V_{CE-\text{sat}}$ are the positive and negative saturation voltages of the amplifier. Because these saturation voltages are low and have temperature coefficients less than 5 mV/°C, they can be neglected for ease of computation. If the charging current is considerably greater than the base current, the frequency f may be expressed as follows [3]:

$$f \simeq \left[2RC \ln\left(\frac{2R_1}{R_2}+1\right)\right]^{-1}.$$

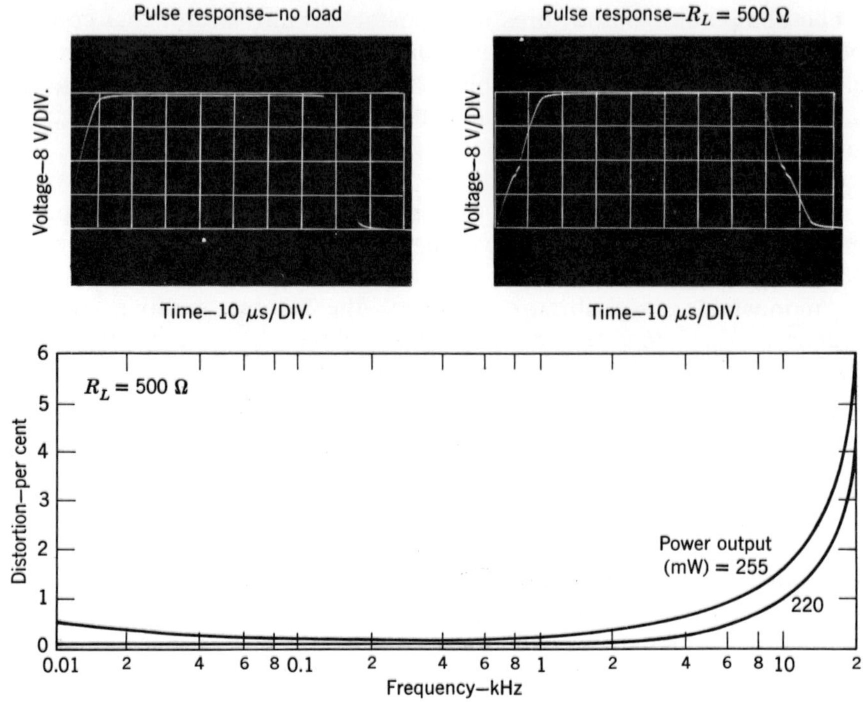

Figure 8.9. Pulse-response waveforms and distortion curves for amplifier of Figure 8.8.

8.8 LINEAR STAIRCASE GENERATOR

The design of a linear staircase generator is nearly identical to that of a linear ramp or sawtooth generator. Figure 8.12*a* shows a linear ramp grounded and a switch S_1 returns the output to the inverting input. When S_1 is closed, the amplifier is in the unity-gain configuration and the output is at ground less the input offset voltage. When S_1 is opened, the output moves in the positive direction when the reference voltage V_{ref} is negative, or in the negative direction when V_{ref} is positive. Because the output under closed-loop conditions tries to maintain the input terminal at zero voltage, the charging current to the capacitor is constant at a rate of $dV/dt = i/C$, where $i = V_{ref}/R$. It is apparent that this analysis is valid as long as the input current to the first stage is considerably less than the charging current.

Figure 8.12*b* shows the circuit for a linear staircase generator. In this circuit a pulse of amplitude *e* couples a charge *Q* to the amplifier input.

The charge Q is equal to $C_1(e-2V_{ak})$, where $2V_{ak}$ is the forward voltage drop across the two diodes. Again, if the amplifier input current is small compared to the effective charging current, capacitor C_2 is incrementally charged in steps of $(e-2V_{ak})C_1/C_2$. If the pulse height is made sufficiently large compared to the expected temperature variations of the diodes, a reasonable degree of temperature independence results.

8.9 COMPARATOR

Regeneration is added around the comparator circuit in this system to accelerate the transition time when the two input voltages are equal. Figure 8.13 shows a complete schematic diagram of the entire system. Waveforms at critical points are shown in Figure 8.14. The 470-pF

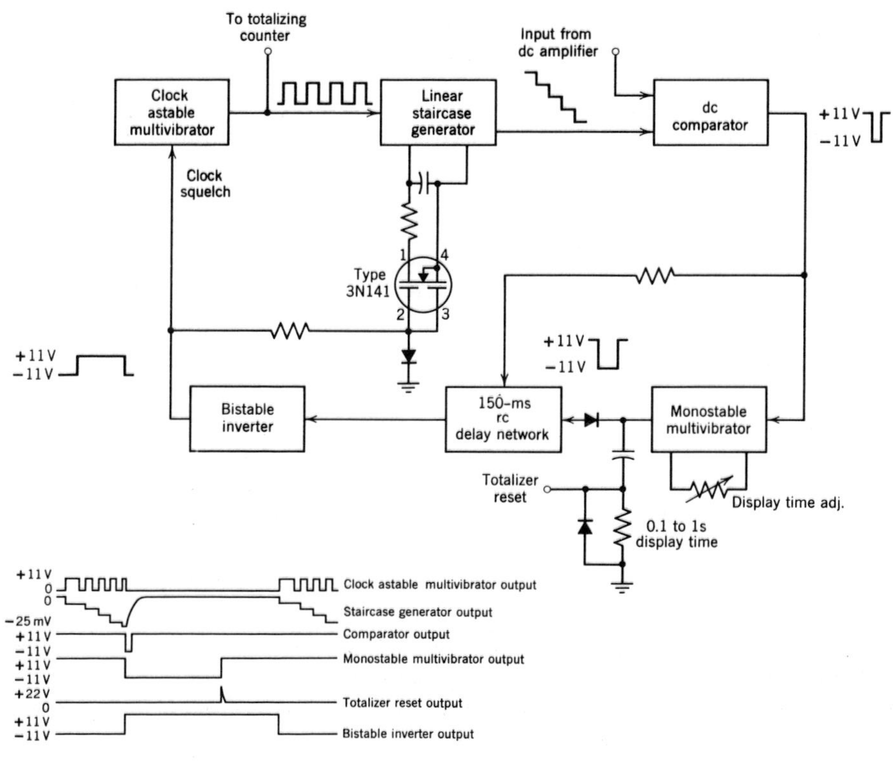

Figure 8.10. Block diagram and waveforms of digital voltmeter system using 5 CA3033s.

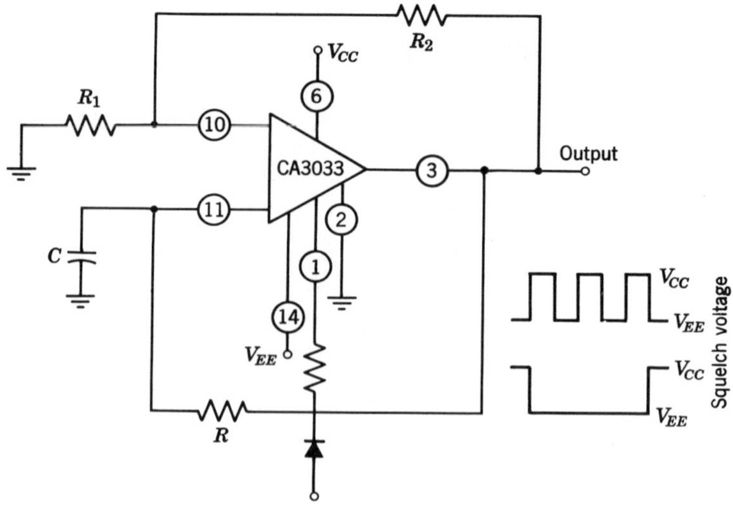

$$f \approx \frac{1}{2RC \ln\left(\frac{2R_1}{R_2} + 1\right)}$$

Figure 8.11. CA3033 squelchable astable multivibrator.

capacitor and 1000-Ω resistor between terminals 3 and 10 of the CA3033 in the comparator circuit provide the regeneration. Two 0.001-μF capacitors on each input filter any externally generated noise.

8.10 MONOSTABLE MULTIVIBRATOR

The monostable multivibrator circuit shown in Figure 8.13 makes use of the extremely low input bias currents of the CA3033 by using small timing capacitors and large timing resistors to obtain the long (one-second) display time. The small timing capacitor in conjunction with the diode D_3 permits the fast recycling time necessary for the one-step case. Triggering of the circuit is accomplished by the 470-pF capacitor to the noninverting input (terminal 10).

An RC timing network is incorporated at the output of the monostable multivibrator to add an additional 150-ms delay that performs two functions. First, the added delay allows more time to complete the timing-capacitor recycling mentioned above. The second and more important function of this network is the generation of a reset pulse at the trailing edge of the monostable multivibrator. Diode D_4 couples the negative output of the multivibrator to the next stage and rapidly charges the 0.22-μF network timing capacitor through the 470-Ω resistor. After the monostable multivibrator completes the cycle, diode D_4 disconnects, and the 0.22-μF capacitor charges to the 1.1-V switching threshold of the next stage through the two 1-MΩ resistors. This cycle is approximately 150 ms.

8.11 BISTABLE MULTIVIBRATOR

The output from the monostable multivibrator is coupled to the next stage, a bistable multivibrator or inverter. The switching thresholds are determined by the 100,000- and 10,000-Ω resistors in the positive feedback loop. The primary function of this stage is to invert the signals from both the comparator and the monostable multivibrator to drive the clock-astable-multivibrator squelch circuit and the staircase capacitor-discharge switch, an RCA-3N141 dual-gate MOS field-effect transistor. Diode D_6 protects the gates of the MOS transistor; the 47-pF capacitor reduces the rise time of the negative-going output of the bistable multivibrator and prevents it from coupling into the beginning of staircase and obscuring the first few steps by this negative transition.

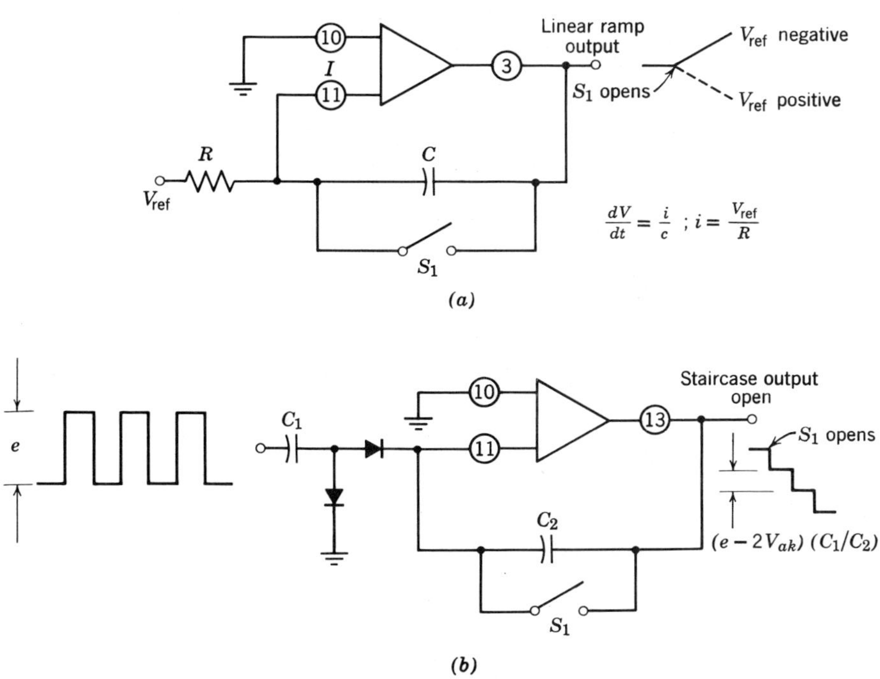

Figure 8.12. Diagrams of (*a*) linear ramp generator and (*b*) linear staircase generator.

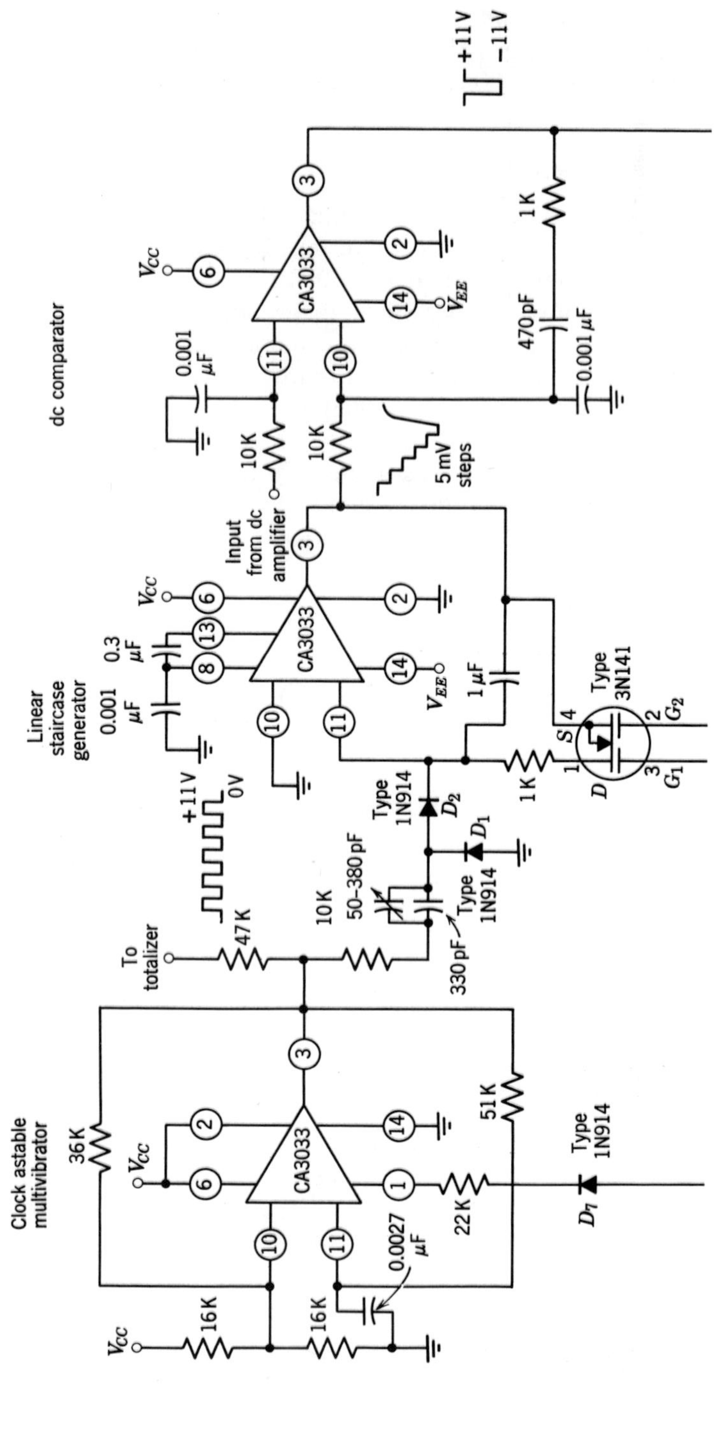

Clock astable multivibrator
Linear staircase generator
dc comparator
CA3033
CA3033
CA3033
V_{CC}
V_{EE}
36K
16K
16K
0.0027 μF
22K
51K
Type 1N914
D_7
To totalizer
47K
10K
50–380 pF
330 pF
Type 1N914
D_1
Type 1N914
D_2
+11 V
0V
0.001 μF
0.3 μF
1 μF
1 K
Type 3N141
D
S
G_1
G_2
Input from dc amplifier
10K
10K
0.001 μF
5 mV steps
470 pF
0.001 μF
1 K
+11 V
−11 V

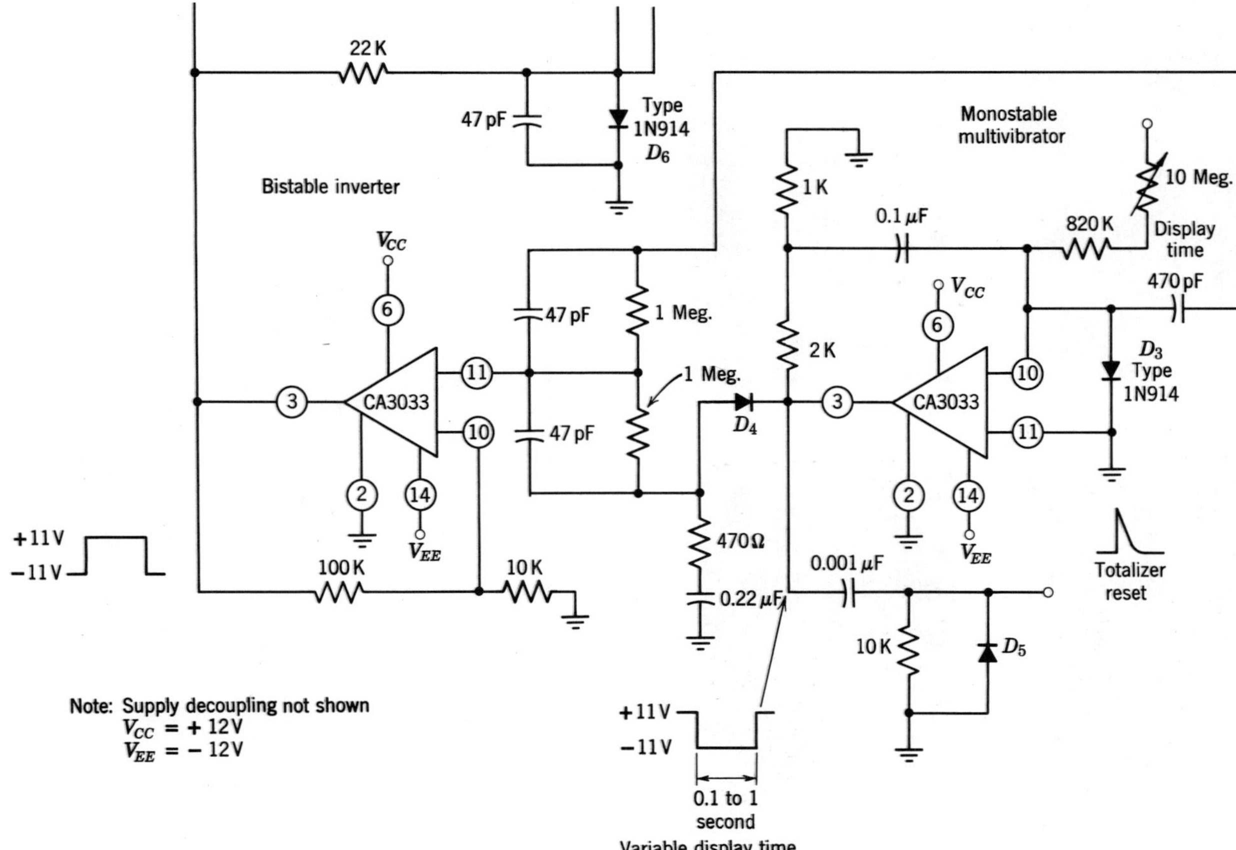

Figure 8.13. Schematic diagram of a digital voltmeter using CA3033s.

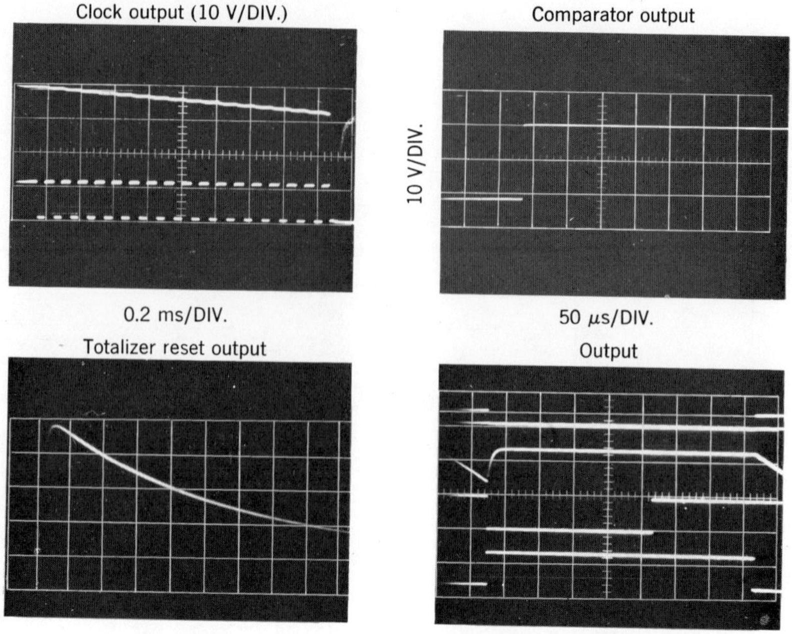

Figure 8.14. Waveforms for the digital voltmeter shown in Figure 8.13.

REFERENCES

[1] Stata, R., "Operational Amplifiers," *Electromechanical Design*, 51 (September 1965).

[2] Valley, G. E., Jr., and H. Wallman, "Vacuum-Tube Amplifiers," Massachusetts Institute of Technology Radiation Laboratory Series, 18, McGraw-Hill, New York, 1948, P. 80.

[3] Engineering Staff of Philbrick Researches, "Philbrick Applications Manual," Second Edition, 72 (June 1966).

ACKNOWLEDGMENTS

A. J. Leidich worked on the design of the CA3033 and CA3033A and the noise-figure curve; J. Klinger breadboarded and evaluated the circuits mentioned.

9

AM and AM/FM Receiver Applications

by J. F. Bogusz, P. E. Hermann and L. H. Hoke, Jr.
Philco-Ford Corp., Microelectronics Division

9.1 INTRODUCTION

Penetration of the consumer-product industry by integrated circuits has been slow in comparison to the acceptance of ICs in other markets. There are several reasons for this very gradual transition to the use of integrated circuitry. The performance/cost ratio is undoubtedly the most important. The transition from vacuum tubes to transistors was relatively easy once it was demonstrated that equal or superior performance could be accomplished at equal or less cost. In the case of battery-operated portable products, the customer was enthusiastic because of lower battery replacement cost and considerably reduced weight and volume.

The advantages of reducing power consumption and overall size and weight through the use of integrated circuits are not as compelling in some portions of the consumer product industry, particularly in the AM and AM/FM receiver field. In this case it is again primarily a matter of economics, the performance/cost factor, that is retarding the widespread use of ICs. Only by reducing cost while maintaining performance can ICs hope to compete with, and replace, the discrete component circuits.

There are several ways to bring the cost of an IC down. As is well known, the cost of the IC chip tends to decrease with decreasing chip size, provided, of course, that the circuit functions remain unchanged. Furthermore, the greater the production volume of any given type of IC chip, the lower its cost will be.

The IC designer is therefore faced with the task of determining the best way to partition electronic circuitry to achieve optimum cost per circuit function. This decision is tempered by the capabilities of the circuit devices obtained by any particular wafer processing technique, the possibility of utilizing the final integrated circuit in several different applications to increase the production volume, and the requirements placed upon the IC by the various applications.

This chapter describes several ICs which were developed for use in AM and AM/FM battery-operated portable receivers. In designing the ICs, special attention was given to the partitioning of the receivers and to the assignment of circuit functions to the various ICs to provide the best solution to the cost/performance/volume dilemma.

In this chapter the basic circuits are described first, stressing simplicity of design, designing to meet performance specifications over wide-supply voltage variation, designing to be compatible with IC processing capabilities, and designing to minimize the use of external components.

The remainder of the chapter is devoted to the application of ICs to two specific types of broadcast receivers. The first is a standard AM broadcast receiver in which two IC chips are used to perform all of the active circuit functions from the rf stage to the audio-output stage. The second is an AM/FM receiver in which three ICs accomplish the functions of AM front-end, FM front-end and FM i-f limiter detector, and audio preamplifier. The three circuits are used in both the AM and FM receiver functions by use of a simple switching arrangement. The partitioning scheme chosen provides the sought-after objective of optimum circuit performance per circuit cost.

9.2 RECEIVER PARTITIONING

The typical AM or FM superheterodyne receiver system consists of an rf stage to achieve minimum noise figure, a frequency converter that can take the form of a mixer stage and separate local oscillator or a self-oscillating autodyne converter, an i-f stage or i-f stages having sufficient gain and power output to drive a particular type of detector, and an audio amplifier section to bring the intelligence to a level sufficient to drive headphones or a loud speaker. Selectivity in such a receiver is usually divided between the receiver input or antenna coupling, the rf mixer interstage, and the i-f interstages.

All the circuitry needed to perform the above functions in a receiver, with the exception of selectivity, can be designed and fabricated by standard monolithic IC processing techniques. One important consideration in the partitioning of a receiver system for integration is the complexity of circuitry to accomplish the above functions and the relationships between complexity, chip size, fabrication yield, and IC cost. A second partitioning consideration arises from the choice of the selectivity elements and the effect upon total system cost and performance. The use of single elements capable of achieving the required selectivity as compared to several interstage bandpass elements must be weighed in making this decision. A third consideration is flexibility: will it be possible to use the IC in various grades of receivers having differing performance levels, combinations of stages, and selective elements? The universal approach is heavily favored in consumer-products applications.

9.3 DESIGN OF INTERNAL VOLTAGE REGULATORS

Monolithic ICs designed for use with widely varying power supplies, such as found in low-cost battery and line-operated consumer electronic equipment, often incorporate internal voltage regulators to provide stabilized bias and temperature compensation.

The simplest of monolithic voltage regulators is the base-to-emitter junction diode of an integrated transistor. During forward conduction the junction presents a voltage offset of about 0.7 V and a relatively low impedance to ground. In the reverse breakdown or avalanche mode the diode voltage is about 6 V. Regulators having higher voltage levels can be formed by stacking a multiple of diode-connected transistors in series, as shown in Figure 9.1.

The output of monolithic voltage regulators is not necessarily limited to integer values of the basic diode voltages. One and two transistor circuits have been developed in which the diode forward conduction voltage is effectively multiplied by the ratio of resistors, as shown in Figure 9.2. The regulated voltage can therefore be selected to have any desired value between 0.7 V and the supply value. The impedance of these circuits has been shown to approach r_e, the impedance of a single base-emitter diode. [1]

An application of the two-transistor regulator in an fm front-end circuit is shown in Figure 9.3. Here, transistors Q_8, Q_9, and Q_{10}, together with the tapped resistor arrangement, comprise the regulation and decoupling network. The regulator circuit has been designed so that transistors Q_8 and Q_9 simultaneously conduct when the minimum voltage is applied on the supply line. Therefore one V_{be} voltage appears across each of the two lower resistors. The voltage at the base of transistor Q_{10} is then simply

$$V_{b10} = V_{be}\left(2 + \frac{R_{CB8}}{R_{BE8}}\right),$$

where R_{CB8} and R_{BE8} are the collector-to-base and base-to-emitter resistors of Q_8. The low impedance of this circuitry in the base of transistor Q_{10} maintains the base at a constant level as the supply voltage is varied from its minimum value. Transistor Q_{10}, connected in the emitter-follower configuration, serves to isolate changes in load current from regulator

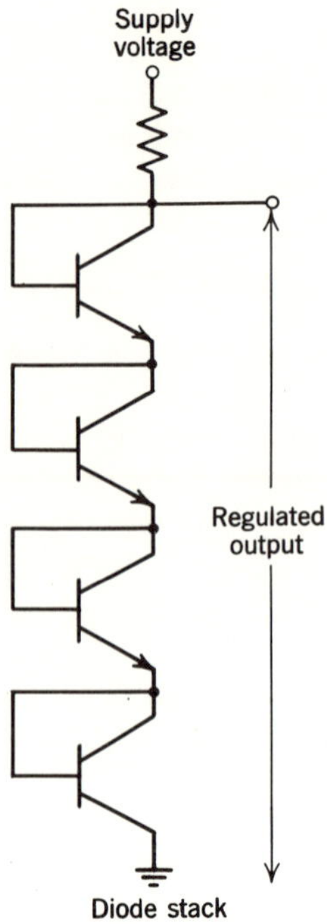

Figure 9.1. Diode stack voltage regulator.

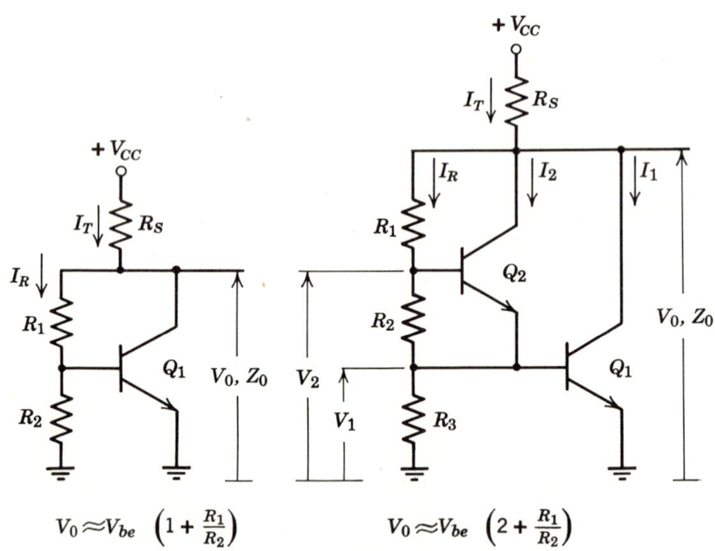

Figure 9.2. V_{be} multiplier regulators.

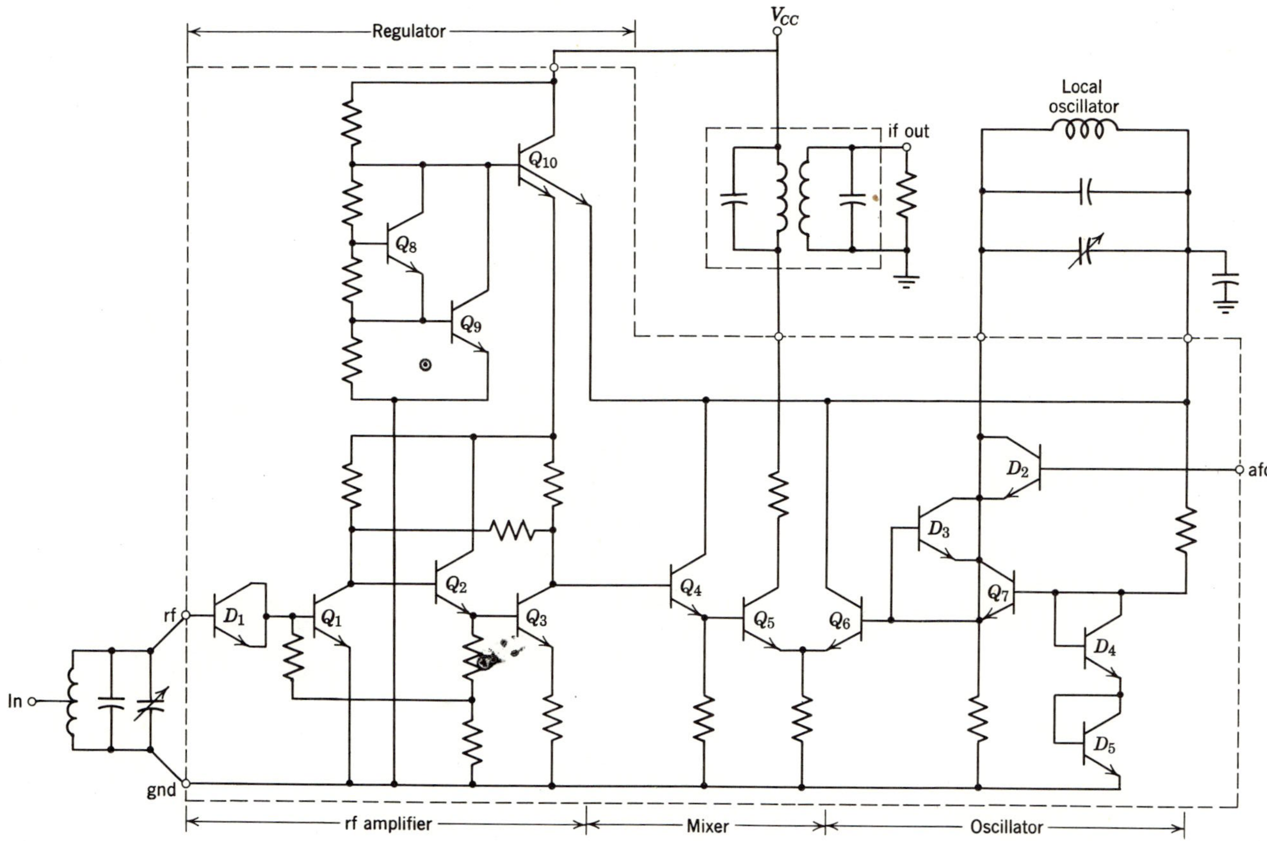

Figure 9.3. FM front-end.

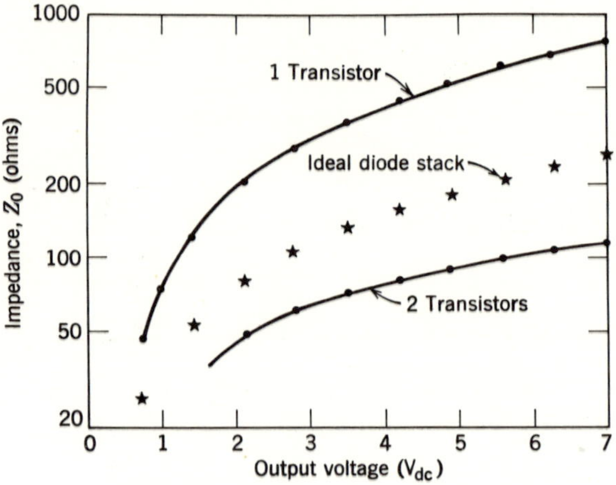

Figure 9.4. Voltage regulator impedance characteristics.

performance. The separate emitters of transistor Q_{10} serve to decouple the rf and mixer circuitry. For the circuit described a plot of output impedance vs output voltage is shown in Figure 9.4 and the voltage transfer characteristic is shown in Figure 9.5.

Another approach to supply regulation is that of feedforward or bridging. A circuit employing this technique is shown in Figure 9.6. This active regulator circuit operates on the principle of a linear amplifier which, upon receiving a disturbance at its input, produces an amplified, inverted replica at its output.

In the regulator circuit the amplifier consists of transistor Q_1 and resistors R_2 and R_3 which primarily establish the gain. A disturbance or change in the supply voltage V_{cc} is applied to the amplifier input through R_1. The inverted output appears at the collector of Q_1 and cancellation with the original V_{cc} change takes place across the summing resistor R_2; that is, the voltage between Q_1 collector and ground remains at a fixed level. An emitter follower Q_2 isolates the load from this reference point. Feedback around the circuit reduces the effects of variations in the transistor characteristics. The gain of the amplifier necessary for perfect regulation is given by

$$\frac{R_2}{R_3'} = 1 + \frac{R_1}{R_4} + \frac{R_1}{R_5},$$

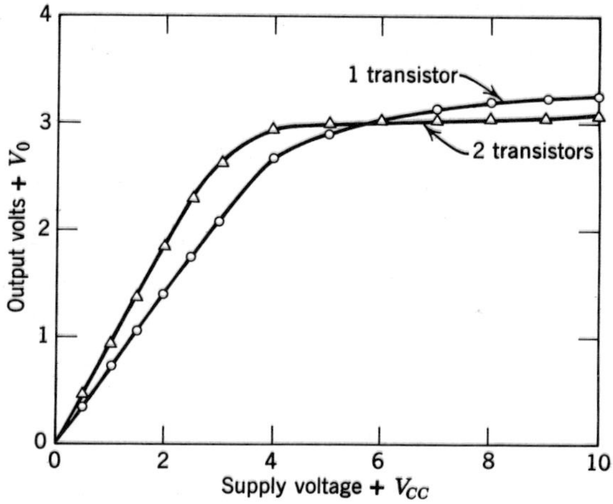

Figure 9.5. Voltage regulator transfer characteristics.

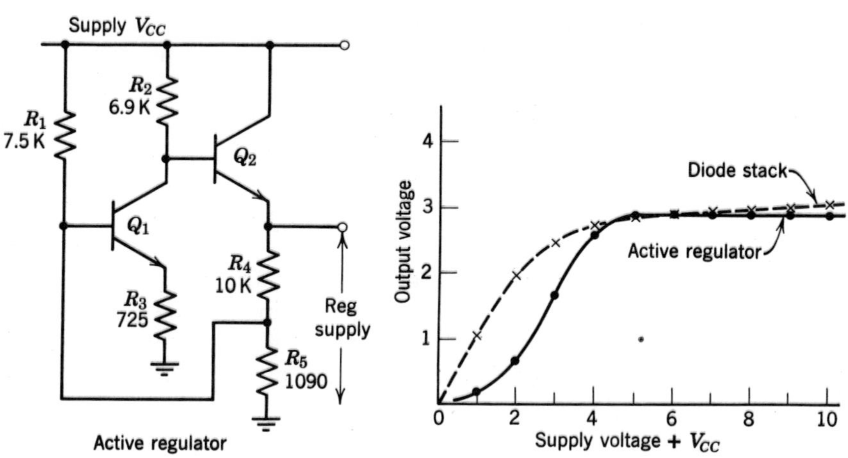

	Active	Diode
Output voltage variation 4.5–10 volts	80 mV	190 mV
Power dissipation @ 6 volts	6.0 mW	6.0 mW

Figure 9.6. Feed-forward voltage regulator.

where R_3' is the sum of R_3 and the Q_1 internal dynamic emitter resistance r_{e1}. The value of the regulated voltage V_r is given by the equation

$$V_r = \frac{V_{be}(R_2/R_3) - 1}{1 + (R_1/R_4)}.$$

The form of these equations indicates that the regulation characteristic is a function of resistor ratios, not absolute values. Experimental data has shown less than 1% change in regulated voltage for a simultaneous change in all resistor values of ±20%. This regulator has found application in connection with an audio amplifier to be described later in this chapter. Here a set-up accuracy of ±30 mV is required to maintain the output stage quiescent collector current within reasonable limits. The negative temperature coefficient of the regulator provides compensation to the amplifier.

9.4 AUDIO AMPLIFIERS

Audio amplifiers that can be incorporated in monolithic-circuit design range from simple one- or two-stage preamplifiers with outputs of a few milliwatts to the more sophisticated power units that can supply several watts. Class A single-ended and class B push-pull are the two primary types. Class D operation has been discussed elsewhere in the literature; however, little use has been made of this technique in commercial products.

The schematic for a two-stage audio preamplifier is shown in Figure 9.7. The emitter follower-common emitter pair is temperature compensated and biased from one of the previously described internal regulators. The input impedance, in this case 10 kΩ, is primarily determined by the bias feed resistor R_1. The output collector terminal can be either RC coupled to a single-ended power output stage or transformer coupled to a push-pull output. This circuit can provide a power gain of 40 dB and an output power of 8 mW to a 1 kΩ load.

A class-B push-pull audio amplifier having 60-dB gain and providing an output power of 100 mW from a 6-V supply and 250 mW from a 9-V supply is shown in Figure 9.8. Referring to this schematic, the output transistors Q_8 and Q_9 are slightly degenerated and biased at a low quiescent current to reduce class B crossover distortion. Voltage sources in the form of emitter followers Q_6 and Q_7 drive the output stage, making the output waveform more linear at high current levels, thus reducing the odd-harmonic distortion due to beta fall-off.

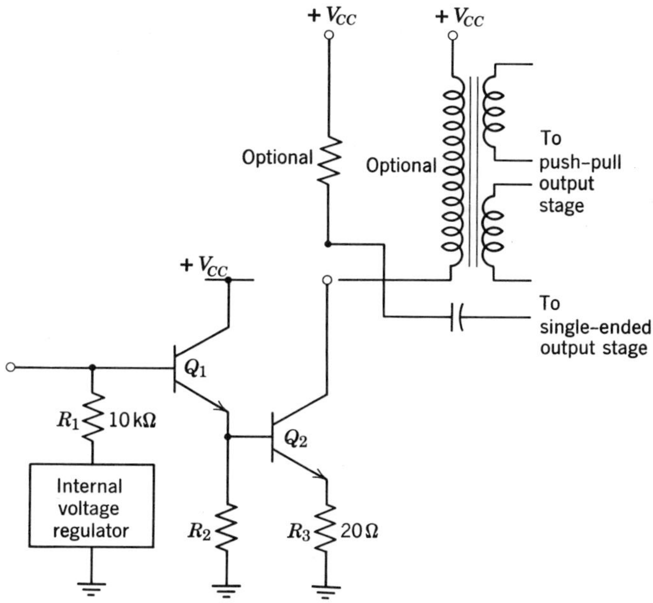

Figure 9.7. Audio preamplifier.

Phase inversion for the upper output transistor is accomplished by the degenerated common-emitter stage containing transistor Q_4 and the dc level-shifting emitter follower Q_5. A tight tolerance is required on the ratio of Q_4 collector resistor to emitter resistor in order to preserve both ac and dc balance to the output devices. In the monolithic IC environment this ratio is held to a tolerance of typically less than $\pm 5\%$.

The preamplifier consists of a self-biasing input stage, transistors Q_1 and Q_2, and a medium-level amplifying stage, transistor Q_3. The input stage achieves an input impedance of 10 kΩ by means of the bootstrapping between Q_1 and Q_2. Bias stability results from the large degree of resistor and transistor matching that can be achieved in monolithic networks.

Temperature compensation for the audio amplifier is primarily accomplished by thermal feedback through the monolithic chip to the internal voltage-regulator circuit whose dc output voltage has a negative temperature coefficient. The effects of this are translated through the circuit to the output stage via the medium-level amplifier, inverter stage, and drivers. Analysis has shown that proper choice of gain in these stages results in correct compensation for the change in output-stage transistor characteristics with temperature.

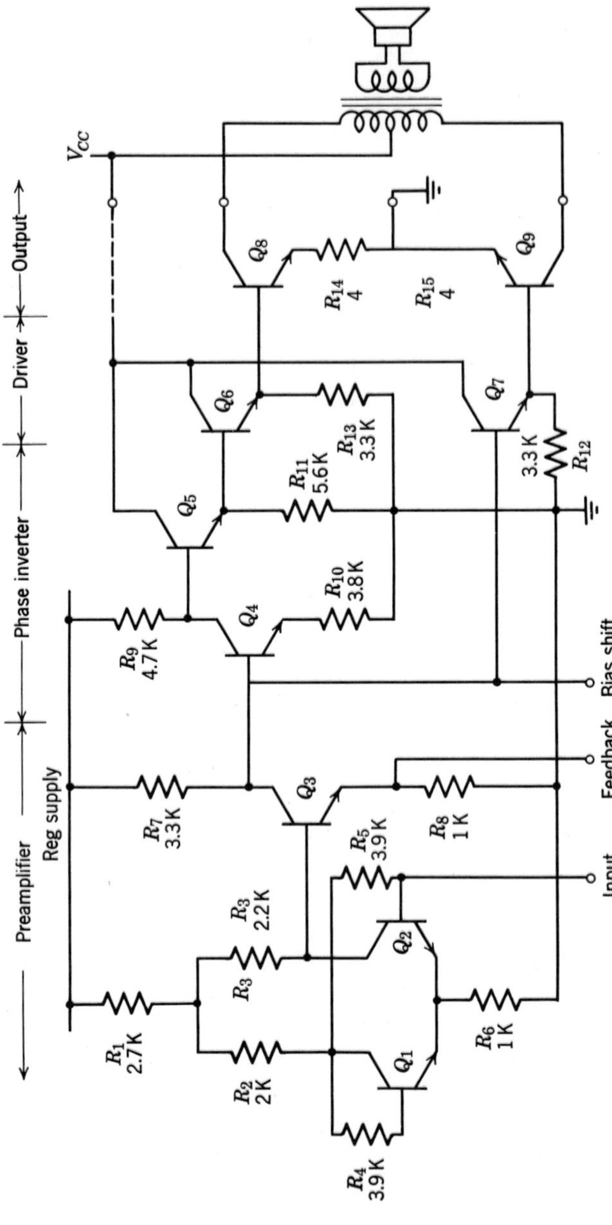

Figure 9.8. Audio amplifier.

Several design options exist in the audio amplifier:

1. The typical power gain of 60 dB can be increased to 67 dB by externally bypassing the resistor in the emitter circuit of Q_3.
2. Feedback from the output transformer primary or secondary can be ac coupled to the emitter of Q_3, resulting in lower audio distortion for the same 60-dB system gain.
3. The audio system can be operated as a class-A driver stage to deliver 10 mW of power to an external output stage. This is accomplished by externally shunting a resistor from the base of Q_4 to ground, thus biasing off one output device while the other handles the full signal.

9.5 AM RECEIVER CIRCUITS [2]

A typical block diagram representation of an AM receiver is shown in Figure 9.9. A particular AM receiver system which utilizes two IC chips is shown in block form in Figure 9.10. Here the first chip contains all of the active elements, most of the passive elements, and the bias and supply regulation for the rf, converter, and first i-f stages. Similarly, the second chip contains the circuitry for the second i-f, detector, and class-B

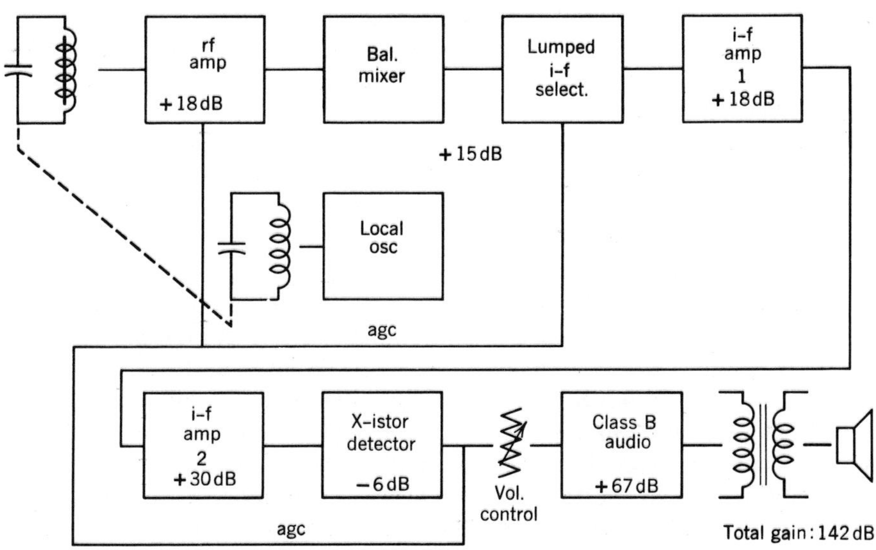

Figure 9.9. Receiver functional block diagram.

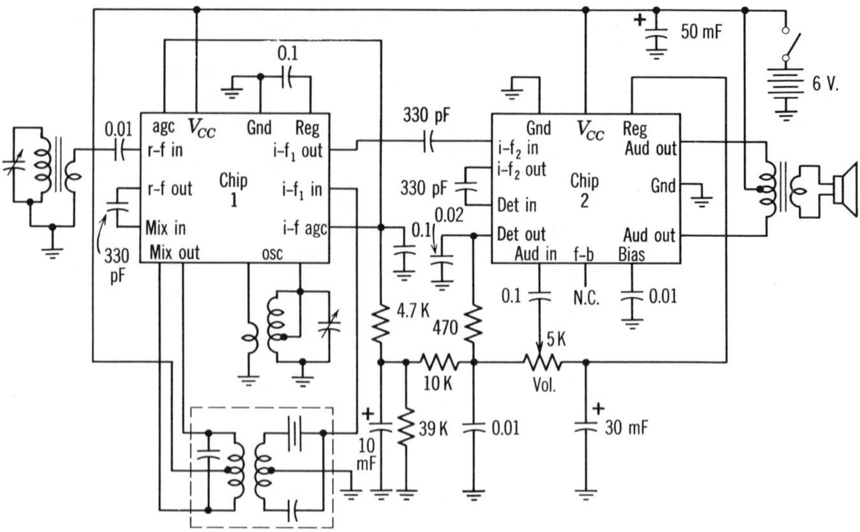

Figure 9.10. Connection diagram of complete AM receiver.

audio stages. I-f selectivity can be lumped between the mixer and first i-f stages as shown or split between that point and the i-f interstage. The system provides a total power gain of 135 to 145 dB depending upon the choice of feedback options in the audio amplifier. Supply voltage for this system is 4 to 10 V.

A second approach is represented by an IC which accomplishes the rf, converter, and i-f functions for an AM receiver and a portion of the i-f function for an FM broadcast receiver. This "AM tuner" IC is used in conjunction with two others to form an AM/FM receiver to be described later.

The performance and flexibility objectives represented by the low-cost consumer portable market can be met with an IC containing the converter, i-f, audio driver, and output stages. Higher performance levels can be achieved by adding to this basic building block a discrete component rf stage, tone control circuitry, and a 117-V line-connected output stage to achieve a receiver for table and clock application.

Numerous ICs to perform individual circuit functions are now on the market. Use of these, however, does not necessarily lead to an economical receiver system. Rather than presenting a survey of the several well-known circuits that can perform an individual function, the following sections will attempt to present designs which have significant advantages, either in simplicity or performance, for each function.

9.6 GAIN CONTROLLED RF AND I-F STAGES

An amplifier stage having a gain control range of 60 dB is shown in Figure 9.11. Amplitude distortion of strong signal levels is considerably less for this configuration than for the more conventional reverse-gain controlled stage. The control mechanism makes use of a degenerated variable emitter resistance in the form of the collector impedance of a transistor. Stage gain has its maximum value of 18 dB when transistor Q_2 is operating in the collector saturation region and is minimum when its collector-base diode becomes reverse biased.

The large value of Q_2 collector resistance greatly degenerates amplifier transistor Q_1 and linearizes its transfer characteristic in the strong signal region, making it possible to achieve satisfactory performance with a broadband rf stage and an antenna Q of 60 to 70. The circuit, however, can be operated as a tuned stage for increased rf selectivity by connecting an external L-C tank, having proper impedance level, across the internal collector resistor.

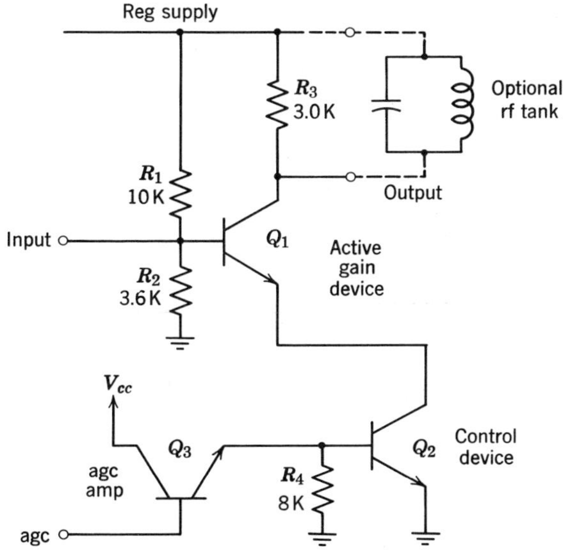

Figure 9.11. Gain controlled rf/i-f stage.

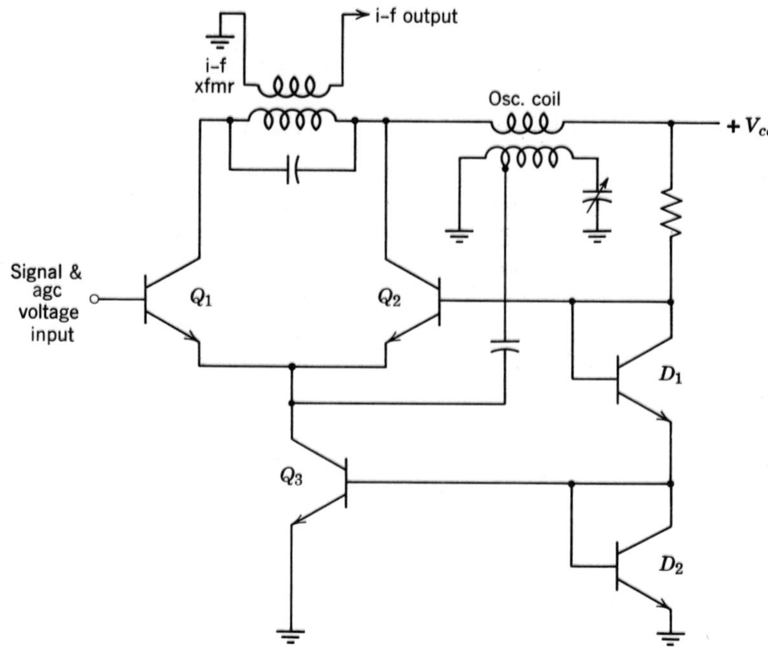

Figure 9.12. Gain controlled converter.

9.7 FREQUENCY CONVERTER STAGES

Figure 9.12 shows a converter stage that can be gain-controlled over a 40-dB range with negligible shift in local-oscillator frequency or level. Mixing action takes place in transistor Q_1, the i-f load being in the Q_1 collector circuit. The local-oscillator operation occurs as a common-mode signal with Q_1 and Q_2 effectively being in parallel. Converter gain control is achieved by reducing the current in Q_1 to a few microamperes, thereby causing a considerable reduction in conversion G_m. At the same time, the current and G_m of Q_2 are increased (differential action). Insofar as the local oscillator is concerned, the net amplification in the loop remains nearly constant.

The mixer-oscillator combination shown in Figure 9.13 has two distinct advantages when compared to a single-ended circuit. First, a reduction of 25 to 30 dB in local-oscillator feed-through to the i-f can be obtained by means of cancellation in the balanced, push-pull matching

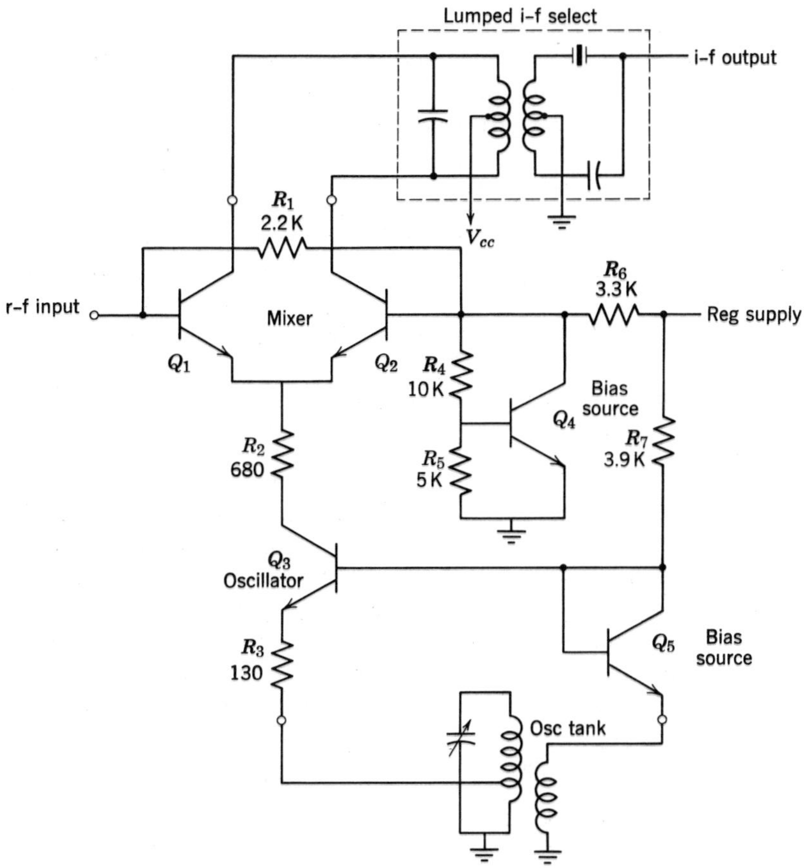

Figure 9.13. Balanced frequency converter.

transformer. Second, spurious responses generated by even-order nonlinearity in the mixer characteristic are greatly attenuated. These can be defined by the equations $n(f_o - f_s) = f_i$ and $n(f_s - f_o) = f_i$, where n is an even integer 2, 4, 6, . . ., f_o is the local-oscillator frequency, f_s is the undesired signal frequency and f_i is the i-f center frequency.

The schematic in Figure 9.13 shows that the rf signal is applied to the base of Q_1 with the base of Q_2 being ac referenced to ground through a low-impedance active feedback network consisting of Q_4 and its associated resistors. Differential-mode operation is achieved by the transistor current sink Q_3 which further serves as the local-oscillator, resulting in common-mode injection and hence its cancellation in the balanced push-pull collector load. Diode connected transistor Q_5 establishes the oscillator dc bias and provides temperature compensation. This stage has a gain of 15 dB, including 10 dB of i-f filter loss.

9.8 FIXED GAIN I-F STAGES

Fixed gain stages can take the form of single-transistor amplifiers, diode biased stages, differential amplifiers, feedback-pair or feedback-triplet amplifiers. Input-output terminal considerations and circuit complexity are factors in determining the proper stage design.

9.9 TRANSISTOR SECOND DETECTOR

Detection of the amplitude modulated i-f carrier can be accomplished in the transistor circuit, shown in Figure 9.14, in which Q_2 is biased near collector cut-off. The power gain of this circuit is of the order of 12 to 15 dB higher than it is for the conventional LC driven diode peak detector operating at the same impedance and power level. Here, use is made of a diode-connected transistor to provide stable bias and temperature compensation. Automatic-gain-control information is derived from the average value of the detector collector voltage. This circuit can deliver a maximum of 25 μW of audio power for a 30% modulated carrier and a 3-V collector supply.

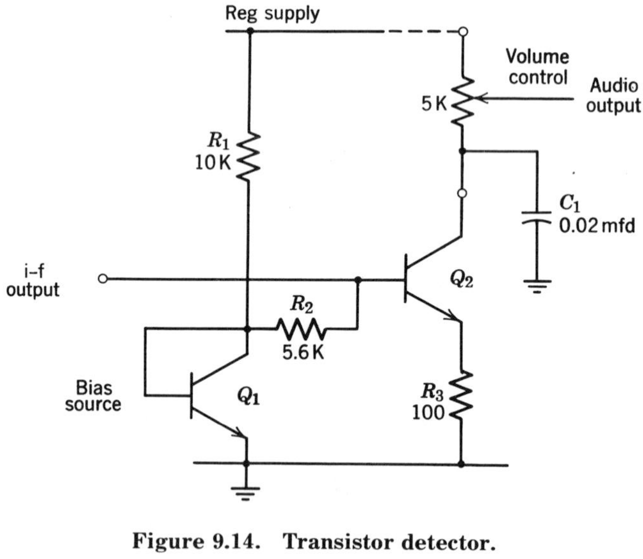

Figure 9.14. Transistor detector.

9.10 FM RECEIVER CIRCUIT DESIGN

The higher operating frequencies used in fm receivers place special requirements upon monolithic techniques. The partitioning of an overall receiver into monolithic circuits is based upon system performance requirements, integrated circuit processing capabilities, maximum utilization of diode area, and minimization of the number of external components. For an FM front-end integrated circuit, performance requirements such as gain and noise figure at high frequencies require good high frequency processing capabilities such as low transistor base-spreading-resistance r_b' and high transistor gain-bandwidth-product f_T. Small geometry transistors combined with high-frequency processing should be utilized for the design of front-end circuitry.

For an i-f amplifier/limiter integrated circuit, system performance requirements such as symmetrical limiting action and higher voltage operation require integrated circuit processing capabilities such as well-matched transistors and higher collector voltage breakdown. Therefore, different integrated circuit processing capabilities can be used for the different functions. Also involved in the partitioning is a trade-off in die area and circuit yield.

In order that each integrated circuit be capable of operating to specifications over a supply voltage range independently of characteristics in the supply line, each section generally has its own internal voltage regulator. In addition to the above, other advantages can be obtained from using an internal regulator, such as their use in ac bypassing, establishing bias levels, providing temperature compensation, and as decoupling elements. These elements are well-suited for integrated circuit design as they require resistor tolerance tracking and transistor matching; these properties are inherent in components fabricated by bulk monolithic processing.

9.11 FRONT-END CIRCUIT

The circuit diagram for an FM front-end monolithic circuit is shown in Figure 9.3. The die has been designed to perform the function of rf amplifier, mixer, and local oscillator with internal reactance control. To minimize the use of external components, dc coupling has been used throughout. An internal voltage regulator circuit is also included.

Referring to Figure 9.3, transistors Q_8, Q_9, and Q_{10}, together with the tapped resistor arrangement, comprise the regulator and decoupling network. This operation of this circuit has been described earlier. Diode D_1 and transistors Q_1, Q_2, and Q_3, along with their associated resistors form the rf amplifier. The input signal travels through diode D_1 to the grounded-emitter transistor Q_1. An interdigitated topology is utilized for the design of transistor Q_1 to minimize r_b' and low-frequency noise figure. As previously mentioned, the processing gives high values for transistor f_T so that the high-frequency noise figure is also low. Transistor Q_2 is used in the emitter-follower configuration as an impedance transformer to provide interstage isolation and stable bias for transistor Q_1. Transistor Q_3 is operated as a degenerated common-emitter stage with bias stabilization via the feedback resistor to Q_1's collector. With a regulated voltage of 2.50 V, an rf power gain of 26 dB with a noise figure of 3 dB is attainable. The overload performance is characterized by the rf transfer characteristic shown in Figure 9.15 where it can be seen that output voltages approaching 0.5 Vrms can be attained.

In reference to Figure 9.3, the converter circuitry is composed of the mixer (transistor Q_5), the local oscillator (transistor Q_7), and the AFC (diode D_2). The rf signal is buffered by emitter follower Q_4 to the base of the mixer, transistor Q_5. The oscillator transistor Q_7 is operated in the common-base mode via the low impedance path offered the base by diodes D_4 and D_5. Extremely stable bias is provided to the oscillator by the double regulation (diodes D_4 and D_5 controlled from the internal regulator). Diode D_3 provides the required feedback for the oscillator. Oscillator injection is accomplished through the emitter follower Q_6 to the mixer transistor. Application of varying dc voltages to the base of diode D_2 causes the junction capacitance to change so that the frequency of the oscillator is controlled.

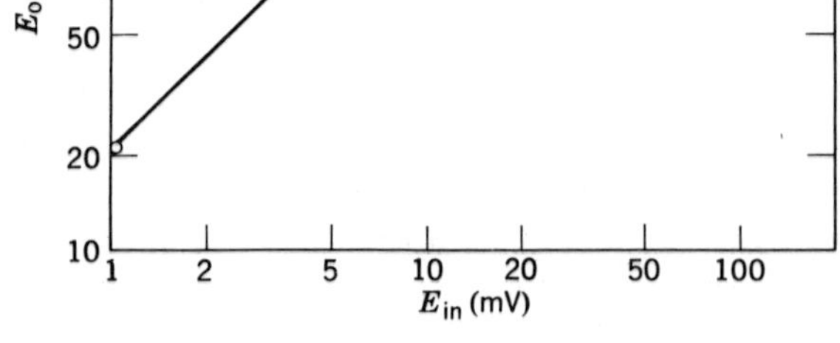

Figure 9.15. Rf transfer characteristic.

To minimize undesired frequency variations in the oscillator circuitry, temperature compensation is provided at the collector of transistor Q_3. The rf circuitry is designed to provide temperature compensation at Q_3's collector equivalent to two base-emitter diodes. The complementary side of the differential amplifier is similarly compensated. With optimized oscillator injection at the emitter of the mixer stage, a conversion gain of 12 dB has been measured. With the front-end circuitry tracked, an AFC sensitivity of ± 350 kHz has been measured with ± 1 V changes at the AFC terminal.

9.12 I-F DETECTOR AND AUDIO PRE-AMP CIRCUITRY

The second FM circuit, shown in Figure 9.16, was designed to perform the functions of i-f amplifier, limiter, detector, and audio preamplifier. As in the design of the front-end circuit, dc coupling has been used throughout, and an internal voltage regulator is included.

As revealed in Figure 9.16 the regulator circuitry is composed of diodes D_1 through D_5 and their series limiting resistor. This diode stack provides stabilized collector and base bias for the i-f/limiter circuitry and base bias for the audio preamplifier. Emitter followers Q_9 and Q_{13} provide decoupling for the first two i-f/limiter stages, and emitter followers Q_{10} and Q_{12} decouple the third high-current limiter stage. The diode-connected transistors D_1 through D_5 maintain a low dynamic impedance, each equal to r_e, so that effective internal bypassing and decoupling is obtained.

The i-f amplifier/limiter circuitry consists of three direct-coupled amplifiers. Referring to Figure 9.16, these amplifier stages are transistors Q_1 through Q_8. Each stage pair is an emitter follower-common base combination with interstage dc level shifting and buffering supplied by transistors Q_3 and Q_6. Transistor Q_{11} dc translates and impedance isolates the degenerative feedback signal from the third stage collector of Q_7 to the input base of Q_1. More than 60 dB of isolation is provided by transistors Q_{12} and Q_{13}, connected base-to-base. Current limiting is utilized in the third stage (determined by the value of the Q_7-Q_8 emitter resistor); in this type of limiter, output level is independent of changes in collector voltage and circuit Q is independent of the degree of limiting. With the design described above, a power gain of 66 dB at 10.7 MHz with a limiting level of 1500 μV (referred to the input) is achieved. A typical limiting characteristic is shown in Figure 9.17.

A pair of matched diodes D_6 and D_7 (refer to Figure 9.16) is included in the integrated circuit die for use in a modified Foster-Seeley discriminator. This circuitry provides AFC voltage referenced to ground for the front-end converter circuitry.

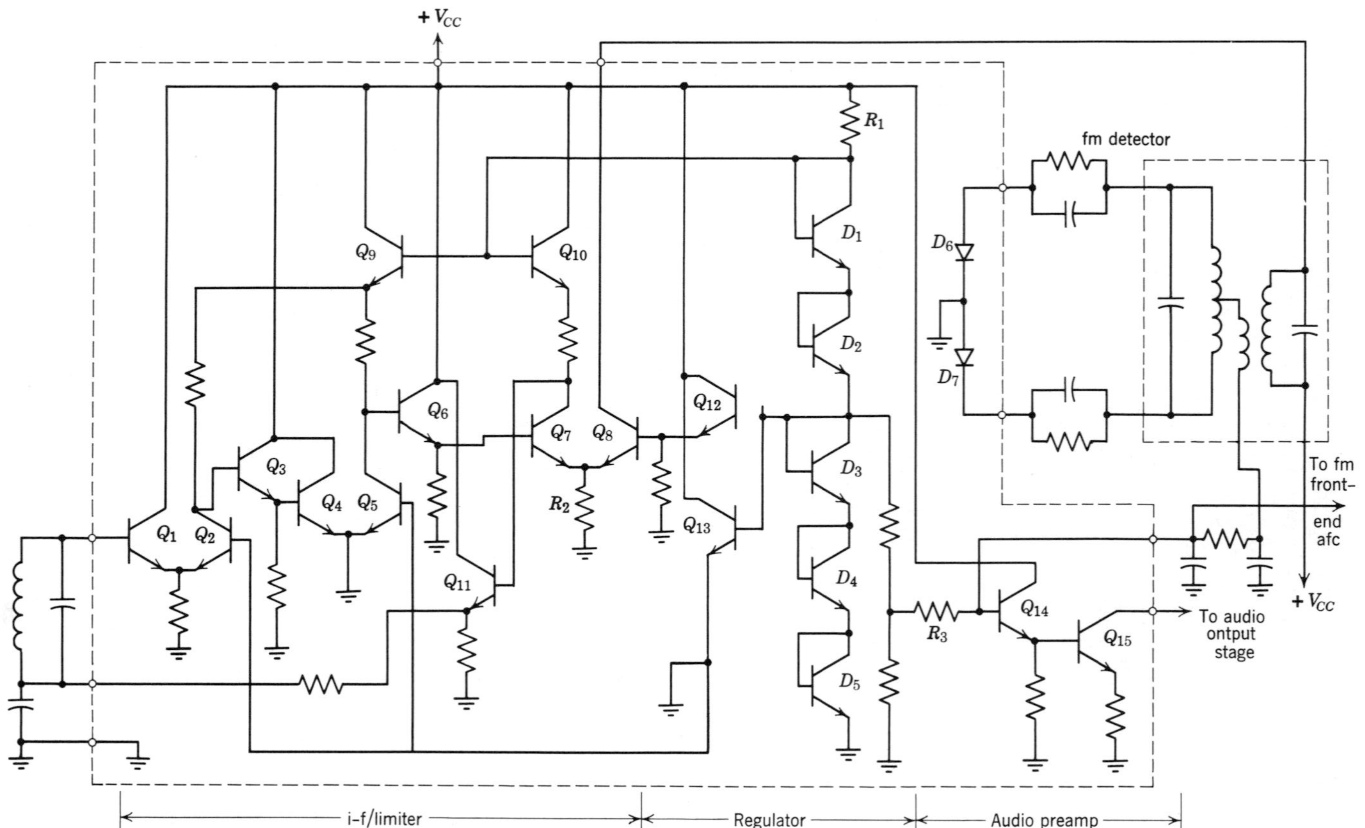

Figure 9.16. FM limiter/audio preamplifier.

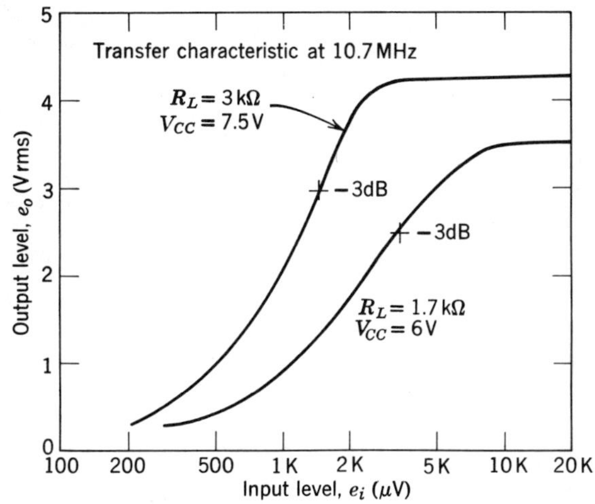

Figure 9.17. FM limiting characteristic.

The audio preamplifier circuit consists of transistors Q_{14} and Q_{15} in Figure 9.16. The input impedance of the circuit is controlled principally by Q_{14}'s bias resistor. Bias temperature stabilization is obtained by diodes D_3, D_4, and D_5 in the regulator circuitry. The input impedance is 10 kΩ, the power gain is 40, dB and the output power is 8 mW to a 1 kΩ load.

9.13 AM RECEIVER DESIGN AND PERFORMANCE

Figure 9.18 shows photomicrographs of the two chips used in the AM receiver described earlier. Chip 1 contains 12 transistors, 21 resistors, measures 46 × 51 mils, and contains the functions of gain-controlled rf converter and gain-controlled first i-f and voltage regulators. A schematic diagram is shown in Figure 9.19. Chip 2 measures 57 × 60 mils, contains 16 transistors and 32 resistors, and includes the functions of second i-f, transistor detector, audio amplifier, and voltage regulator. A schematic appears in Figure 9.20.

The advantages of system integration are depicted in Figure 9.21. Here, the printed-circuit boards of the conventional eight-transistor, two-diode receiver and the production-IC receiver are compared. Thirty-six components, as shown in Figure 9.22, have been replaced by the two monolithic chips bonded face-down to a carrier board. A schematic diagram of the required external components and the inter-chip connections of the complete receiver are shown in Figure 9.10.

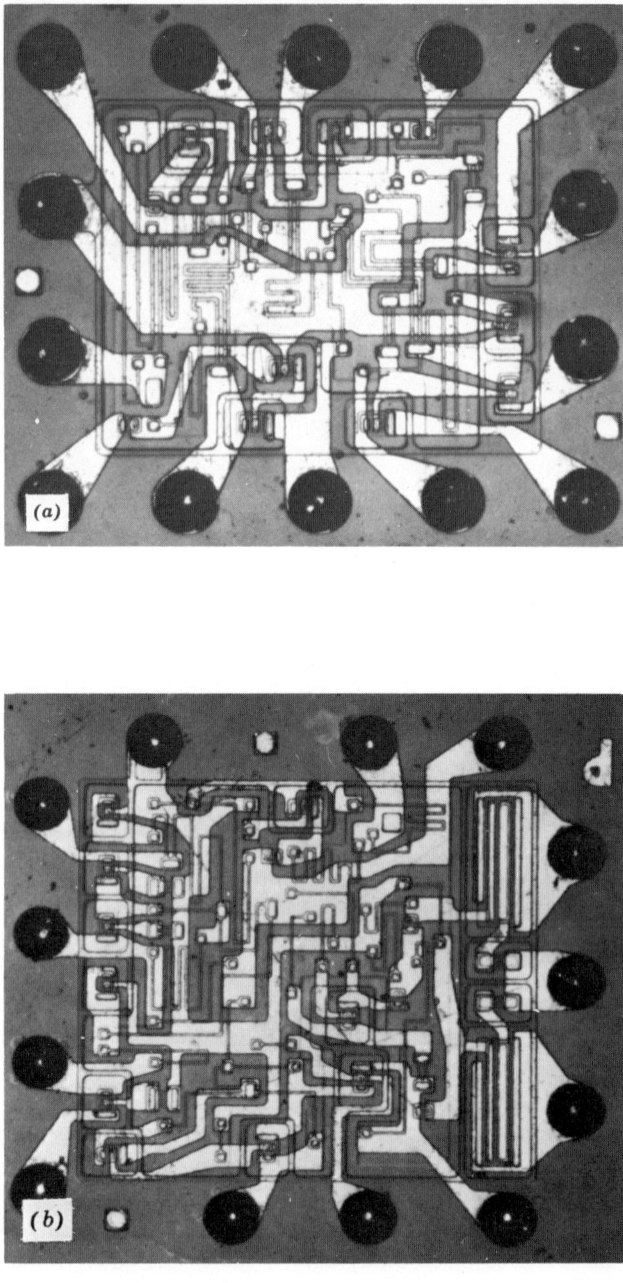

Figure 9.18. Photomicrographs of AM monolithic radio chips: (*a*) chip 1; (*b*) chip 2.

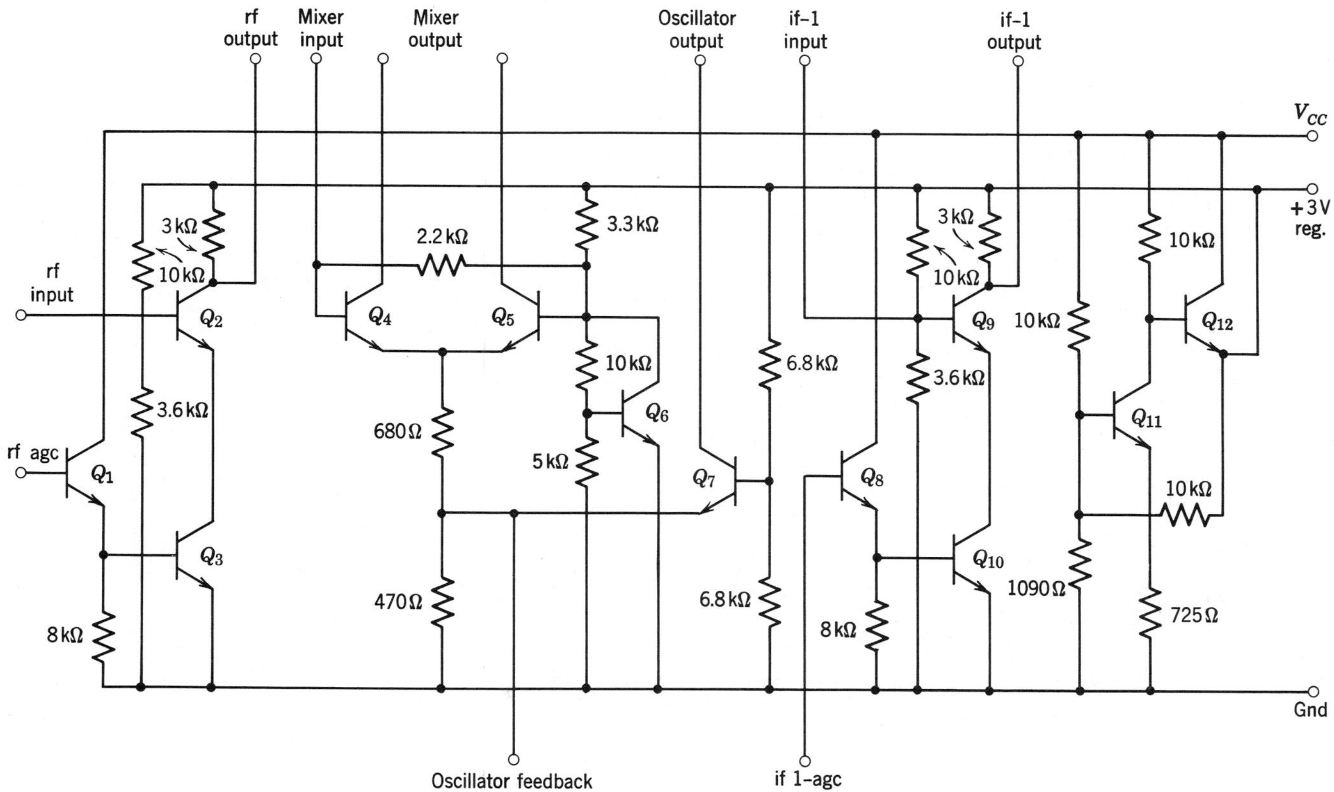

Figure 9.19. Schematic diagram of integrated first section of AM receiver.

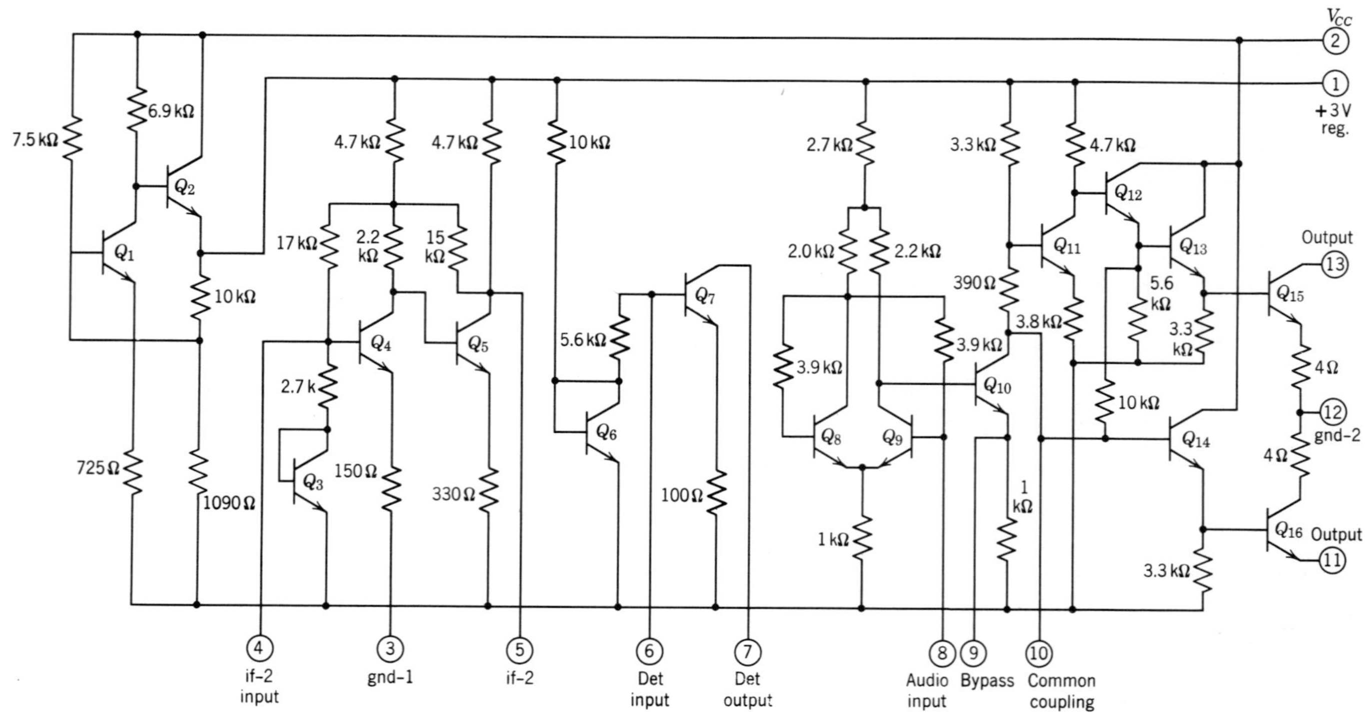

Figure 9.20. Schematic diagram of integrated second section of AM receiver.

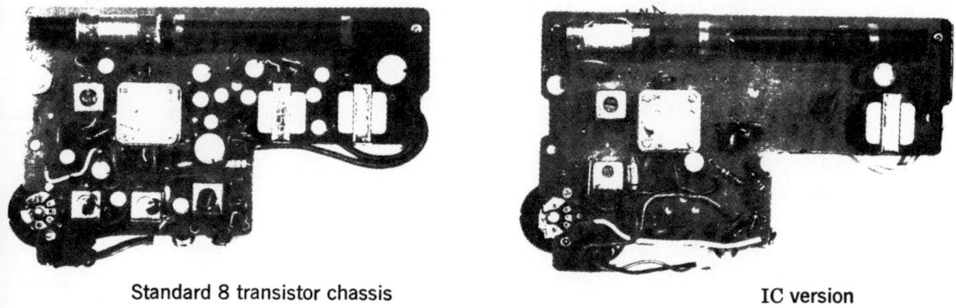

Figure 9.21. AM receiver comparison.

Listed in Table 9.1 is a comparison of the performance of the two receivers of Figure 9.21. The sensitivity, selectivity, and overload performance meet the requirements necessary for large portable and table model receivers. Note that the ceramic filter used to lump the i-f selectivity gives better passband characteristics than three single-tuned transformers.

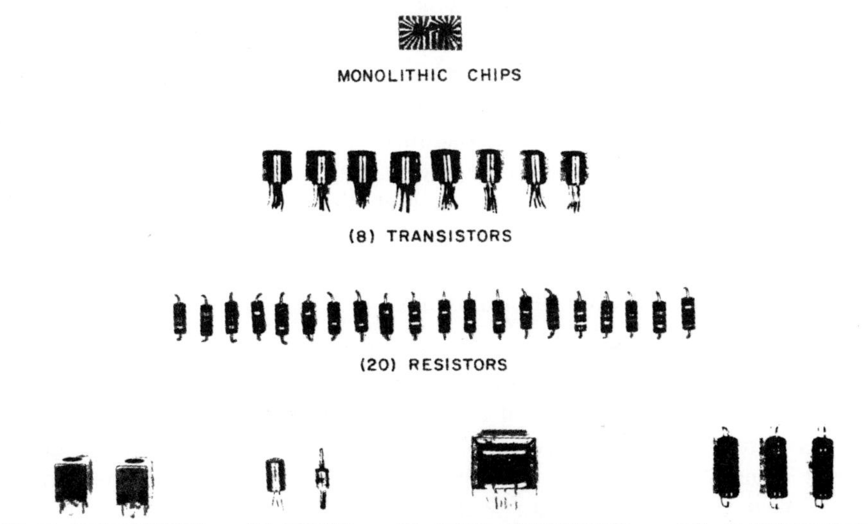

Figure 9.22. AM receiver integrated circuit carrier board and the components no longer needed.

TABLE 9.1 Performance Comparison

	IC Receiver	8-Transistor Receiver	Units
Sensitivity for 50 mW	50	100	μV/m
6 dB bandwidth	7	7	kHz
I-f ACA at 10 kHz	26	20	dB
80% modulation overload	2	0.2	V/m
Agc figure of merit*	80	40	dB
Power output for 10% distortion	140	140	mW

9.14 AM/FM RECEIVER DESIGN AND PERFORMANCE

FM Front End.

A photomicrograph of the FM front-end die is shown in Figure 9.23. The circuits were described in the discussion of Figure 9.3. Good high-frequency layout has been utilized to maintain rf stability both internal and external to the die. To reduce internal coupling, only associated resistors are biased to a common isolation region. In die assembly care is taken to minimize the lead length of the critical externally assessable terminals such as ground, the rf input, the oscillator-tank circuitry, etc. This same care must be taken when laying out the external circuitry.

This chip performs the functions of rf amplifier, mixer, and local oscillator and has been designed primarily for use in both portable and console frequency modulation broadcast receivers (88 to 108 MHz). The signal output from the chip is at the conventional i-f frequency of 10.7 MHz. Total conversion gain from rf to i-f is 34 dB. Tuned LC networks are utilized at the rf input, in the oscillator section, and at the i-f output. An optional LC tank can be connected at the rf to mixer interstage coupling point, thereby providing improved high-signal-level spurious response performance.

Design features include the following:

1. Rf stage noise figure in the 3-to-4-dB range.
2. Direct coupled local-oscillator-mixer circuit which includes a voltage-sensitive reactance element for AFC operation.
3. An internal voltage regulator which provides temperature compensated bias to each stage, thereby minimizing gain variation as a function of both temperature and supply.

*The Agc figure of merit is the ratio of input reduction for an audio output change of 10 dB, referenced to the overload level of the receiver.

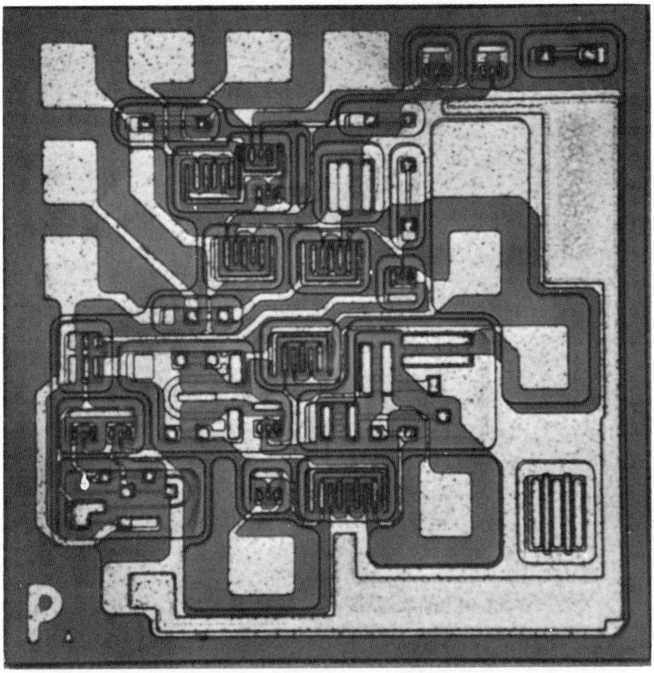

Figure 9.23. Photomicrograph of FM front-end integrated circuit.

FM Limiter, Detector, and Audio Preamplifier

A photomicrograph of the i-f amplifier/limiter die is shown in Figure 9.24. The same principles discussed in conjunction with Figure 9.16 were used to design the integrated-circuit mask set and assembly of this die.

This circuit can be used in FM broadcast-receiver or television sound systems to provide the functions of intermediate-frequency amplifier, limiter, frequency modulation detector, and audio preamplifier. LC tuned networks required with this chip include a 4.5/10.7-MHz input network and a double-tuned discriminator transformer. The i-f amplifier/limiter circuitry consists of three direct-coupled differential amplifier stages with dc feedback to the input.

A pair of matched diodes for use in a modified Foster-Seeley discriminator circuit are included on the chip. This FM detector circuit provides AFC voltage, referenced to ground for center frequency. The audio circuit consists of an internally-biased amplifier which has 10-kΩ input impedance and 40-dB power gain, and provides 8 mW of power to drive an external audio power output stage. The audio pre-amp can be utilized for both the AM and FM functions of an AM/FM receiver.

Performance features of this design include the following:

1. Stabilized gain over a wide variety of supply voltage (4-to-10 V) and temperature.

2. Low supply current drain makes the design applicable for use in battery-operated portable receivers.

AM Tuner

The circuit diagram for the AM tuner integrated circuit is shown in Figure 9.25; a photomicrograph of the die is shown in Figure 9.26. This circuit has been designed to perform the AM receiver functions of agc'd rf, agc'd converter, and first and second i-f amplifiers. The i-f amplifiers can also be used at the FM i-f frequency of 10.7 MHz. This circuit also

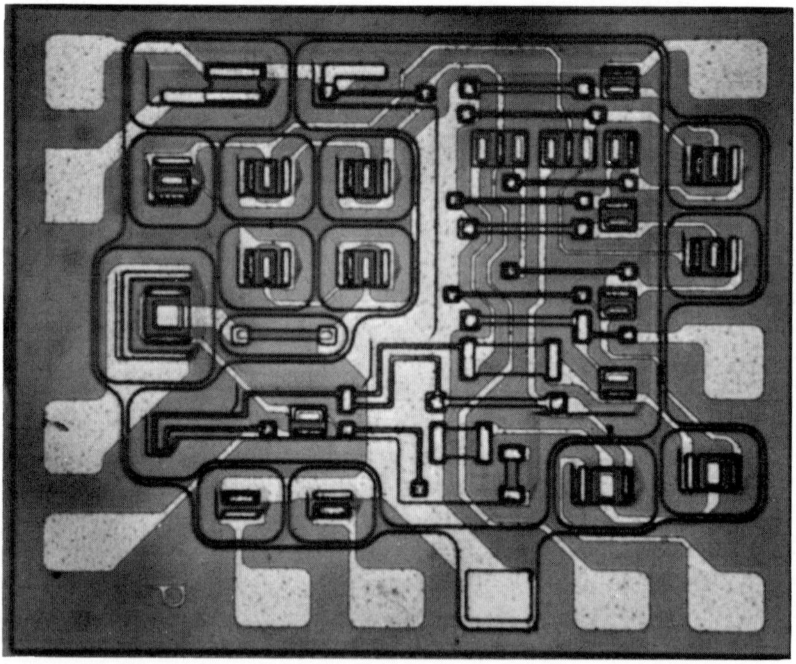

Figure 9.24. Photomicrograph of FM limiter/audio preamplifier integrated circuit.

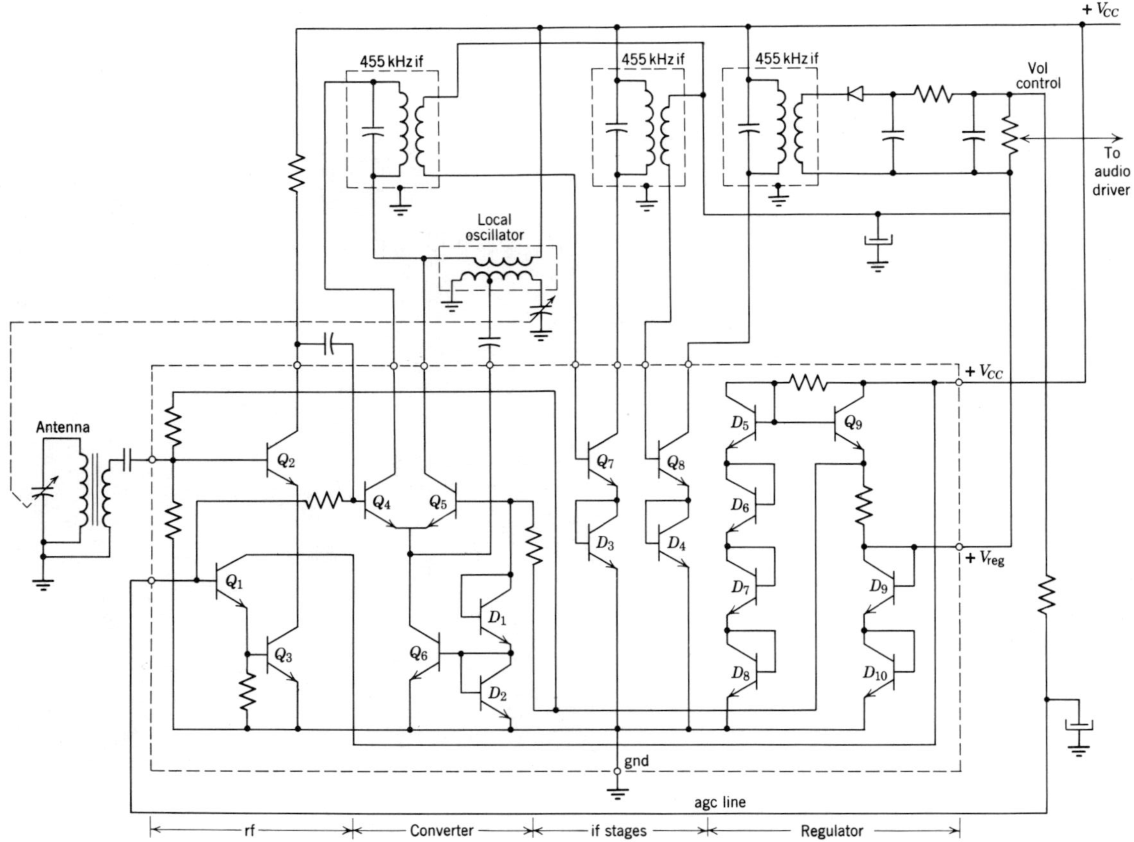

Figure 9.25. Schematic diagram of AM front-end integrated circuit.

combines with the previous circuits to provide AM/FM receiver functions from antenna to audio stage. In this design, emphasis has been placed upon circuit simplicity, both on the chip and external to it, in an effort to be both cost and performance competitive with conventional discrete component designs. Use of standard i-f transformers, in contrast to lumped i-f filtering, has been selected for reasons of circuit simplicity and cost, although some of the miniaturization possibilities are lost.

Noteworthy performance features of the design are the following:

1. An rf stage having an agc range of 50 dB and capable of handling signals in the range of several volts per meter when used with a 4-to-6 inch ferrite-core antenna.
2. A converter stage having an agc range of 40 dB with negligible shift in local oscillator frequency or level.
3. An internal voltage regulator which maintains the bias to each stage constant and supplies a fixed reference for the agc system over a wide range of battery supply voltages.

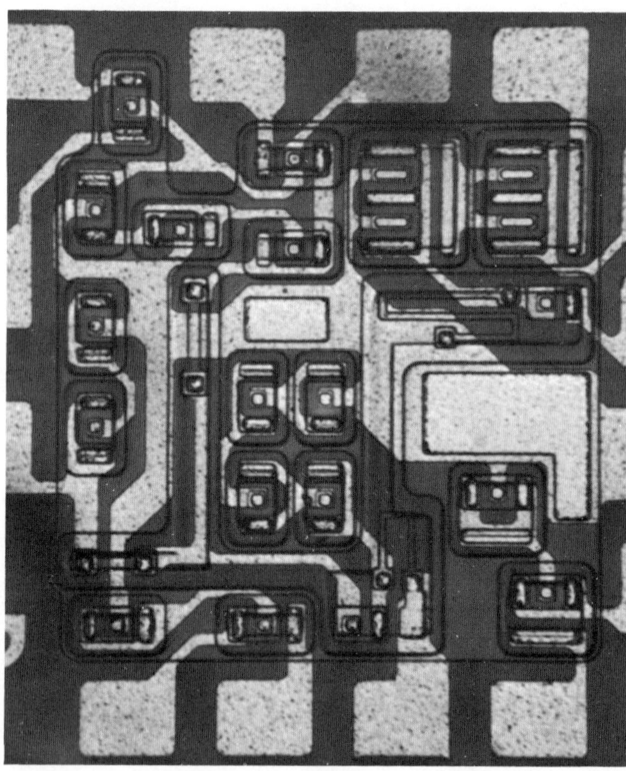

Figure 9.26. Photomicrograph of AM front-end integrated circuit.

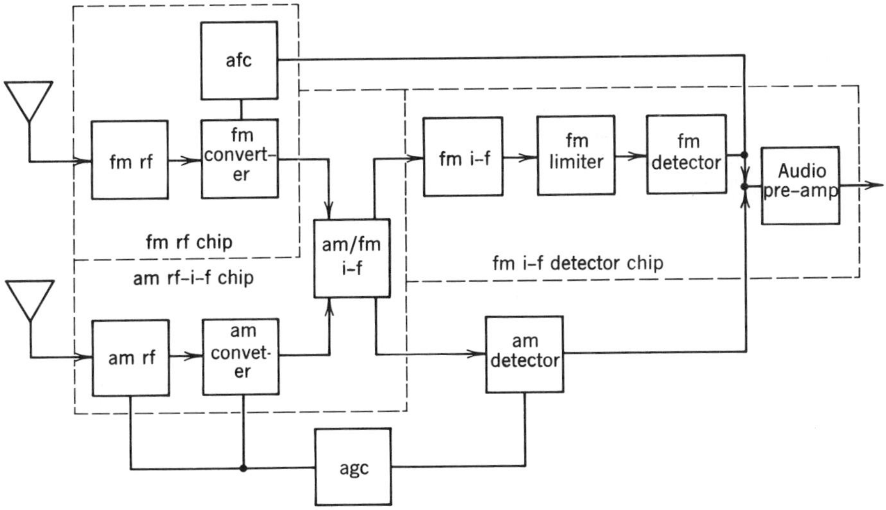

Figure 9.27. Block diagram of integrated AM/FM receiver.

4. Rf, converter and i-f stage designs that require identical bias or agc set-up voltage and therefore can be referenced to the regulator output, that is, single bias source for all stages.

AM/FM System Performance

With the integrated circuits assembled in ceramic dual-in-line packages, and the appropriate external circuitry utilized, a system designed to perform AM and FM reception is shown in the block diagram in Figure 9.27.

Specifications for each of the three integrated circuits are given in Table 9.2. Table 9.3 shows a comparison of significant performance specifications for the multichip monolithic receiver vs the discrete receiver design [3].

9.15 CONCLUSIONS

If ICs are to compete successfully with discrete-component circuits in consumer-product applications, designs must be used that do not place undue burdens upon the device capabilities of a given monolithic wafer-processing technique. In addition circuit functions must be assigned to an integrated circuit chip that permits the IC's utilization over a broad range of applications. This chapter has discussed ICs suitable for AM and AM/FM receiver applications and the partitioning of receiver systems to allow the use of devices having high volume potential with a good performance/cost tradeoff.

TABLE 9.2 Integrated Circuit Specifications

	Am Rf, I-f	Fm Rf	Fm I-f, Detector
Conversion power gain (dB)	45	34	
A-m, i-f power gain (dB)	60		
Fm, i-f power gain (dB)	40		60
Audio power gain (dB)			40
Noise figure (dB)	3.0	2.5	
AFC (kHz/V)		150	
Oscillator drift ($\Delta T = 70°C$) (kHz)		100	
V_{CC} range (V)	4-10	4-10	4-20
Overload (V)	1.0	0.5	
Agc range (dB)	90		

TABLE 9.3 Comparison Chart

(a) Performance

	A-m Monolithic	A-m Discrete	Fm Monolithic	Fm Discrete
Sensitivity for 10 mV recovered audio	75 μV/M	95 μV/M	1.0 μV	1.7 μV
12 dB S/N	50 μV/M	60 μV/M	0.75 μV	1.5 μV
30 dB S/N	750 μV/M	500 μV/M	2.0 μV	3.75 μV
Limiting threshold (−2 dB)			2.5 μV	6.5 μV
Overload	10 V/M	10 V/M	>0.2 V	>0.2 V

(b) System economics

	Discrete Components Removed by Monolithic Design
High power hi-fi	90
Low power hi-fi	96
Table radio	69

It is difficult to predict how various consumer product electronic devices will be partitioned in the future. Of course, the goal is to use a single monolithic chip to accomplish all of the functions for any receiver system.

REFERENCES

[1] Allen, W. F., et al., "Voltage Regulators as Circuit Elements in the Design of Integrated Circuits," *Proc. NEC,* 273-278 (1967).

[2] Hoke, L. H., et al., "Monolithic Circuits for Radio Receivers," *IEEE Trans. on Broadcast and Television Receivers,* **BTR-13**, No. 2, 64-70 (July 1967).

[3] Hermann, P. E., et al., "Multi-Chip A-m/Fm Receiver," presented at International Solid State Circuits Conference, February 16, 1968.

10

Wideband-Amplifier ICs

by L. A. Pope
Sylvania Electric Products

10.1 INTRODUCTION

The availability of general-purpose wideband-amplifier ICs provides the designer with a flexible tool. This will be illustrated in this chapter using the SA-20 as an example. The SA-20, shown in Figure 10.1, is a general-purpose wideband amplifier offering high-gain bandwidth, large signal swing, and excellent linearity. Parasitic capacitance has been minimized to achieve wideband performance through the use of one-half mil resistors and small-geometry transistors. To maintain good ac stability, the number of stages has been purposely minimized, thus reducing phase shift. Special processing techniques have been developed to provide transistors with high breakdown voltage, thereby permitting the use of high supply voltages to achieve large-signal swings. Feedback loops are employed to reduce distortion and maintain uniform gain over a wide range of operating conditions. External access to every component of the device enhances the versatility of the SA-20. Shunting of internal elements with discrete components permits operation of the circuit under diverse conditions. A photomicrograph of the SA-20 die is shown in Figure 10.1.

10.2 CIRCUIT ANALYSIS

The circuit diagram for the SA-20 in Figure 10.2 reveals that three direct-coupled npn transistors are used in a single-ended configuration. Bias for the input transistor is established internally by the resistive divider consisting of R_1 and R_2. A zener diode D_Z is used to raise the bias level of Q_2 to permit operation of Q_1 in the linear range. The emitter follower Q_3 buffers the output load circuit from the intermediate stage Q_2 thereby giving a higher total open-loop circuit gain even with low-value loads. Series type dc negative voltage feedback is applied from the output

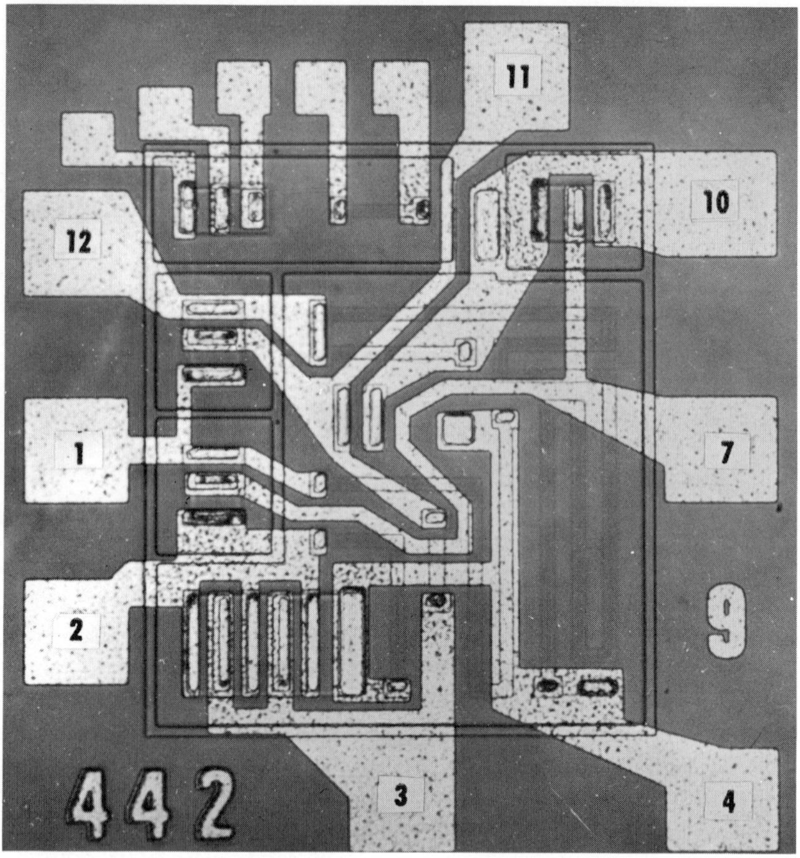

Figure 10.1. Photomicrograph of SA-20 chip.

of the emitter follower to the emitter of Q_1 by the resistive voltage divider consisting of R_4 and R_6. The operation of the circuit is such that the closed-loop gain is equivalent to the relationship $(R_4 + R_6)/R_4$.

From Figure 10.1 it is evident that power supply current flows in four current paths: I_{R1}, I_{R3}, I_{R5} and I_{CQ3}. The summation of these currents determines the device current drain. The current of each of these paths may be expressed as follows:

$$I_{R1} = \frac{(V_{CC} - V_{Q\text{in}})}{R_1}, \tag{10.1}$$

where $V_{Q\text{in}}$ = input quiescent voltage;

$$I_{R3}=\frac{(V_{CC}-V_Z-V_{BE2})}{R_3}, \tag{10.2}$$

where V_Z = zener voltage;

$$I_{R5}=\frac{(V_{CC}-V_{Q\text{out}}-V_{BE3})}{R_5}, \tag{10.3}$$

where $V_{Q\text{out}}$ = output quiescent voltage; and

$$I_{CQ3}=\frac{(V_{Q\text{out}}-V_{Q\text{in}}-V_{BE3})}{R_6}. \tag{10.4}$$

Examination of the typical component and voltage values given in Figure 10.2 shows that the major current path is through the output transistor Q_3. Total device current drain is therefore strongly related to the quiescent output voltage and the value of R_6. Direct coupling to a resistive load will, of course, also affect total current drain. Very serious consideration must be given to several factors before adjustment of either of these parameters can be considered. The limitations that must be considered are the maximum junction and maximum case temperature of

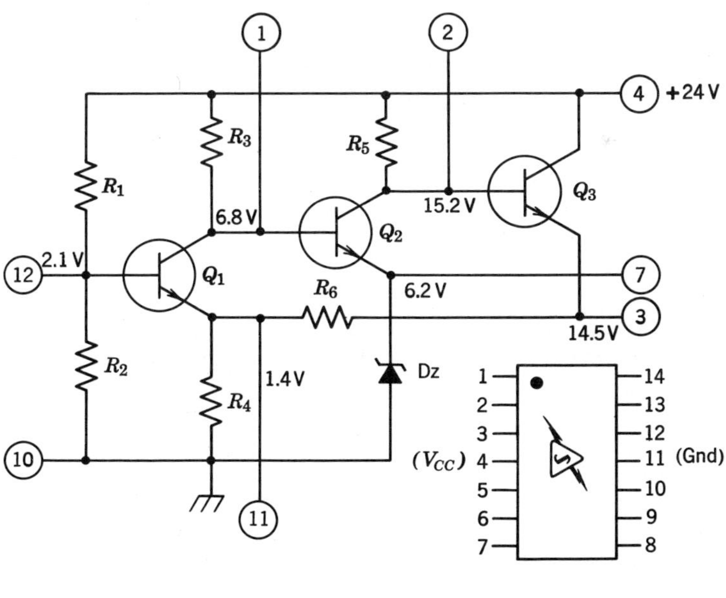

Figure 10.2. Circuit diagram for SA-20.

the device and the maximum current capability of the output transistor Q_3. The output transistor has a maximum current capability of I_{C3} max = 50 mA. Maximum junction and case temperature for the device are 200 and 155°C, respectively. Worst-case thermal gradients of the dual inline package used for the SA-20 are $\theta_{JA}=0.15$°C/mW and $\theta_{JC}=0.075$°C/mW.

Let us assume that the circuit is to be operated directly coupled to a resistive load of 1.0 kΩ. Under these conditions it is apparent that $I_{C3}<$ 50 mA. At 24 V for V_{CC}, a typical unloaded device consumes 450 mW. If the load is applied, the device dissipation increases to

$$450\ \text{mW}+(V_{CC}-V_{Q\text{out}})\frac{V_{Q\text{out}}}{1\ \text{k}\Omega}\simeq 600\ \text{mW}.$$

The temperature rise of the junction is 90°C, which dictates a maximum ambient temperature, without safety factors, of 110°C.

The expression for input quiescent voltage $V_{Q\text{in}}$ may be determined as follows: Assume $I_{C1} \gg I_{B2}$ and $\beta > 50$; then $I_{B1} \simeq (V_{CC}-V_{BE2}-V_Z)/R_3B_1$.

Applying the Thevenin Theorem to the input circuit,

$$V_{Q\text{in}}=\frac{V_{CC}R_2}{R_1+R_2}-\frac{V_{CC}-V_{BE2}-V_Z}{R_3B_1}(R_1//R_2). \tag{10.5}$$

In most instances the base current of the input transistor is small relative to the current in the input divider and the expression simplifies to

$$V_{Q\text{in}}=\frac{(V_{CC}R_2)}{(R_1+R_2)}. \tag{10.6}$$

It is apparent that the bias level of the input transistor may be varied merely by shunting resistors R_1 or R_2. Consideration must be given, however, to the effect of a change in $V_{Q\text{in}}$ on such factors as current drain and output quiescent voltage.

The output quiescent voltage $V_{Q\text{out}}$ may be determined by the loop equation

$$\frac{V_{CC}R_1}{R_1+R_2}-\left[\frac{V_{CC}-V_{BE2}-V_Z}{R_3B_1}\right]R_1//R_2-V_{BE1}=\frac{V_{CC}-V_{BE2}-V_Z}{R_3}(R_6//R_4)+ V_{Q\text{out}}\frac{R_4}{R_6+R_4}.$$

Collecting terms,

$$V_{Q\text{out}}=\frac{R_6+R_4}{R_4}\left[\frac{V_{CC}R_2}{R_1+R_2}-V_{BE1}-\frac{V_{CC}-V_{BE2}-V_Z}{R_3}\left(\frac{R_1//R_2}{\beta_1}+R_6//R_4\right)\right]. \tag{10.7}$$

Since the last two terms of the expression are normally $\ll (V_{CC}R_2)/(R_1+R_2)$, the expression simplifies to

$$V_{Q\text{out}}=\frac{R_6+R_4}{R_4}\left(\frac{V_{CC}R_2}{R_1+R_2}-V_{BE1}\right). \tag{10.8}$$

It is apparent that the output quiescent voltage is a function of the product of two resistor ratios. $V_{Q\text{out}}$ can be changed by external resistive shunting of any of the four resistors in the expression.

The maximum voltage swing of the device with an $R_L=\infty$ is essentially a function of the supply voltage and the zener-diode voltage. In the positive direction the voltage is limited by the voltage drop across R_5 and the base emitter voltage of Q_3. The voltage swing in the negative direction is limited by the zener voltage plus the saturation voltage of Q_2 minus the junction drop of Q_3. With a high R_L the device will normally swing between 22.5 V and 6.5 V. If the quiescent output voltage is optimized, the ac swing at the output, assuming negligible resistive loading, is ± 8.0 V. In processing single polarity signals, the output quiescent voltage may be offset by resistive shunting as shown in (10.8), and output swings approaching 16.0 V may be obtained.

It is apparent that the negative-going output voltage swing of the amplifier as well as other characteristics of the device can be modified by clamping the emitter of Q_2 at some voltage lower than that of D_Z. This may be accomplished with resistive dividers, power supplies, or zener diodes.

The maximum output voltage may be determined by

$$e_o=V_{CC}-V_{BE}-R_5(I_{C2}+I_{B3}). \tag{10.9}$$

By establishing the expression for I_{B3} in terms of $V_{Q\text{out}}$, e_o max, and R_L, and assuming that the maximum positive swing occurs when Q_2 is cut off, $I_{C2}=0$, the expression for e_o max becomes

$$e_{o\max}=\frac{V_{CC}-V_{BE}+(R_5/(B_3+1))\,[\,(V_{Q\text{out}}/R_L)+(V_{Q\text{in}}-V_{BE})/R_6\,]}{1+(R_5/R_6(\beta_3+1))+(R_5/R_L(\beta_3+1))}. \tag{10.10}$$

In the normal case when R_L is low,

$$\frac{(V_{Q\text{in}}-V_{BE})}{R_6}\ll V_{Q\text{out}}/R_L$$

and

$$1\gg\frac{R_5}{(R_6(\beta_3+1))}<\frac{R_5}{(R_L(\beta_3+1))}.$$

Equation 10.10 can be simplified to

$$e_{\text{omax}} = \frac{V_{CC} - V_{BE} + [(R_5 V_{Q\text{out}})/((\beta_3+1)(R_L))]}{1 + (R_5/(R_L)(\beta_3+1))}.$$

Assuming a typical device with $V_{CC} = 24\ V$, an ac-coupled 50-Ω load, and $\beta_3 = 100$, which is normally the case, the maximum absolute voltage level as seen at the emitter of Q_3 is $e_{\text{omax}} = 16$ V.

The maximum positive swing as seen at the load is therefore 1.5 V. The negative-going swing from the quiescent voltage level is a function of the quiescent current of the device. In terms of absolute voltage the minimum output voltage e_{omin} of the device with low value of ac coupled load can be expressed as

$$e_{\text{omin}} = V_{Q\text{out}} - \left[\frac{V_{Q\text{out}} - (V_{Q\text{in}} - V_{BE1})}{R_6}\right] R_L.$$

Again, assuming a typical device and an ac-coupled load of 50 Ω, $e_{\text{omin}} = 14.5 - 0.675 = 13.8$ V. The maximum linear swing above and below the quiescent voltage with an ac-coupled 50-Ω load is therefore +1.5 to −0.67 V.

It is possible to increase the swing capability of the device in the negative direction when driving low-resistance loads by increasing the quiescent current of the device. This may be accomplished by a dc resistive load or current source. A 1000-Ω dc resistive load connected to the output of a typical SA-20 increases the negative-going voltage swing across an ac coupled load from 0.67 V to 0.67 + [(14.5 V) 50 Ω/1000 Ω] or 1.4 V. Because of the current limitations of the device, the maximum possible voltage swing cannot be obtained with ac coupled resistive loads of less than approximately 600 Ω.

The transfer characteristics of the SA-20 as shown in Figure 10.3 indicate that linear reproduction of input voltages can be maintained over a very wide range of output quiescent voltage.

10.3 AC CONSIDERATIONS

Although the betas of Q_1 and Q_2 are typically > 70, the voltage gain of the first stage is relatively low due to the shunting effect of the input impedance of Q_2. If it is assumed that feedback resistor R_6 is temporarily removed, the total emitter resistance $R_E + r_e$ of $Q_1 \simeq 125$ Ω. If the total resistance in the emitter of Q_2, $r_Z + r_e$, is 10 to 15 Ω, it is clearly evident that the gain of Q_1 is the range of 5 to 7. The emitter-follower stage serves to buffer the load from the collector resistor of Q_2. The gain of the second stage is $\simeq$ 150 with $R_L = 50$, giving a total three-stage open-loop gain

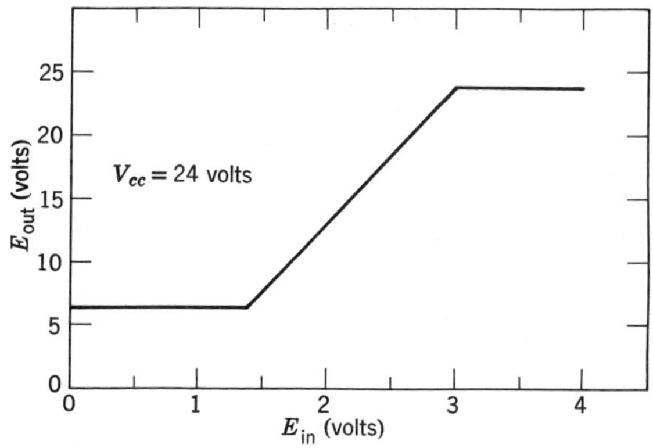

Figure 10.3. SA-20 transfer characteristics.

$A_{VOL} \simeq 900$. By applying conventional feedback theory, the closed-loop voltage gain A_{VCL} can be expressed as

$$A_{VCL} = \frac{A_{VOL}}{1 + A_{VOL}\beta}, \tag{10.11}$$

where $\beta = R_4/(R_6 + R_4)$.

Referring to Figure 10.2, it can be seen that the values of resistors R_4 and R_6 are 100 Ω and 1000 Ω, respectively. By shunting R_4 or R_6 with an external resistor, the gain of the circuit can be varied. As shown previously, however, other characteristics such as $V_{Q\text{out}}$ and associated parameters are also changed by padding either of these resistors. To change the mid-frequency gain without affecting the quiescent level, it is necessary to shunt the internal resistor with a series *RC* network.

Since the quantity $A_{VOL}\beta$ at low and medium frequencies is significantly greater than 1, the expression for closed-loop gain simplifies to $A_{VCL} = 1/\beta = 11$.

To maintain ac stability an additional feedback loop is required with the SA-20. This feedback loop is of the shunt type and uses an external capacitor that is connected between the base and collector of Q_2. Referring to Figure 10.4, it can be concluded from the shape of the amplifier frequency-response curve that the phase margin of a typical device is $>45°$ when the feedback capacitor is 3.6 pF.

The gain dependency of the SA-20 on resistor ratios rather than absolute values of resistors indicates good unit-to-unit uniformity in gain values. IC manufacturers using conventional processing techniques can hold resistor ratios to less than ±5%, whereas absolute values of resistors may vary by ±20%.

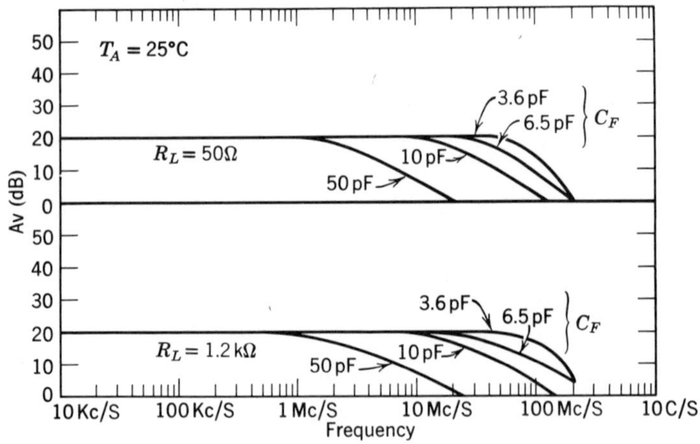

Figure 10.4. Typical gain versus frequency for the SA-20.

The use of negative feedback offers additional benefits in the form of increased input impedance and reduced output impedance. The input impedance of Q_1 without feedback is βR_E, whereas with feedback the input impedance increases by the factor $A_{VOL}\ [\ R_4/(R_4 + R_6)\ ]$. A high input impedance is most desirable in reducing distortion. Nonlinearity of the base current vs V_{BE} is a major contributor to distortion when Z_{in} is low, particularly when the device is driven by a voltage source which is frequently the case. Intermodulation distortion measurements on the SA-20 show second-order distortion of better than -73 dB with fundamental-frequency output-voltage swings of 5.0 V peak to peak.

Since the input resistance of Q_1 is high with respect to R_2, the input impedance of the device is determined essentially by $R_1//R_2$. The value of $R_1//R_2$ is typically 1.6 kΩ. It is apparent that a 20% increase in the value of R_1 and R_2 would result in a 20% increase in input resistance. Consequently the unit-to-unit variation in input resistance is essentially due to variation in resistor values.

The output impedance of the amplifier is also significantly affected by feedback. The open-loop output resistance of Q_3 is $\sim R_5\ (1 - \propto_3)$ or $\sim 60\ \Omega$. By using conventional voltage feedback theory it can be shown that the output impedance will be reduced to $[\ R_5(1 - \propto_3)\ ]/[\ A_{VOL}(R_4)/(R_4 + R_6)\ ]$. Since $A_{VOL}\,R_4/(R_4 + R_6) \simeq (900)\ (1/11)$, it is apparent that output impedance can be as low as 1.0 Ω.

10.4 STANDARD APPLICATION

The SA-20, when connected in the standard configuration as shown in Figure 10.7, exhibits the characteristics shown in Figures 10.4, 10.5, and 10.6. Referring to Figure 10.4, it can be seen that by varying the value of the capacitor C_f used in the shunt feedback path, the bandwidth of the device can be varied considerably.

10.5 MODIFIED APPLICATIONS

In addition to the modifications described previously, the circuit can be connected as shown in Figures 10.8, 10.9 and 10.10. In the maximum-gain configuration shown in Figure 10.8, ac series feedback is removed from the circuit by shunting R_4 with a capacitor. With the device connected in this manner the shunt feedback capacitor which is normally connected between pin 1 and pin 2 is not required. The gain of the amplifier with all forms of ac feedback removed is typically 60 dB up to 5 MHz. From 5 MHz to 100 MHz the amplifier rolls off at an average rate of ~ 8 dB/octave.

In the selective-amplifier configuration of Figures 10.9 and 10.10, the circuit functions as bandpass and notch amplifiers. Series or parallel tuned circuits are used in the shunt feedback loop to achieve the desired response.

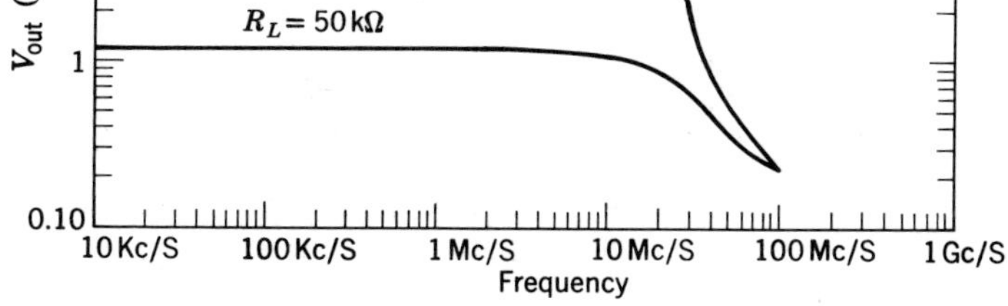

Figure 10.5. Maximum available output voltage curves for the SA-20.

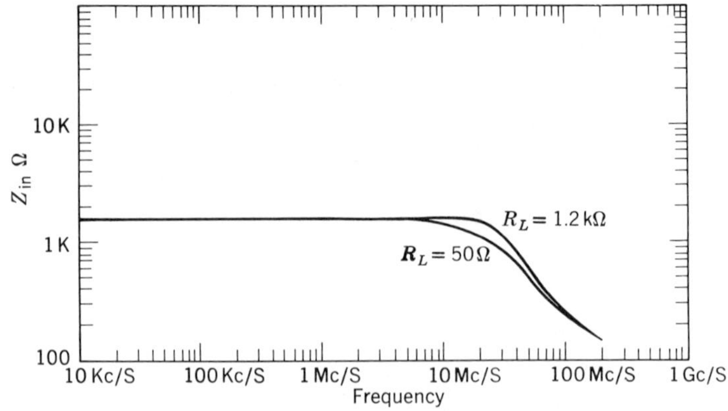

Figure 10.6. Z_{in} as a function of frequency for the SA-20.

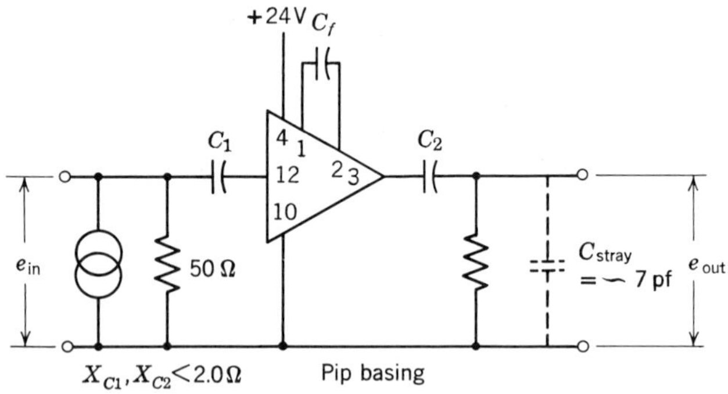

Figure 10.7. SA-20 connected in standard configuration.

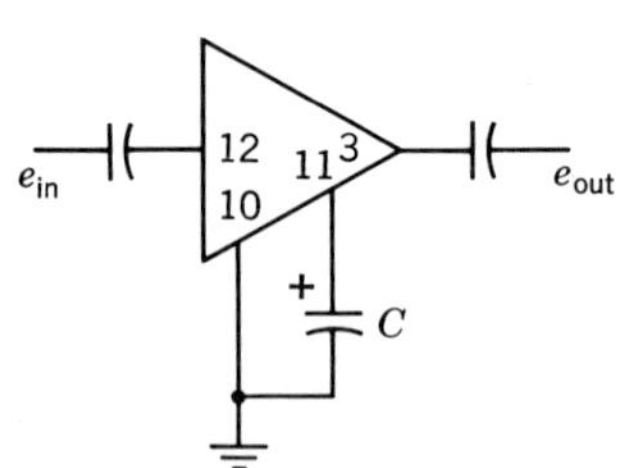

Figure 10.8. High-gain configuration.

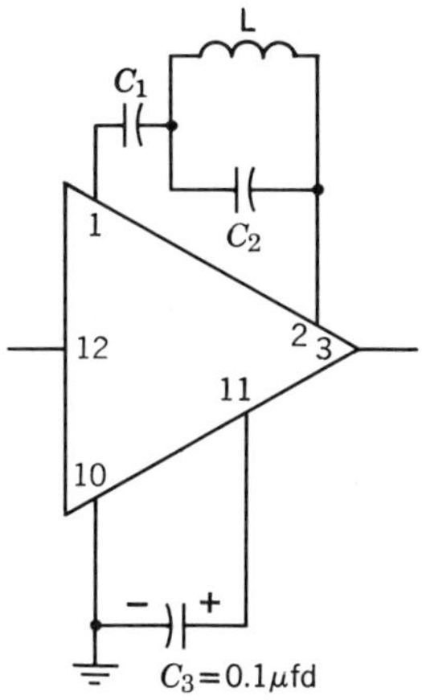

Figure 10.9. Selective amplifier.

10.6 CONCLUSION

Linear integrated circuits may in many instances be operated at conditions other than those used by the manufacturer for characterization of the device. Frequently the characteristics of off-the-shelf linear ICs can be modified to meet special requirements. The flexibility of general-purpose wideband amplifiers is enhanced by providing access to every internal component, when possible, as in the case of the SA-20.

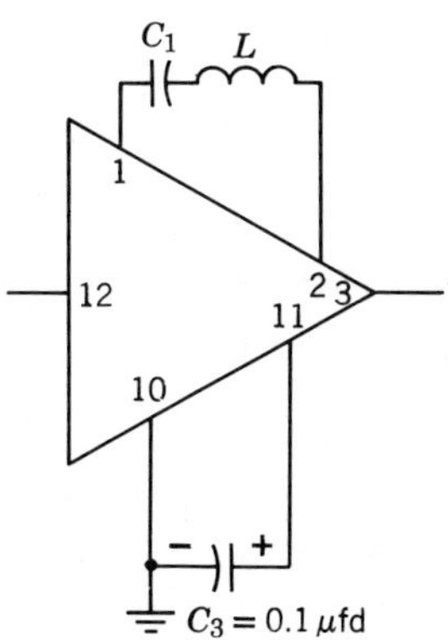

Figure 10.10. Selective amplifier configuration with C_1 and L series connected.

11

Analog Integrated Systems

by J. A. Livezey
Raytheon Company, Semiconductor Operation

11.1 INTRODUCTION

Since the introduction of semiconductor technology, analog systems have been moving from discrete-device designs to designs with fewer and more compact microcomponents and circuits. Milestones have been transistors, printed circuit boards, welded modules, thick and thin film components and networks, monolithic circuits, and finally hybrids combining various technologies. The phases of this trend in semiconductor device development may be described by the various levels of integration shown in Figure 11.1. As components become more complex, system size is reduced, while reliability, as a result of the smaller number of component interconnections, is improved. The trend in integration is depicted in Figure 11.2.

The discrete-component level of integration has existed throughout most of the history of electronics. While retaining essentially the same terminal parameters, devices have become progressively smaller as materials technology has developed. Passive components are primarily classified along lines of lumped parameters and power-handling capability. Active components used for amplification and other nonlinear functions require a great deal of special characterization and bias circuitry for application because of their nonlinear nature. Discrete system interfaces are often greatly complicated by level shifting and other matching problems. However, usually only a few different discrete components with limited parameter variations are required for analog and digital systems.

After the introduction of planar diffusion technology, components appeared that contained circuit terminal characteristics. These were the first components that could be described as ideal amplifiers. For digital applications this element is the gate with all of its active and passive components. These nonlinear amplifiers may be combined to provide all

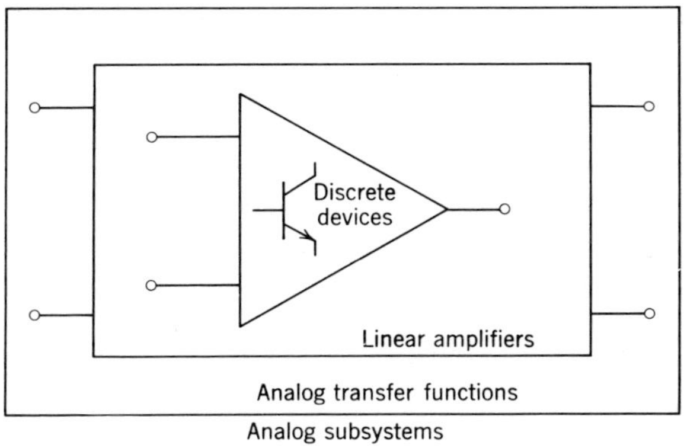

Figure 11.1. Levels of analog integration.

the transfer functions required for digital systems design. Progressively more of these amplifiers were combined to provide further integration. For analog circuits the linear amplifier is the basic building block for most transfer functions and is presently being widely used with external conditioning components for large portions of analog systems. There are still many secondary parameters and level shifts involved at various interfaces.

Based on these concepts, the next level of integration may be defined as the analog transfer function level. Devices such as sense amplifiers, rf strips, and voltage regulators may be classified in this manner since they perform a complete system transfer function with a minimum of

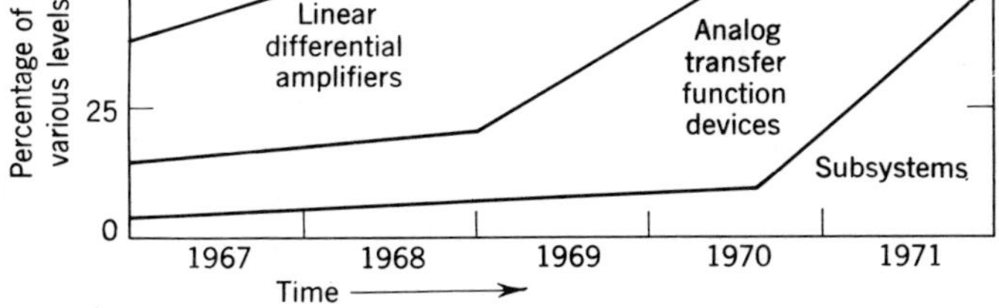

Figure 11.2. Relative number of component types at various levels of integration.

external conditioning. We are only now beginning to move onto this level with a large number of developmental types such as modulators and oscillators. The order of development will, of course, be closely tied to the generality and volume usage that may be expected from each type.

Beyond this come the subsystem and system levels which are harder to define. Although some subsystems have been developed using bipolar and MOS technologies, power and access limitations still require hybrid elements in most of these.

11.2 MONOLITHIC TECHNOLOGY TRENDS

Monolithic technology is progressing on several fronts and is providing greater design flexibility and better performance. The diode isolation concept is under attack. In the past, other techniques have involved complicated additional process steps and low yields, but the air-gap isolation technique first reported by Bell Labs is being developed into a reliable production technique by various equipment, systems, and component houses in the industry; Raytheon is presently contributing to these developments. This technique involves the use of wafers grown and cut to take advantage of the etch-rate characteristics of the (100) one-zero-zero plane of the crystal structure. The diffusion, etching, metallurgical, and testing technologies required to take advantage of this technique have been worked out to the point of providing production prototypes for evaluation. The 703 circuit configuration, Figure 11.3, was used as a vehicle to demonstrate the advantages of removing the parasitics associated with diode isolation. The f_T of this device, with no circuit or geometric modification, was moved from 200 kHz to 1 GHz. Improved geometries and a redesigned circuit will continue the improvement to several GHz. Frequency performance of such a device should be equal to or better than discrete counterparts because of the small and controlled interconnection geometries that are possible.

Several new components are becoming compatible with monolithic technology, and many new and novel components are evolving from the teaming of circuit and process engineers. As circuit designers become more familiar with the design criteria of various semiconductor components, innovations are occuring in the structures to provide unique functions. The ultimate manifestation of this effect is a functional device that uses geometry and diffusion techniques almost entirely to provide transfer functions that have formerly been accomplished with relatively complex circuit configurations. Design analysis techniques are, of course, much more complex for these devices and often require either computer or empirical analog aids. Junction FETs and epitaxial and pinched resistors are a few examples of the components that integrated circuit

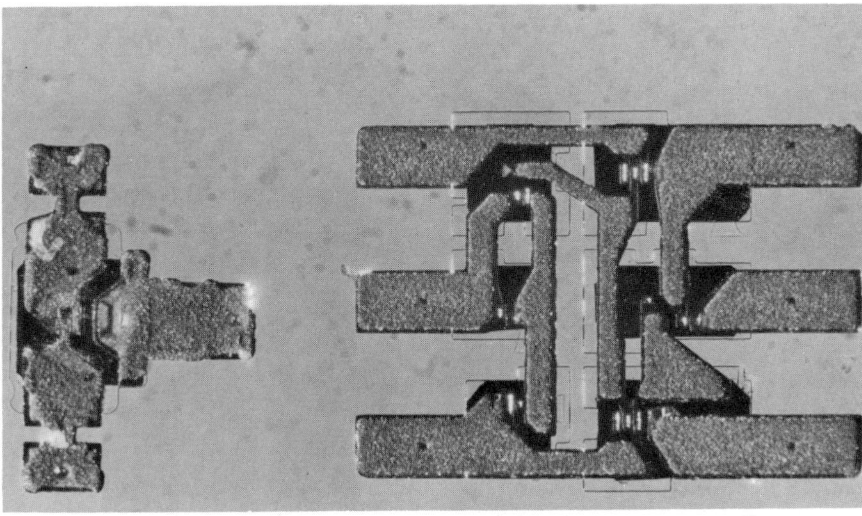

Figure 11.3. Beam-lead construction.

designers are learning to use. The (100) plane material used in air-gap isolated processing also produces transistors with superior noise performance.

11.3 TRANSFER FUNCTION DESIGN AND FABRICATION

As ability to integrate semiconductor devices has increased, other micro-components have become smaller, while maintaining their performance characteristics. Although their ultimate size, cost, and reliability cannot compete with integrated circuits (except in small quantities) their performance can be much better controlled and tailored because of the broad range of materials that may be used. Thick and thin film resistors and capacitors have temperature coefficients in the hundreds of parts per million and a range of resistance values comparable to those of discrete components. Although special external components are required in some critical locations, it is possible to achieve almost any performance level with micro-circuit designs.

Most transfer functions are presently fabricated by hybrid techniques because of the lower initial cost and faster development turn-around time. Monolithic circuits cost between 30,000 and 60,000 dollars to bring into production and have developmental periods of six months to a year because of the several weeks or months required to make a single change.

Hybrid prototypes may be fabricated in a week, but large-volume production costs cannot compete with monolithic circuits that will eventually be produced with fully automated equipment.

Hybrid transfer-function performance characteristics are typically an order of magnitude better than present monolithic designs. Hybrid differential amplifiers achieve picoamperes of input bias and drifts of 1 μV/°C because of the ability to use as yet monolithically incompatible components or selective matching.

For these reasons it is being found by the industry that it is most economical to limit monolithic product developments to standard products and manufacture special-purpose circuits in hybrid form. There are, of course, special cases requiring the size and reliability available only in monolithic devices.

Active component micro-packages with tabs are being used for hybrids, and in some cases leadless integrated devices have been produced by die attaching and bonding the chip into a channel of ceramic for handling ease. Chips and wafers are being sold directly to hybrid-circuit manufacturers in the same manner that discrete transistors and diodes have been in the past. The components are then die-attached and wire-bonded directly into the hybrid module.

Raytheon is in the final production development stages for a Flip Tab* device (see Figure 11.4) that uses air-gap isolation technology to produce individual integrated circuits with weldable tabs that can be etched chemically directly from the wafer. This eliminates completely the need for wafer scribe, break, and die-attach steps, and provides a unit which may be welded directly into a package or hybrid circuit without additional heat cycling of the chip. All tabs may be welded in one step without excessive heat reaching the chip. The tabs are made of thick beams of gold directly attached to the oxide over the diffused junctions and thereby provide very high thermal conductivity paths away from the junctions. It would appear at this time that power dissipation ratings for flip-tab chips in conventional packages will be significantly higher than those of existing devices.

As compatible thin film, flip-tab, and beam-lead technologies are applied to monolithic products, they will make up progressively larger portions of the hybrid circuits in which they are contained, and the trend will be toward the use of more standard transfer function devices. Even if this increases the system complexity somewhat, the total cost will be reduced. Ultimately, the only discrete components remaining outside monolithic structures will be special tailoring and power-dissi-

* Raytheon trademark

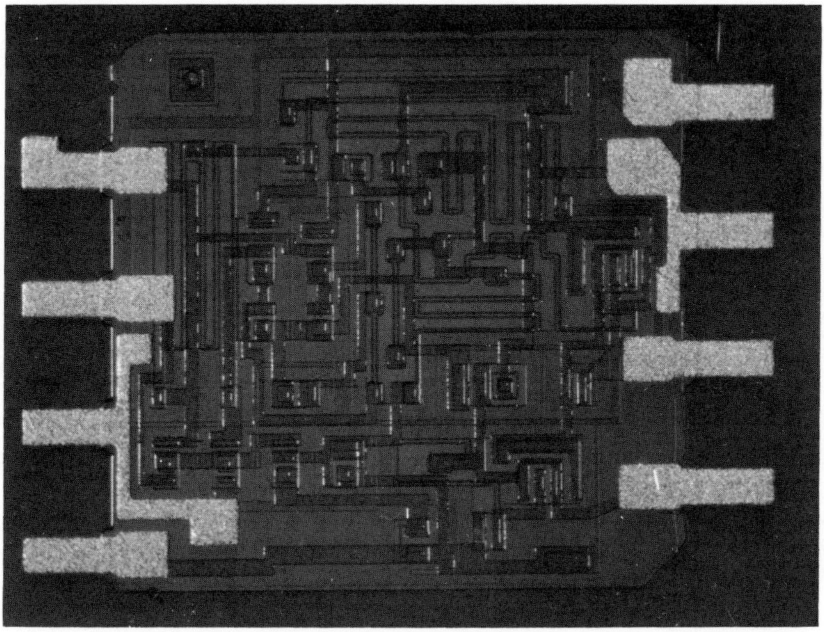

Figure 11.4. Beam-lead IC.

pating devices. A current example of this is the use of high-gain monolithic differential amplifiers in many locations where they could be replaced by much simpler discrete circuits. Cost is reduced by the use of this standard component.

At present, circuit configurations, operating frequencies, and levels within analog systems can be quite varied, since discrete components define few constraints. For this reason, when systems houses request integration of their circuits, they do not conform to any standard functional blocks or specifications, and from this some have concluded that total analog integration is not possible. To counter this argument, it should again be noted that at present relatively complex differential amplifiers are being used in areas for which much simpler circuits would be adequate but actually more expensive, and this trend will continue with broad usage standard devices. Present units, such as the 709, are defining power supply and signal levels. Although larger signal swings may become possible with analog process improvements, the need for them will probably be limited to higher power transfer devices. The decoupling required for small signal stages will bring signal supply voltages down to the existing levels. It is quite conceivable, therefore, that standard 12-to-18 volt systems will be common for some time. This very conveniently allows scaling of dynamic signal levels to ± 10 V. Although one order of magnitude smaller than standard analog computing systems, this does not seem to represent a severe limitation to other than the most sophisticated computing systems. Standardization on a balanced supply voltage and idealization of input and output impedances of each transfer function will greatly reduce the large number of level-shifting operations common to analog systems. This approach increases the circuit complexity of each device but is of small economic consideration compared to the volume advantages. Circuit designs that are relatively independent of supply voltages, as long as they are balanced, allow successive decoupling to lower voltages within the system using the same device.

As numerous transfer function devices are introduced, standard terminologies are going to become increasingly necessary. Somewhat standard multi-port transfer function terminology should be adapted for specifications and data sheets to include nonideal interaction with power supplies and grounds and to make complete characterization possible with limited special definitions.

As an example, the symbology for the common-mode rejection ratio of differential-in differential-out amplifiers presently requires extensive definition. In another chapter (see Chapter Five), the use of a term CDMRR for "Common to Differential Mode Rejection Ratio" is proposed. On first encounter one must study the definition carefully to ensure that all its implications are understood. It would be much cleaner if a multi-port transfer function notation were adopted in the industry. Figure 11.5 shows a DIDO amplifier and a suggested notation for various CMRRs.

It is obviously unnecessary to characterize all the possible interactions, but as various parameters become relevant to particular transfer functions, their definitions will, using a standard format, be much more understandable on first exposure to the user.

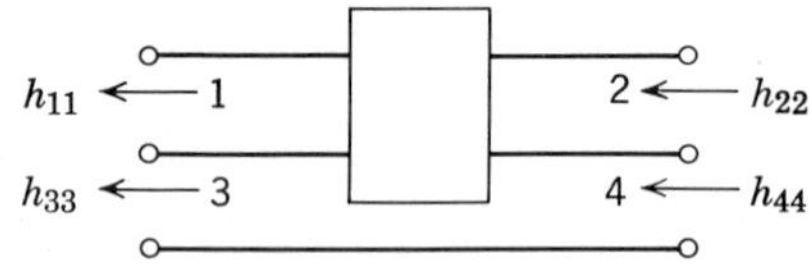

Differential input impedance = h_{11}
Common-mode input impedance = h_{33}
Common-mode to differential-mode coversion = h_{23}

Figure 11.5. DIDO amplifier.

12

A Ferrite-Core-Memory Sense Amplifier

by V. Sarkissian and H. B. Grutchfield
Signetics Corporation

12.1 INTRODUCTION

The performance required from a sense-amplifier IC and its characteristics vary widely depending on the application. There are some parameters, however, that are important in all ferrite-core-memory applications. These are threshold uncertainty, propagation delay, preamplifier overload recovery, pattern insensitivity, and input common-mode characteristics. In general, monolithic sense amplifiers should be designed to optimize these key performance characteristics.

12.2 FERRITE-CORE MEMORIES

Some insight into the performance required from a sense amplifier can be gained by looking at the basic organization and performance of a ferrite-core memory. A simplified block diagram for a typical digital computer memory organization is shown in Figure 12.1. The appropriate command to the address register selects the appropriate X and Y address lines. As illustrated in Figure 12.2, two wires are threaded through each core in the memory. One is common to all cores in that column and the other is common to all cores in that row. Thus, any X address line and any Y address line have one core in common.

Both reading and writing are accomplished by driving current through the wires linking the cores. If this current exceeds a minimum value I_m, the core is reset to a "zero." Any one of the cores may be reset by driving $I_m/2$ through one of the X address lines and through one of the Y address lines. Only the core common to both lines will receive the sum of the currents and will be reset. However, each of the other cores on the two lines will be half-selected by the current $I_m/2$. If a core is to be set to the "one" state, the current may be reversed in the two selecting wires,

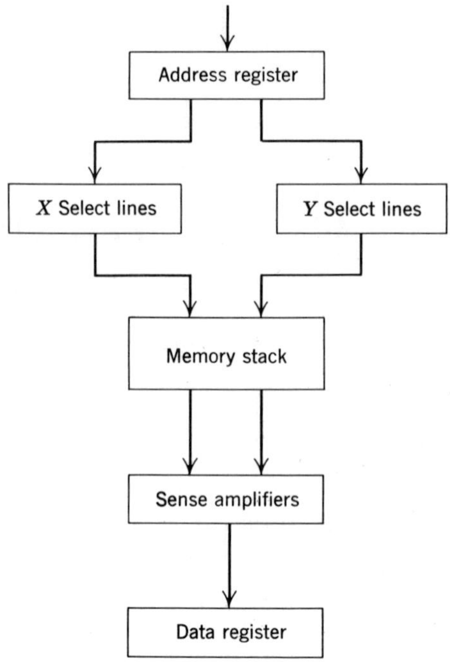

Figure 12.1. Typical digital computer memory organization.

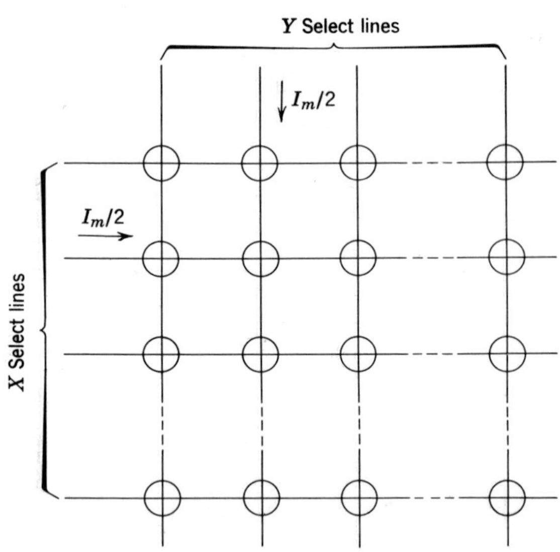

Figure 12.2. Core select wiring.

or additional wires may be provided that link the cores in the opposite direction.

Reading is accomplished by providing a sense-line that links all the cores in the plane. To read a particular core, a "zero" is written into that core. If the core is initially in a "one" state, it is reset and generates a pulse on the sense line. If the core is in a "zero" state, it will not be changed and no output pulse is produced. If the core is in a "one," it is necessary to re-insert a "one" into the core. This is accomplished by writing back into the core the information that it originally contained. The sum of the read and write times is defined as the memory cycle time and is a measure of the speed of the memory.

When a core is half-selected by the driving current $I_m/2$, a noise voltage is generated in the sense line. Other sources of noise in a memory system are electrostatic and magnetic interference within the memory stack itself. The noise appears at the sense-line terminals as either common-mode noise (the potential of both sense-line terminals changes in the same direction and with the same magnitude) or as differential-mode noise (the potential of one sense-line terminal changes with respect to the other). A large differential signal is generated on the sense line when writing into a core. A smaller differential noise is generated when reading a core. However, the noise generated when reading is more critical.

If the amplitude of the noise generated is sufficient, the noise pulse may be recognized as a "one" by the sense amplifier, or conversely, the amplitude of a "one" pulse may be degraded to the point that the sense amplifier detects a "zero." A large common-mode noise pulse may result from electrostatic coupling between the sense line and the activated address line. Similar noise can result from inductive pickup by the sense line.

The sense amplifier must be able to detect the difference between the "disturbed zero" voltage and "disturbed one" voltage. The threshold voltage of the sense amplifier should be adjusted to approximately the mid-point between the sum of the "disturbed zero" voltages and the minimum "disturbed one" voltage. This is illustrated in Figure 12.3. The region on either side of the threshold voltage in which the sense amplifier cannot detect the difference between a "zero" and a "one" is called the "uncertainty region." The "uncertainty region" of a sense amplifier, which is the sum of the variation of the input threshold voltage and the differential offset voltage, must be minimized. Ideally, it would be zero; however, it should at least be smaller than the difference in voltage between the minimum "one" and the maximum "zero."

In most core-memory organizations the disturbed "zero" noise is minimized by threading the sense winding in such a manner that the sense

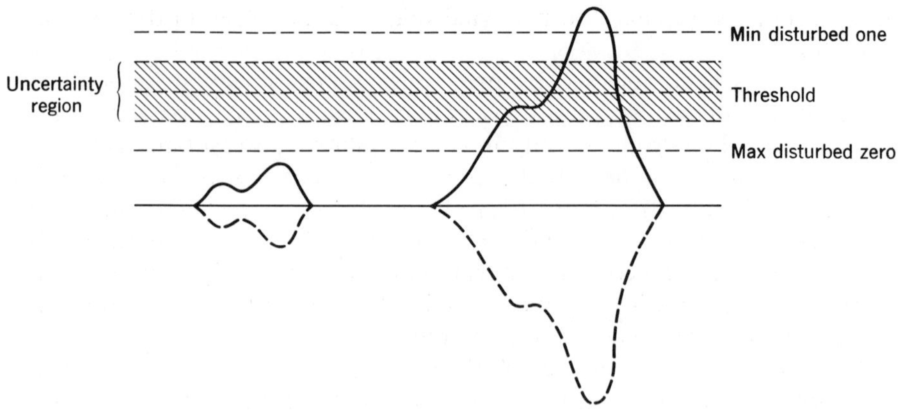

Figure 12.3. Typical core disturbed zero and disturbed one signals.

winding passes through half of the cores in one direction and the other half in the opposite direction, as illustrated in Figure 12.4. This results in partial cancellation of the noise generated by half-selected cores. However, this results in "one" signals of either polarity on the sense line. Therefore, the sense amplifier must respond to bipolar signals indiscriminately, being sensitive only to their amplitude.

Additionally, the sense amplifier must be capable of withstanding large common-mode or differential signals and, in order to minimize the memory cycle time, must recover quickly from overload. These signals can be categorized into two types:

1. Those that the preamplifier should withstand and recover from without damage.
2. Those that the preamplifier should be operative during the presence of the signal.

A large differential signal, for example, appears in a 3D memory just prior to the read time; this is the inhibit noise. This high-amplitude signal normally saturates the preamplifier. The amplifier must recover from this state fast enough to be ready to detect the actual core turnover signal. The time between the trailing edge of the high-amplitude differential pulse and the time when the amplifier is settled to within one percent of its steady state value is called the differential-mode recovery and is a measure of the preamplifier performance. This recovery time is an important factor in determining the memory cycle time.

A similar saturated, or disturbed, state can occur during a high-amplitude common-mode signal. The maximum common-mode signal which the amplifier can tolerate and still detect a small differential signal is called the common-mode range.

In order that the sense amplifier produce an output only during the time when the core is being "read," the output is gated with a strobe line.

12.3 DESIGN CONSIDERATIONS

In order to meet the requirements of state-of-the-art ferrite-core memories, the following were considered:

1. The device had to be truly monolithic. No external components were to be used.
2. In order to avoid pattern sensitivity* or dc restoring circuit, the device had to be dc coupled throughout the circuit. This would increase dc drift and consequently threshold variation. These variations had to be minimized.

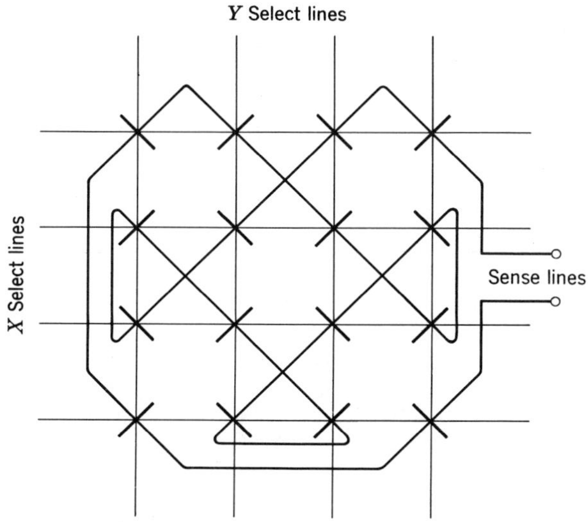

Figure 12.4. Core sense-line wiring.

* An ac coupled amplifier blocks the dc average of the signals appearing at its inputs and, hence, the threshold can vary with the signal pattern. This problem is eliminated if a dc coupled amplifier is used.

3. In order to accommodate high common-mode variations the design had to provide a common-mode range of at least ±3 V, and a 60-dB common-mode rejection at 50 MHz.

4. The faster recovery time needed dictated another constraint on the preamplifier; less than 40 ns both for differential and common-mode overloading signals.

5. The faster memories also require shorter propagation delay, approximately 30 ns.

6. The device had to be easily interfaced with existing logic families.

7. The device had to be versatile to fit as many applications as possible.

Therefore single shots or flip-flops at the output would not be considered. Some method had to be devised, however, to make these functions realizable with simple external connections.

12.4 DESIGN PHILOSOPHY AND CIRCUIT DESCRIPTION

A dc coupled two-stage differential amplifier was considered. The two stages are isolated with an emitter follower and a classical common-mode feedback is provided as shown in Figure 12.5. This circuit provides the

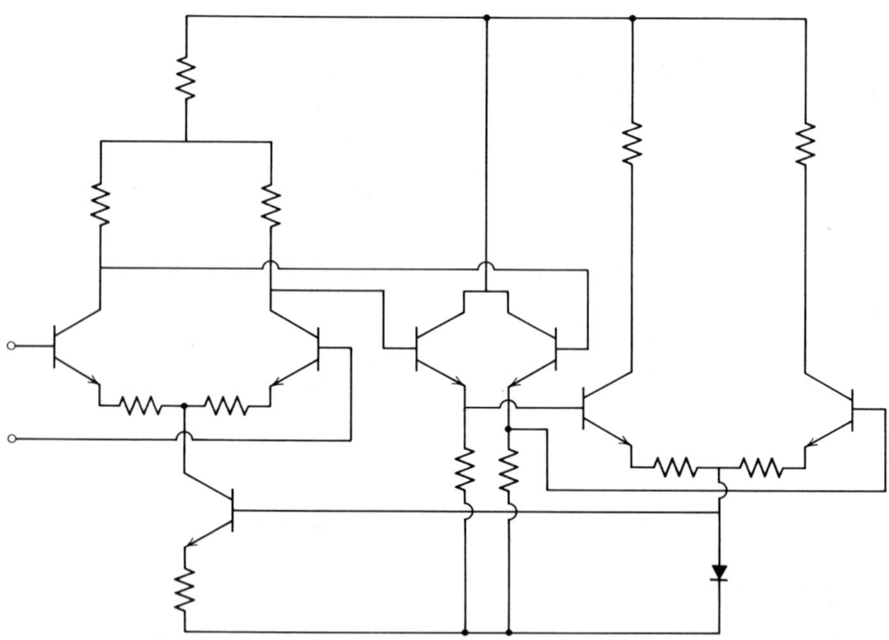

Figure 12.5. Classical two-stage differential amplifier with common-mode feedback.

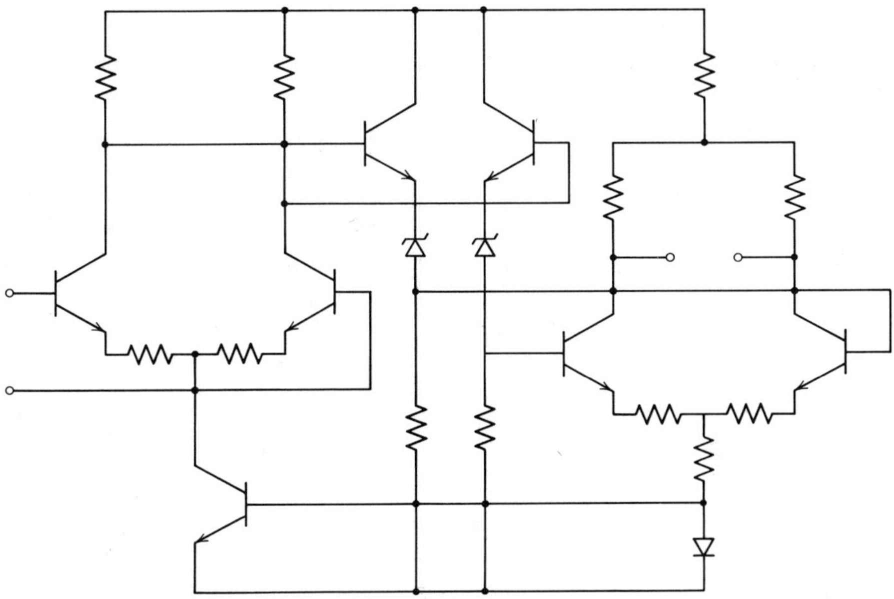

Figure 12.6. Two-stage differential amplifier with zener diode level shifting.

necessary gain and frequency response with relatively good common-mode rejection. However, it suffers from two important factors, that is, common-mode range and overload recovery. The first stage collectors have to be biased at a relatively low voltage to allow coupling to the second stage and provide the necessary voltage swing. This does not provide sufficient collector-to-base voltage on either stage. The low collector voltage of the first stage restricts the input common-mode range. Additionally, high differential-mode signals may saturate both states. In either case, the amplifier has to recover from a saturated condition which creates long overload recovery times. For this reason, the basic amplifier was modified as described in the following paragraphs.

The first stage collectors were designed to be at approximately +4 V and the level was shifted down by zener diodes to couple to the second stage as shown in Figure 12.6. This provides sufficient voltage so that both stages will current limit before saturation. For instance, if a large differential signal is applied at the input of the differential amplifier $+A$, $-B$, the collector voltage of the input $+A$ will decrease from $V+$ by $R_L X$ (total current supplied to both emitters). The base voltage will increase by $V_{in}/2$ (assuming a differential signal); therefore, non-saturation can be guaranteed as long as $V_{in}/2 \leq R_L(I_{1\,total})$. The second stage has similar limiting characteristics; the maximum negative collector swing is less than the maximum positive base swing. Thus, nonsaturating characteristics are ensured throughout the amplifier. Consequently, the recovery time from large differential signals is minimized.

The zener diodes, however, do present another problem. Since the absolute zener voltage variations from unit to unit affect the preamplifier output level, a method had to be devised to minimize this effect. For this, an additional common-mode feedback network was employed as shown in Figure 12.7.

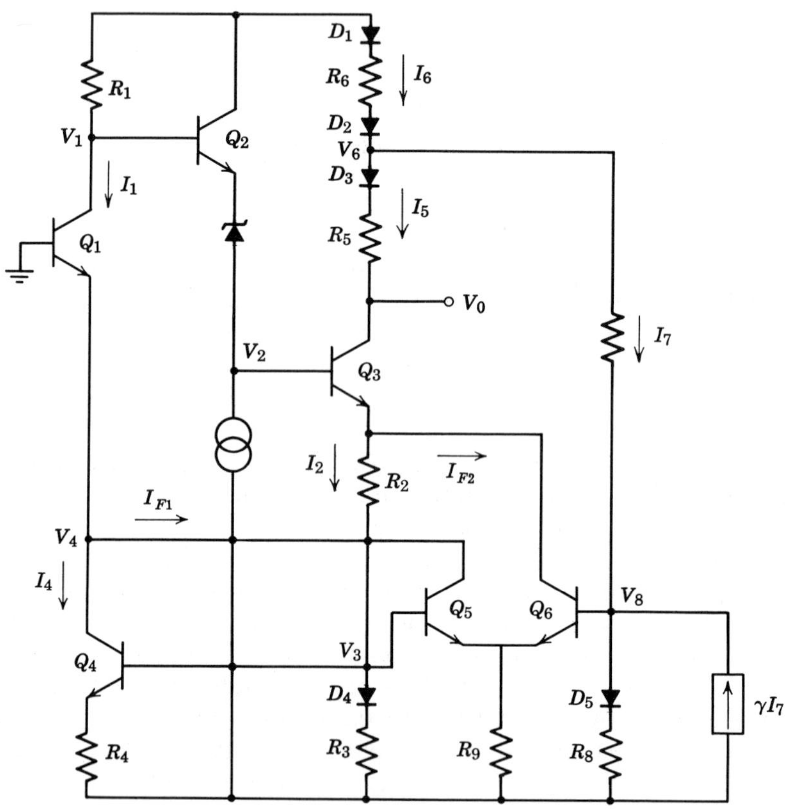

Figure 12.7. Simplified common-mode feedback network.

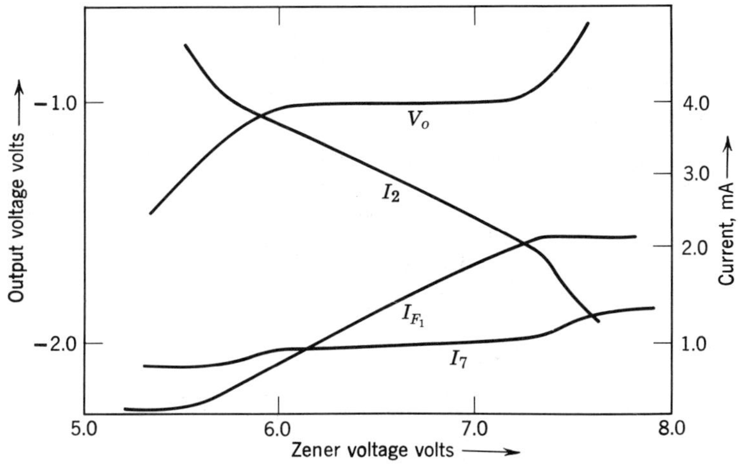

Figure 12.8. Output voltage and currents I_2, I_F, and I_7 versus zener voltage variations.

If V_0 is to remain constant, the current I_5 must also remain constant with zener-voltage variations. The feedback amplifier, Q_5 and Q_6, detects variations in current I_2 resulting from zener voltage variations and compensates for it by varying the feedback currents I_{F1} and I_{F2} in such a manner than I_5 remains unaffected. The generator γI_7 simulates the biasing network for the following current-mode circuit. Curves for V_0, I_2 and I_7 vs zener voltage are shown in Figure 12.8. The diodes D_1, D_2, and D_3 provide temperature tracking. It is evident from Figure 12.8 that the dc output voltage is almost invariant for zener voltages between 6.0 and 7.2 V.

The complete schematic of the sense amplifier is shown in Figure 12.9. The output of the preamplifier is fed into a current-mode circuit that performs both rectification and threshold detection. The preamplifier output is designed to be at approximately −1 V. With a preamplifier gain of about 70 and the reference of the CML circuit at ground potential, the input threshold voltage will be set at approximately 15 mV. Since either of the collectors can turn the CML circuit on, rectification, or detection of bipolar signals, is achieved. The threshold voltage can be changed, as desired, by varying the reference voltage (threshold set). The output of the emitter follower drives a TTL gate that has another input used for strobing. The output gate is a regular Signetics 8000-family type gate with the exception that it has an external input that makes other functions possible without extra propagation delay.

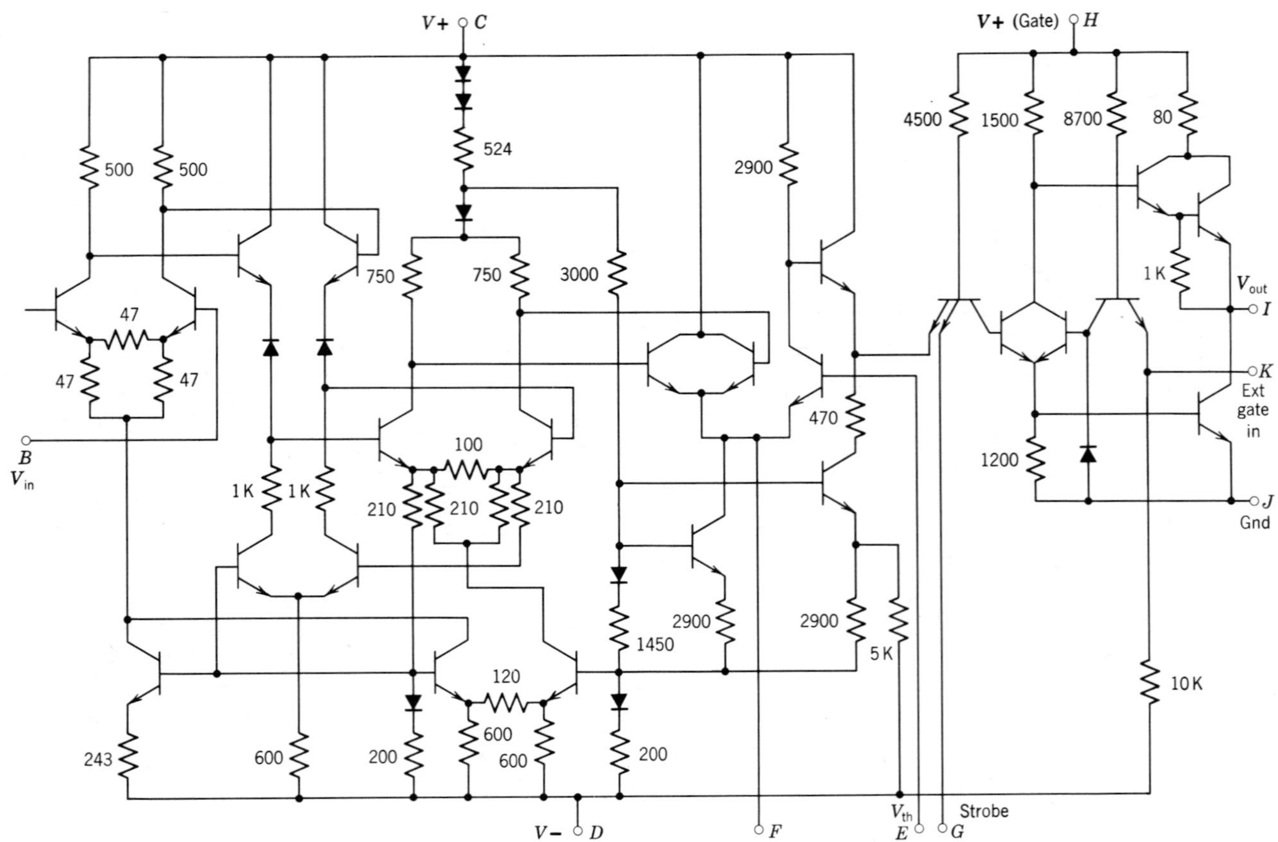

Figure 12.9. Schematic of SE525 sense amplifier.

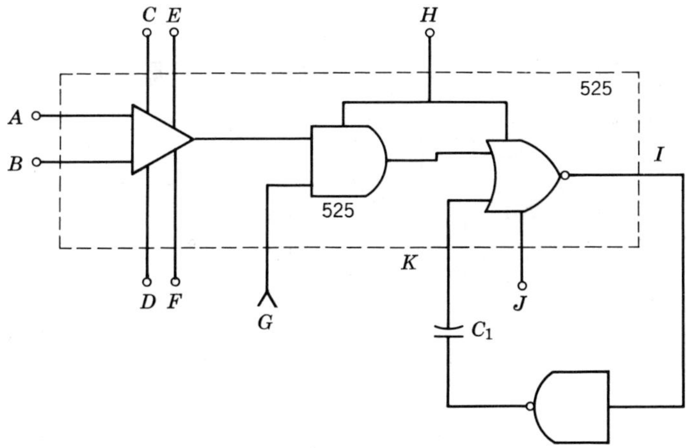

Figure 12.10. Pulse stretcher.

One such possibility is pulse stretching. If the output of the sense amplifier is fed back to the extra input through an inverter and a capacitor as shown in Figure 12.10, the output pulses can be stretched as required by varying the size of the capacitor. Another possibility is the use of an external NOR gate, as shown in Figure 12.11, to construct a simple latch circuit at the output.

Separate power supply connections have been provided for the preamplifier and the output gate. This allows preamplifier strobing prior to the read cycle for power reduction.

12.5 CONCLUSIONS

The monolithic sense amplifier was characterized at ±5 V and the following results were obtained:

1. The preamplifier has a gain of approximately 70 and a frequency response of about 20 MHz. These results were very close to those obtained from an earlier computer analysis.
2. The dc level variation of the preamplifier output was within 200 mV of the design target value (−1.0 V). This, combined with preamplifier offset, gain variation, and temperature drift, defines an uncertainty region of 6 to 8 mV within a temperature range of 0 to 70°C.
3. Propagation delay from input to output was typically 25 ns with a 25 mV input signal and a load consisting of 300 Ω to $V+$ and 22 pF to ground. The typical strobe release time with the same load is 12 ns.

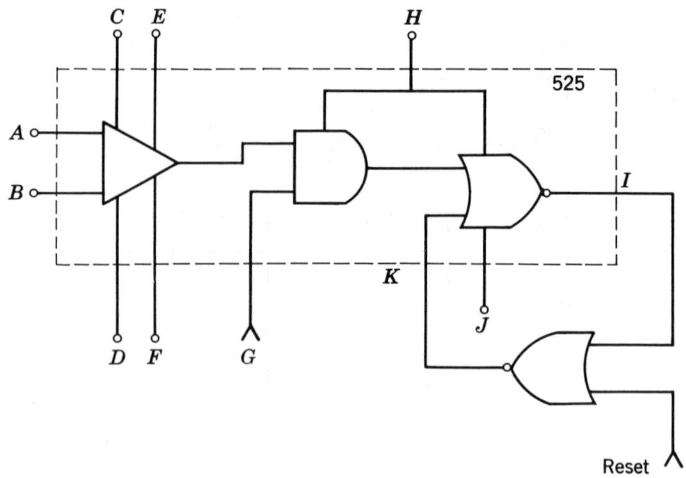

Figure 12.11. Latch.

Figure 12.12 is an oscillogram of the input and output waveforms. The spikes (marked by *X*) on the input pulse are due to current spiking by the output gate. This can be minimized by using a low impedance supply or by using separate supplies for the preamplifier and the gate. Figure 12.13 is an expansion of Figure 12.12 to illustrate the propagation delay in more detail.

4. The input common-mode range is a minimum of +4 and −3.5 V and the common-mode rejection ratio of the preamplifier is better than 60 dB at 20 MHz.

5. The device was operated with both the latch and the pulse stretching output configuration with very satisfactory results. Oscillograms of the pulse stretcher output together with the input signal and the strobe are shown in Figures 12.14*a* and 12.14*b*. Figure 12.14*a* illustrates the pulse stretching capability with a 68-pF capacitor connected as shown in Figure 12.14*c*. The strobe is less than 40 ns, the limit of the pulse generator used. The theoretical limit of the minimum strobe pulse width is approximately 25 to 30 ns. Figure 12.14*b* illustrates the possibility of operating at very high duty cycles since the capacitor can be easily discharged.

6. As mentioned earlier, the preamplifier can be strobed in some applications where power consumption is a prime consideration. This is accomplished by applying a pulse of amplitude equal to the V+ supply

and duration long enough to allow settling of the amplifier, the read portion of the signal, and the strobe. Figure 12.15 shows oscillograms of the output power supply strobe, input signal, and signal strobe pulses all superimposed. In Figure 12.15*a* the input signal is just below the threshold voltage; hence, the output goes low for the duration of the input signal. It is evident that the settling time of the amplifier is shorter than 30 ns.

It should be clear also that this technique can be used with a pulse-stretching configuration at the output, and thus longer output pulses can be obtained even though the preamplifier will not be on after the signal strobe time.

12.6 SUMMARY

A monolithic integrated sense amplifier which meets the requirements imposed by state-of-the-art ferrite-core memory systems was designed and built. Test results show that all design objectives were realized. The finished device is a versatile, high performance sense amplifier, useable in a wide variety of ferrite-core memory sense amplifier applications.

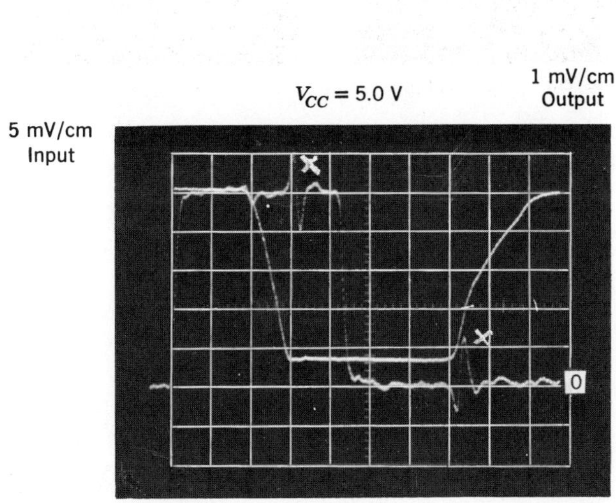

Figure 12.12. Input/output characteristics.

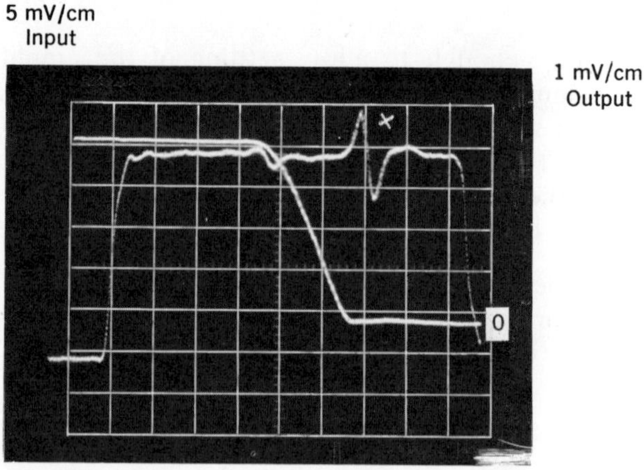

Figure 12.13. Propagation delay.

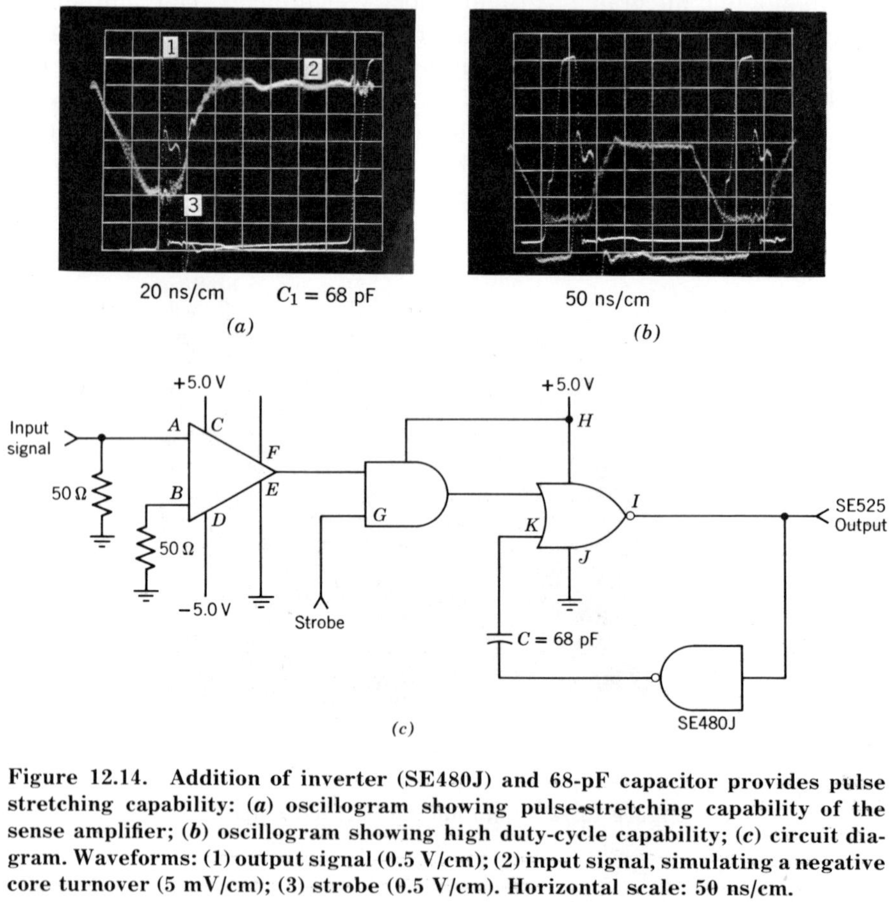

Figure 12.14. Addition of inverter (SE480J) and 68-pF capacitor provides pulse stretching capability: (*a*) oscillogram showing pulse-stretching capability of the sense amplifier; (*b*) oscillogram showing high duty-cycle capability; (*c*) circuit diagram. Waveforms: (1) output signal (0.5 V/cm); (2) input signal, simulating a negative core turnover (5 mV/cm); (3) strobe (0.5 V/cm). Horizontal scale: 50 ns/cm.

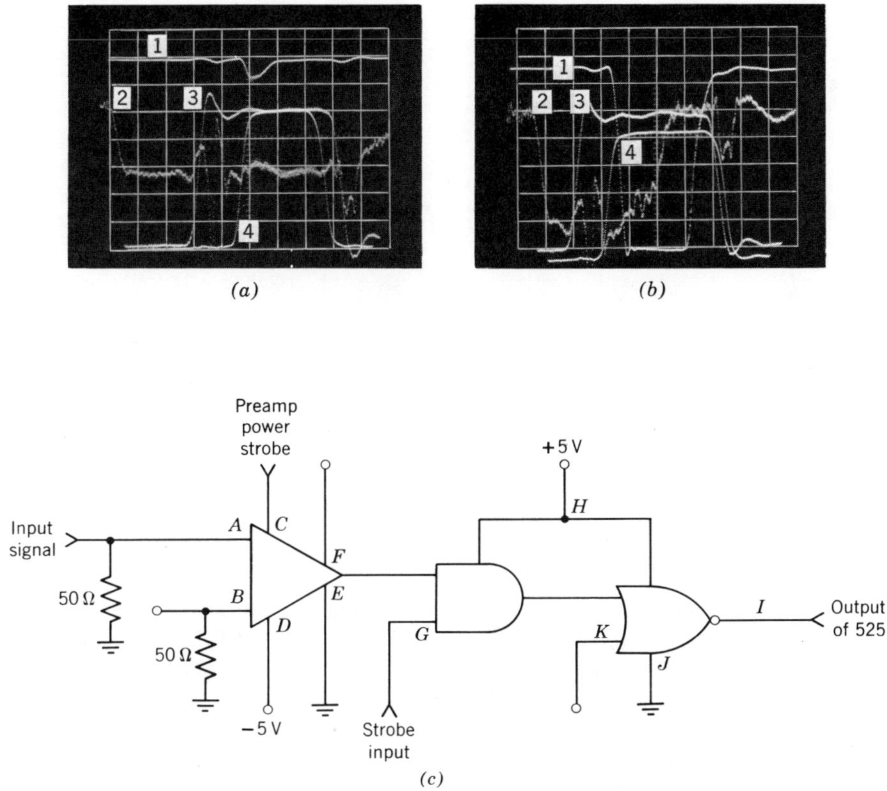

Figure 12.15. Operation with a strobed preamplifier: (*a*) input signal below threshold; (*b*) input signal above threshold; (*c*) circuit diagram. Waveforms: (1) output signal (0.5 V/cm); (2) input signal (5 mV/cm); (3) preamp power strobe (1 V/cm); (4) gate strobe (0.5 V.cm). Horizontal scale: 20 ns/cm.

13

A Monolithic Voltage-Regulator Building Block

by W. H. Williams
Westinghouse Electric Corporation, Molecular Electronics Division

13.1 INTRODUCTION

The series voltage regulator is one of the more popular discrete voltage regulators. An integrated version of the circuit, the WM330, can be used in applications that require up to 2 A of output current. Within the range of 1 to 43 V output, this device performs as well as or better than its discrete counterpart. A significant saving in space is also realized because the regulator IC is housed in the low-profile TO-3 package. Equally important is the connection of the substrate to the circuit ground which permits mounting of this device directly to a chassis for heat sinking.

The simplicity of the circuit and the availability of several terminals make the WM330 an excellent basic building block. The circuit may be modified to satisfy the performance requirements of most systems by adding only a few external components. This chapter shows how improved regulation, overload protection and extension of the unregulated input voltage range are accomplished. The regulation of voltages above and below the rated output values ($1.0 < V_0 < 60$ V) are treated, along with means for increasing the output current capability. The connection for producing a negative output voltage is also covered.

Line regulation can be achieved to within 0.01% while load regulation is maintained to less than 0.2% variation. These values are not exceeded with output voltages up to 60 V. All measurements were made in free air at room temperature; thus, heat sinking will provide improved performance.

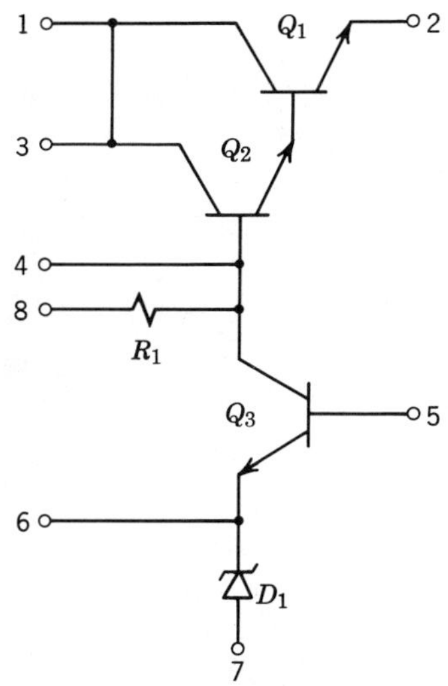

Figure 13.1. Circuit schematic for the WM330 voltage-regulator IC.

13.2 THE BASIC CIRCUIT

The circuit schematic of the WM330 is shown in Figure 13.1. The Q_1 and Q_2 compound-connected Darlington pair function is the series control element. This compound connection has a dc current gain (h_{FE}) in excess of 10,000 at 1 A of collector current. The transistor Q_3 is the amplifying stage, while D_1 serves as the internal reference element. D_1 is an avalanche diode having a breakdown of about 6.2 V and an impedance of 14 Ω when 1 to 5 mA of current is supplied. The bias resistor R_1 has a value of 1000 Ω. The temperature coefficients of D_1 and V_{BE3} are equal and opposite, which results in excellent tracking over temperature.

The integrated circuit chip, shown in Figure 13.2, has an overall die dimension of 140 × 120 mils. The device is mounted in an 8-pin version of the TO-3 package. The thermal resistance of the package in free air is 20°C per watt. This can be reduced to 3°C per watt when a reasonable sized heat sink is used.

The simplest circuit connection for using the WM330 is shown in Figure 13.3. Pins 1, 3 and 8 are connected to the unregulated input. The sensing potentiometer is connected between the output, pin 2, and the return with the wiper attached to pin 5. Pin 7 is connected to the return. Pin 4 accommodates the compensating capacitor which suppresses internal oscillations.

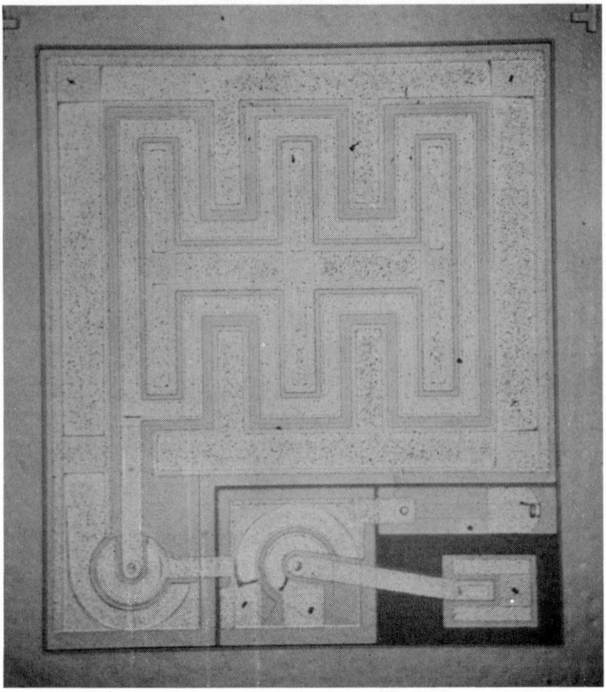

Figure 13.2. Photo of the voltage-regulator IC chip.

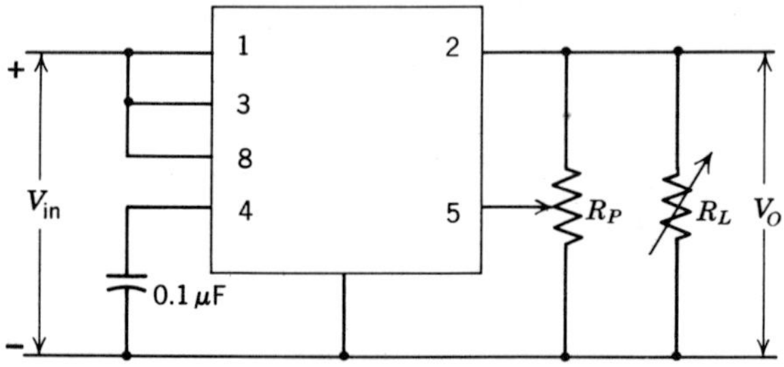

Figure 13.3. Basic connections for the regulator.

Typical operation of the regulator requires that the potentiometer, R_p, be adjusted to give the desired voltage at the load terminal. A portion of the variation in the load voltage is applied as an error signal to the base of Q_3. The error signal is amplified by Q_3 and applied to the base of the series control transistor; this results in an increase or decrease in the load current according to the need. This action causes the load voltage to be returned to the desired value. In this arrangement line and load regulation are maintained to within 2% for $0 < V_0 < 43$ V.

The voltage setting of the sensing potentiometer is determined by the voltage reference element. The overall value of the bleeder resistance selected for the potentiometer is dictated by the drive current required into the base of Q_3. Since the total bleeder current should be at least 10 to 20 times that supplied to the base of Q_3, it must also be low enough to insure regulation stability over temperature. For the WM330, the sensing element is normally selected for a total resistance of at least 100 Ω per V of the output.

13.3 PREREGULATION (METHOD 1)

Improvement in line regulation can be achieved by using a preregulator circuit. Preregulation is accomplished by connecting three resistors and an avalanche diode as shown in Figure 13.4. Since a 1-kΩ resistor already exists on the chip, only two resistors and a diode are necessary for some applications.

The zener diode D_2 has a sharp breakdown voltage which is relatively constant with zener current. This device maintains a constant voltage drop across R_2 which insures that bias current is constant for changes in the line voltage. The breakdown voltage of D_2 is selected according to the expression $V_{Z2} < V_{in}\text{min} - V_0$ and is normally chosen, as a rule of thumb, to be about four times the base-emitter drops of the series control element.

The resistor R_2 is selected according to the current flow in the collector of Q_3. This assumes that current flow into the compound-connected Darlington is negligible because of the high current gain associated with these transistors. Thus $R_2 = (V_{Z2} + 2V_{BE})/I_{C3}$ provides a reasonable means for selecting R_2. If the internal 1-kΩ resistor is used, some adjustments may be required in the selection of other components.

I_{C3} is selected primarily according to the beta vs I_C characteristic of Q_3. Since the internal reference element of Q_3 is operated in the 1 to 5 mA range, the current flow in R_4 is selected accordingly and in relationship to the value of I_{C3}. For instance, if I_{C3} is chosen to be 1 mA, R_4 may be

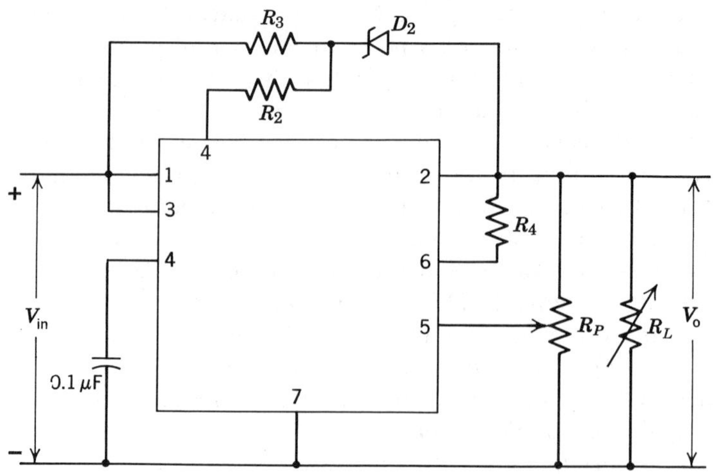

Figure 13.4. Preregulation using Method 1.

selected to provide an additional 2 mA of current into the reference element. Therefore $R_4 = (V_0 - V_{Z1})/I_{Z1}$, and since V_0 is the regulated output, the established value of I_Z will be constant for line and load variations.

The value of R_3 must be selected so that it supplies the required bias current for the zener D_2 and bias resistor R_3 under the worst case conditions. In so doing, R_3 is selected to be as follows:

$$R_3 \leq \frac{V_{\text{in}}\,\text{min} - V_0 + V_{Z2}}{I_{Z2}\,\text{min} + I_{R2}}.$$

This circuit provides an excellent means for obtaining improvement for irregularities of the line voltage when relatively constant loads are being supplied.

13.4 PREREGULATION (METHOD 2)

Figure 13.5 shows another way that the basic regulation characteristic of the device may be improved with only two additional resistors. Resistor R_3 provides practically all of the current necessary to maintain the zener reference element D_1. R_2 serves only to provide a minimum of bias current to the series control element and the collector of the amplifier Q_3. As stated previously, the current gain of the compound Darlington is so high that its bias current can be neglected. The use of this configuration actually depends on the low-current beta characteristics of the amplifier stage Q_3. This transistor has a typical dc current gain (h_{FE}) of about 40 at a collector current of 100 μA. As a result the circuit functions adequately as long as sufficient base drive is available. To insure that sufficient base drive is available, the bleeder resistor should be selected so that the bleeder current is a minimum of ¼ the collector current of Q_3. The approximate design equations are the following:

$$R_3 \leq \frac{V_0 - V_{Z1}}{I_Z \min},$$

$$R_2 \leq \frac{V_{\text{in}} \min - V_0 + 2V_{BE}}{I_{C3} \min},$$

$$I \text{ bleeder} \leq \frac{I_{C3} \min}{4}.$$

13.5 PREREGULATION (METHOD 3)

A great improvement in the regulation characteristics of the basic device can be obtained by using an external transistor as a constant-current source. This can be accomplished by supplying the bias current to the control element and amplifier stage through pin 4. Figure 13.6 shows how a pnp transistor may be connected to the regulator and function as a constant-current source. The resistor R_3 and diodes D_2 and D_3 provide bias current and voltage at the base of Q_4 to feed 1 to 5 mA of current through R_4 and into the biasing terminal. Since the base voltage of Q_4 is very nearly equal to its emitter voltage, changes in the input

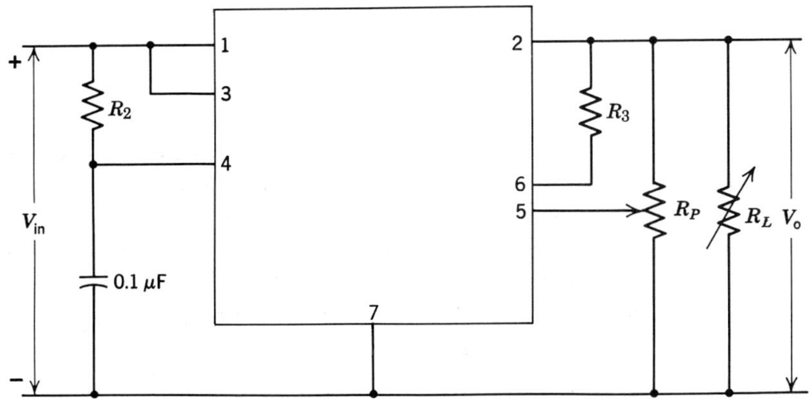

Figure 13.5. Preregulation using Method 2.

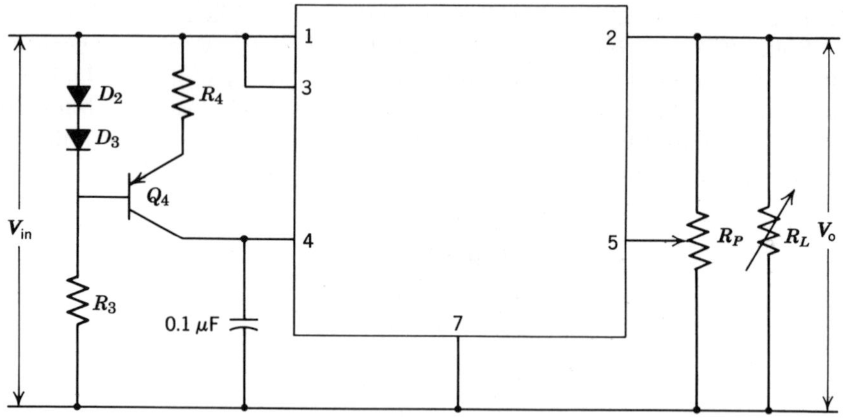

Figure 13.6. Preregulation (Method 3) using a constant-current source.

voltage do not reflect a significant change in collector current flow. With 1 to 5 mA flowing into the junction point, pin 4, Q_3, and D_1 are operated in their optimum modes simultaneously with Q_3 having high voltage gain and D_1 in a stable breakdown mode. As a result the overall regulation capability of the device is considerably improved.

The results of the methods just discussed are given in Table 13.1.

TABLE 13.1 Comparison Chart

Method	Circuit	Line Regulation	Load Regulation
1	Figure 13.4	0.05%	0.5%
2	Figure 13.5	0.5%	1.0%
3	Figure 13.6	0.02%	0.2%
4	Figure 13.11	0.01%	0.1%

13.6 OVERLOAD PROTECTION

Nearly all applications involving the use of voltage regulators require some form of overload protection. For some systems overload protection may be adequately achieved by using fuses, circuit breakers, or series resistance. With semiconductor devices, however, be they discrete or integrated, the above methods are prohibitive. The following two methods overcome the inadequacies of fuses, circuit breakers, and resistance as protective elements for the WM330.

Method 1: the SCR Lockout or Crowbar Circuit

Figure 13.7 shows a configuration which is commonly referred to as a crowbar circuit. In this arrangement R_2 is the current sensing element and R_3 provides current to initiate turning on the SCR when an overcurrent condition exists. The SCR, once turned on, provides a path which is very nearly a short to the fuse and causes the fuse to become open circuited from excessive current flow. The current demanded by the SCR is so much in excess of the rated value of that required for normal fuse action that the fuse will blow in a very short time. This time is short compared to the thermal time constant (which may cause destruction) of the transistor in the series element. When a value of overload current has been selected, R_2 and R_3 may be determined as follows:

$$R_3 \geq \frac{V_0'}{I_{OL}\min} \text{ or } V_0' = I_{OL}\min R_3,$$

and

$$R_2 \leq \frac{V_0' - V_g \min}{I_g \max}.$$

These values ensure that the circuit will function for worst-case conditions of the load and SCR characteristics.

Method 2: Overload Current Limiting

A transistor and its associated biasing resistors, connected as shown in Figure 13.8, constitute another popular technique for overcurrent protection. This arrangement is generally referred to as the fold-back method because when a fault is sensed, the output voltage and current are simultaneously reduced. The output voltage is reduced to zero while the current flow is decreased to a low sustaining value. This sustaining current is generally much less than the value of the overload current that initiates

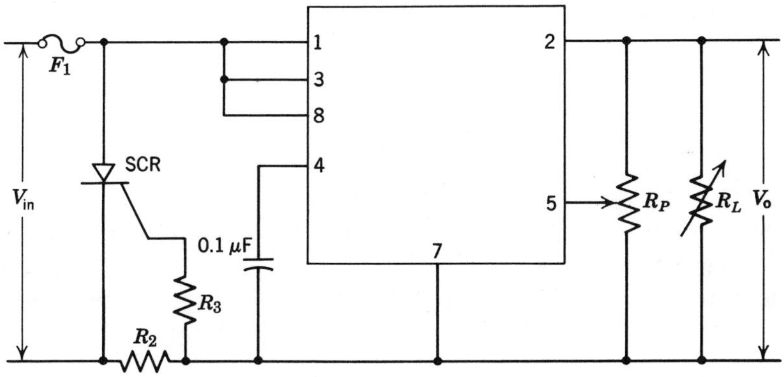

Figure 13.7. Over-current protection using an SCR.

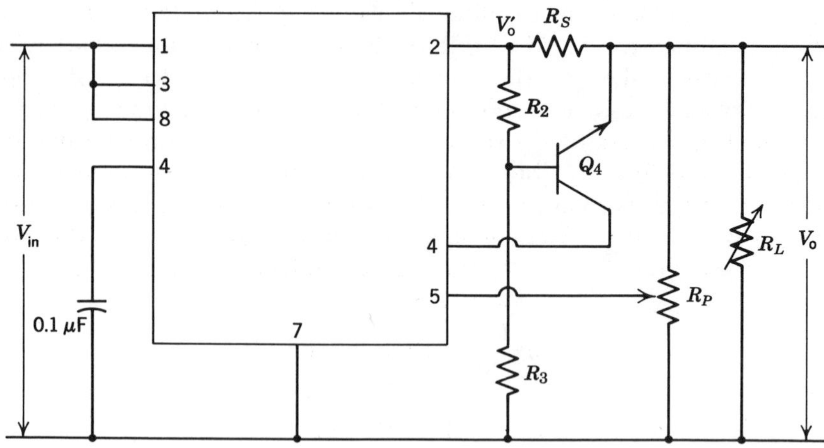

Figure 13.8. Over-current protection using fold-back current limiting.

the circuit operation. Since the sensing circuit is within the basic feedback loop of the regulator, the overall regulation will not be significantly affected.

Resistor R_S is the current sensing element and is usually of small value ($1\,\Omega$ to $10\,\Omega$). The resistors R_2 and R_3 constitute the bias network for Q_4 and should be selected to provide sufficient drive capability during fault conditions. The transistor Q_4 should be selected to have a power capability in excess of that required to handle the maximum current supplied through R_1.

In the normal mode, Q_4 is operated in a non-conducting state such that the following conditions are satisfied:

$$I_L R_S < V_{B4} + V_{BE4},$$

where

$$V_{B4} = \frac{V_0' R_3}{R_2 + R_3}.$$

Thus $V_{B4} < V_0$, and a reversed biased condition exists.

When a fault is experienced, $I_L R_S > V_{B4} + V_{BE4}$ so that Q_4 conducts. In the conducting state, Q_4 removes the bias current from the series control element and the amplifier stage. When Q_4 becomes saturated, the maximum collector current that it must handle is given by

$$I_{C4} \simeq \frac{V_{\text{in}}\,\text{max} - V_{CE4}}{R_1}.$$

After Q_4 becomes fully saturated, the current flow in R_S is reduced to a small value such that the power dissipation of the series element is not exceeded. Even though the series element is supposedly cut off, it does provide a small but sufficient bias current to maintain the on condition of Q_4 and a trickle current through R_4.

Figure 13.9 shows the results obtained for a specific application requirement, while Figure 13.10 presents a graphical means for selecting R_S based on the sense current to be detected for this special arrangement.

13.7 EXTENDING THE UNREGULATED INPUT VOLTAGE RANGE

The useful range of the unregulated input voltage may be extended to a value beyond the device rating with the circuit shown in Figure 13.11. In this circuit the zener reference D_2 must be chosen so that the maximum rated input voltage of the regulator is not exceeded. Similarly, transistor Q_4 must be selected with a rating greater than the maximum power it must dissipate. Thus $V_{CE4}I_L$ max $\simeq (V_{\text{in}}$ max $- V_{Z2} - V_{BE4})\, I_L$ max $\leq P_D$, where P_D is the power handling capability of Q_4. The resistor R_2 must provide a sufficient bias current for the zener D_2 when the minimum input exists:

$$R_2 \leq \frac{V_{\text{in}}\,\text{min} - V_{Z2}}{I_{Z2}}.$$

The above expression neglects the base bias current needed to maintain Q_4.

Even though the primary intent of this modification is to permit the handling of larger inputs, some improvement in regulation does result. This occurs because D_2 is stable for reasonable excursions of the input and maintains a constant input to the basic regulator.

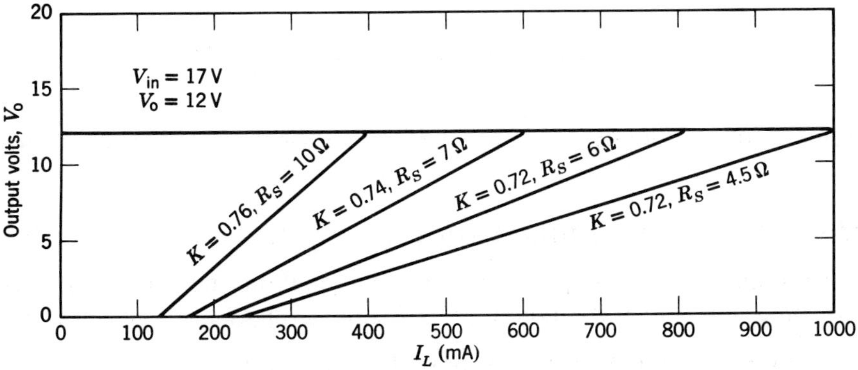

Figure 13.9. Fold-back current limiting.

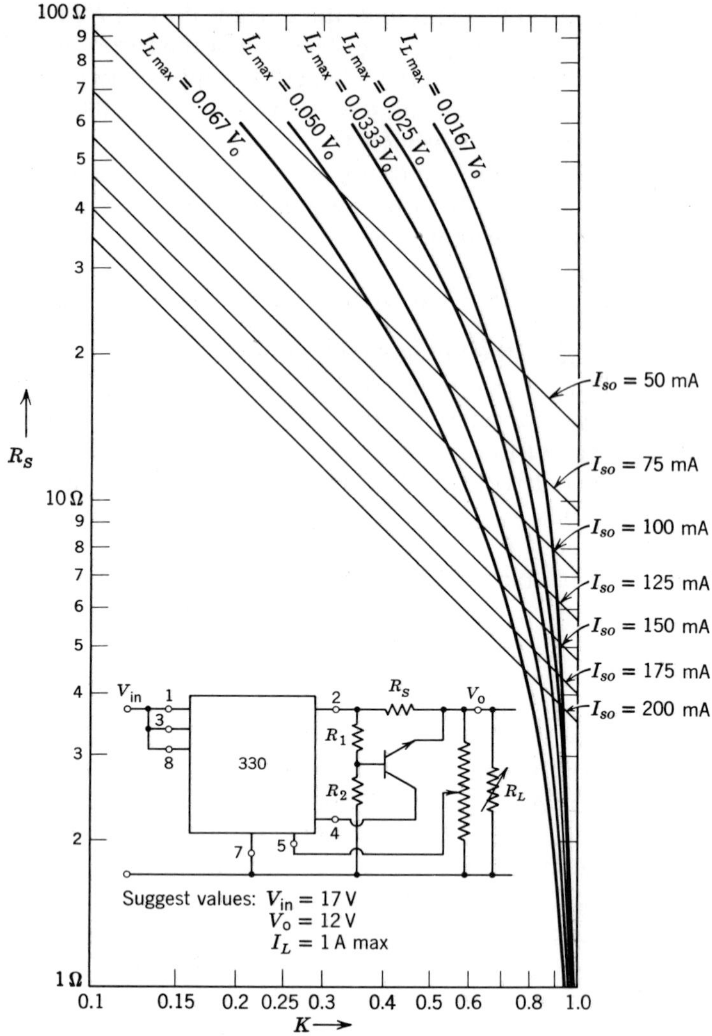

Figure 13.10. Curves for selection of current limiting resistor and bias networks.

13.8 REGULATING HIGH-OUTPUT VOLTAGES

To regulate high output voltages, two methods may be employed: elevating the substrate potential or using a piggy-back arrangement.

Method 1: Elevating the Substrate Potential

It is possible to obtain regulated output voltages which are above the device rating. This may be accomplished with a method similar to the one used for regulating low outputs from high-voltage inputs. The procedure is to elevate the substrate above ground potential by an amount that is consistent with overall device limitations. For instance, if the maximum output is specified as 50 V and it is desired to have a 100-V output, the substrate must be elevated by 50 V. If this is done, the reference element will be approximately 56.2 V. This circuit arrangement is illustrated in Figure 13.12.

Method 2: Piggy-back Arrangement

An extension of the above approach can be used to achieve two or more regulated outputs from a single input. This is done by connecting the regulators in a piggy-back fashion as shown in Figure 13.13. Note that the zener reference in the upper regulator is used in a preregulator arrangement for the lower one. If used as shown, the upper output is regulated to within 2% while the lower one is regulated to within 1%.

One precaution which must be observed is that whenever pin 7 is not operated at ground, the package should be isolated from the mounting surface with a mica washer or by other means. Also, the need for overload protection dictates the use of a method equivalent to the SCR lockout circuit.

13.9 REGULATING LOW-VOLTAGE OUTPUTS

Two methods that can be used to regulate low-voltage outputs are lowering the substrate potential and employing a resistor divider network.

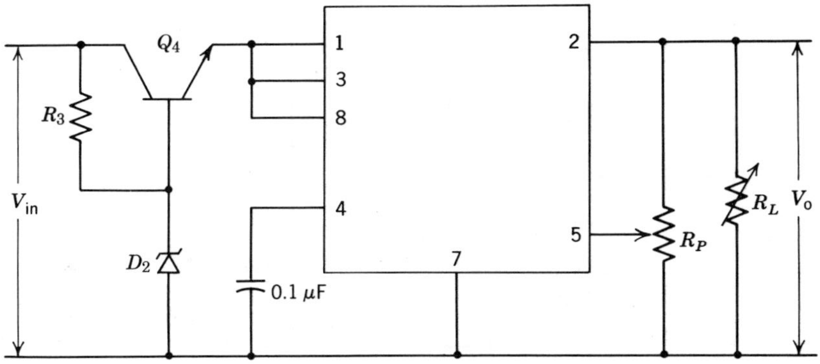

Figure 13.11. Method for extending the input voltage range.

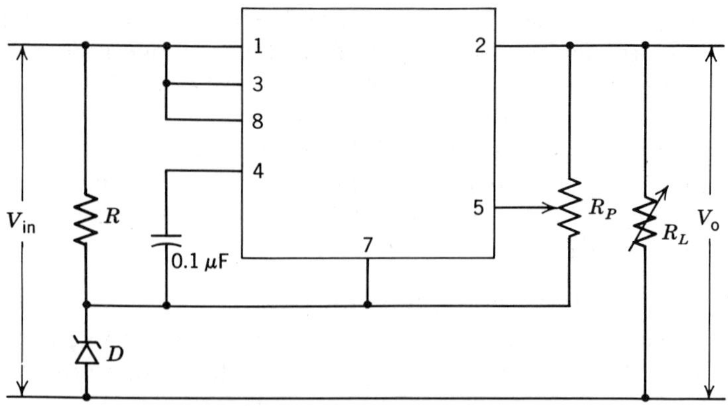

Figure 13.12. Circuit for elevating the substrate potential to obtain large output voltages.

Method 1: Lowering Substrate Potential

Just as the substrate potential was increased to permit the regulating of higher output voltages, it may be lowered to allow lower outputs to be regulated. In so doing, regulated outputs of 6 V and lower can be obtained. Figure 13.14 illustrates a circuit arrangement for a 6-V supply. Regulation to within 1% can be achieved while tracking over temperature is excellent.

The reference element D_2 should be exactly equal to or less than the internal reference D_1. Thus $V_{Z2} \simeq V_{Z1}$, and the resistor R_2 is determined

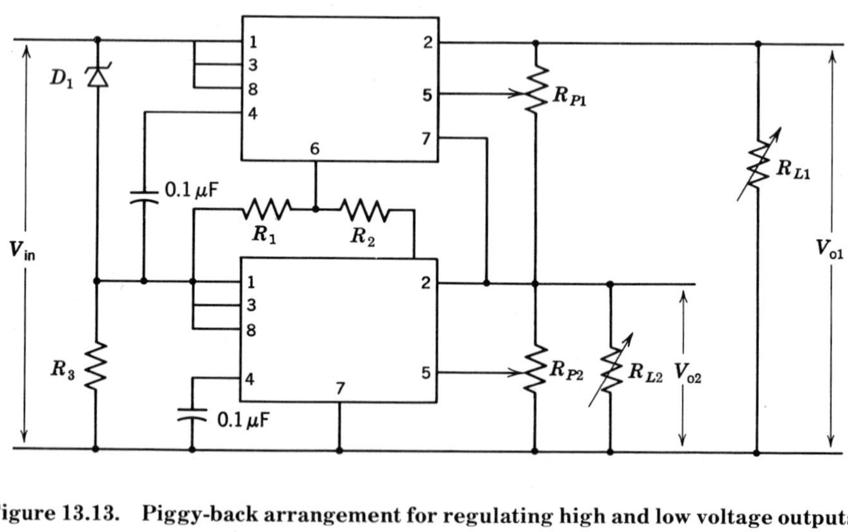

Figure 13.13. Piggy-back arrangement for regulating high and low voltage outputs.

by $R_2 = (V - V_{Z2})/(I_{Z2} + I_0')$, where V is the negative source voltage and I_0' represents the current flowing in D_1 and the sensing potentiometer. V_{Z2} and I_{Z2} are the voltage and current of D_2; D_3 is a temperature compensating diode used to counteract the change in V_{BE3} over the temperature range.

Method 2: Resistor Divider Network

Another approach to obtaining a lower reference voltage than the internal element affords is to use the connection shown in Figure 13.15. This method does not use an additional voltage source as required by the first method. The reference zener D_2 in this case is selected to be 6 V or greater to take advantage of its positive temperature coefficient. This offers some measure of stability over temperature by counteracting the change in V_{BE3}. Of course, this stability is very much dependent on the quality and composition of the biasing resistors R_2, R_3, and R_4. The approximate design equations stipulate that

$$R_2 \leq \frac{V_{\text{in}}\min - V_{Z2}}{I_Z + 1},$$

$$I = \frac{V_{Z2}}{R_2 + R_3},$$

$$R_2 = \frac{V_{Z2} - V_{\text{REF}}}{I},$$

$$R_3 = \frac{V_{\text{REF}}}{I + I_0},$$

where I_0 is the current available from the emitter of Q_3.

13.10 BOOSTING THE OUTPUT CURRENT CAPABILITY

The easiest means for boosting the output current capability of the WM330 is to add another transistor to the Darlington arrangement. Since the current capability of the basic device is large, a very powerful transistor can be selected for the added current requirements. For example, the output of Q_1 is sufficient to drive 10 to 50 A power transistors to outputs from 5 to 25 A. The only precautions to observe are that the discrete power transistor's heat sink must be isolated from the monolithic circuit's heat sink and the voltage limitations of the IC must not be exceeded. Figure 13.16 illustrates the appropriate connection.

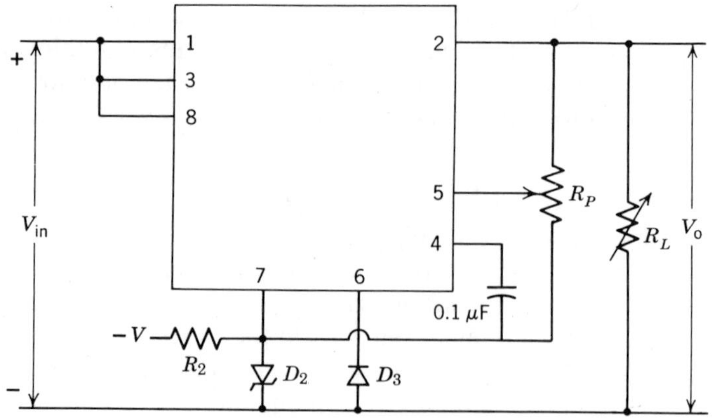

Figure 13.14. Low-voltage (under 8.0V) outputs using negative power source and zener.

13.11 REGULATING NEGATIVE OUTPUT VOLTAGES

To regulate negative voltages is simply a matter of grounding the normal positive output and connecting the load to pin 7. Such a connection is illustrated in Figure 13.14 and the degree of regulation is equivalent to that achieved with all the methods previously discussed.

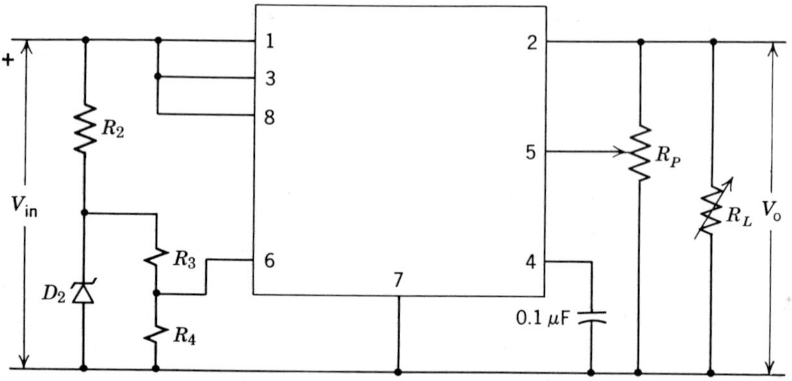

Figure 13.15. Low-voltage (under 8.0V) circuit using resistor divider network.

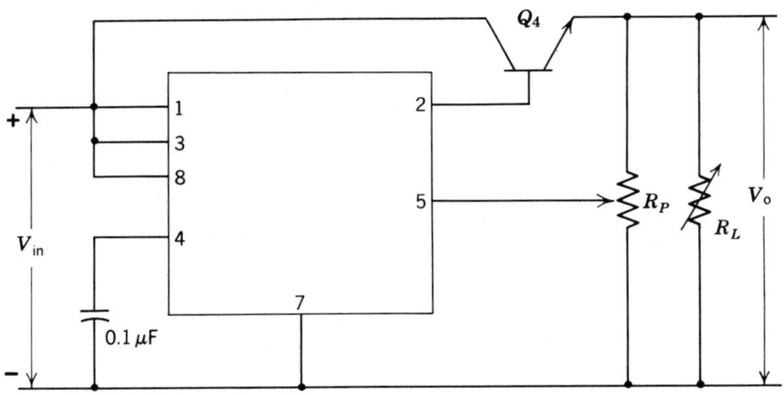

Figure 13.16. Use of an external transistor to extend output current capability.

13.12 CONCLUSION

The WM330 voltage regulator is a nearly universal basic building block. Since it is impossible to have a single IC which satisfies on one chip all the requirements of the design or systems engineer, the building block approach offers the best compromise for achieving specific design goals.

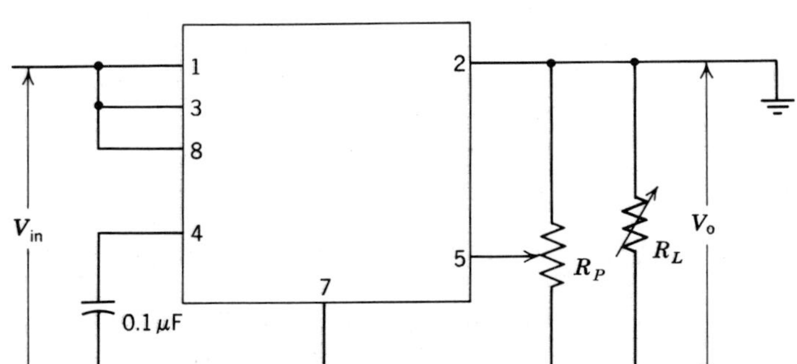

Figure 13.17. Connection for regulated negative output voltages.

14

Integrated Low-Voltage Breakdown Diode

by Robert J. Widlar
National Semiconductor

14.1 INTRODUCTION

Conventional breakdown diodes in the 2 V to 6 V range, presently made by the alloy process, exhibit a fairly soft breakdown characteristic with the voltage changing by about 0.7 V for a factor-of-ten change in reverse current. The reverse punchthrough of a graded-base transistor has a constant-voltage characteristic which is superior to that of an alloy diode. In this case the breakdown voltage is determined by depletion of the base region rather than by an abrupt junction, so that breakdown voltages below 1 V can be made using standard diffusion techniques [1,2]. Transistors and resistors can be fabricated simultaneously with this punchthrough diode with present integrated circuit processes, so that active circuitry can be put on the diode chip to provide even better performance.

The results of using this approach are presented in Figures 14.1 and 14.2. Figure 14.1 shows a normal curve-tracer display comparing an alloy zener with an active regulator using a punchthrough diode. Figure 14.2 is a logarithmic plot and includes a punchthrough diode by itself.

14.2 CIRCUIT DESCRIPTION

A schematic representation of the diode and regulator circuit is given in Figure 14.3. In this circuit, Q_1 is the punchthrough diode, and Q_2 is a field-effect transistor using the base diffusion of the *npn* transistor as a channel and the *npn* emitter as a gate. Its function is to establish the operating current of Q_1 at a level greater than the base current of Q_3. As can be seen from Figure 14.4, Q_1 and Q_2 are actually fabricated as one device. Q_3, which provides most of the gain for the regulator, is an ordinary *npn* transistor. Q_4 is a vertical *pnp* transistor made using the

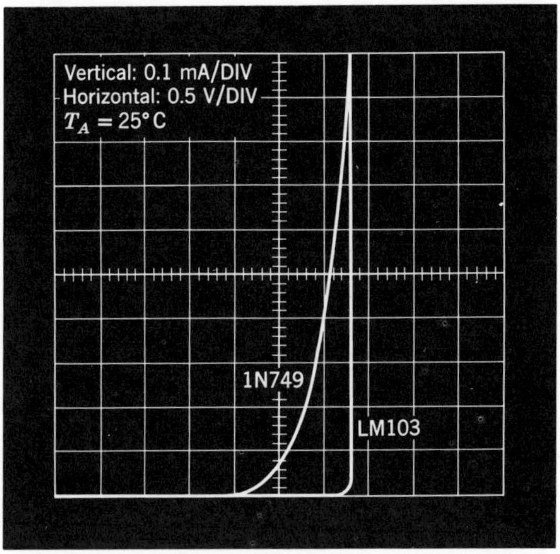

Figure 14.1. Comparison of breakdown characteristic for a LM103 active-punch-through device and a typical alloy-junction diode.

npn base as an emitter, the *npn* collector as a base, and the *p*-type substrate of the integrated circuit as a collector. Its purpose is to provide additional current gain. It is evident that the *pnp*, which is required to handle currents up to 10 mA, must be made quite large because of the low emitter efficiency of the *npn* base diffusion. Finally, R_1, a pinch resistor, sets the operating collector current of Q_3. The value is chosen so that Q_3 sees a minimum variation of collector current for operating currents from 100 μA to 1 mA, giving best regulation in this region.

In operation, as voltage is applied across the device, the emitter-base junction of Q_1 is reverse biased. When a voltage is reached at which the emitter depletion region penetrates the base to the collector, Q_1 begins to conduct, turning on Q_3 which in turn regulates the voltage across the device terminals.

14.3 PERFORMANCE

For reverse voltages up to about 1 V of the nominal breakdown voltage, the reverse current is primarily junction leakage current in the low-nanoampere range. Above this voltage Q_1 breaks down, conducting into the "on" resistance of Q_2. Any further increase in voltage causes Q_1 to conduct more, but, at the same time, it pinches off Q_2, giving rise to the initial hump in the reverse characteristics shown in Figure 14.5. This

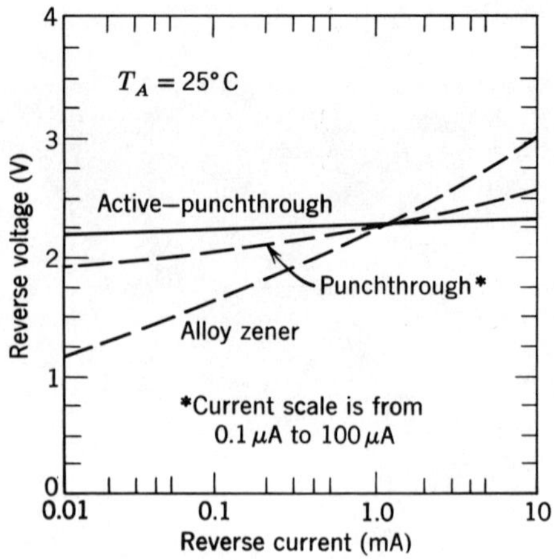

Figure 14.2. Comparison of active-punchthrough diode with small-geometry punchthrough and alloy-junction diodes.

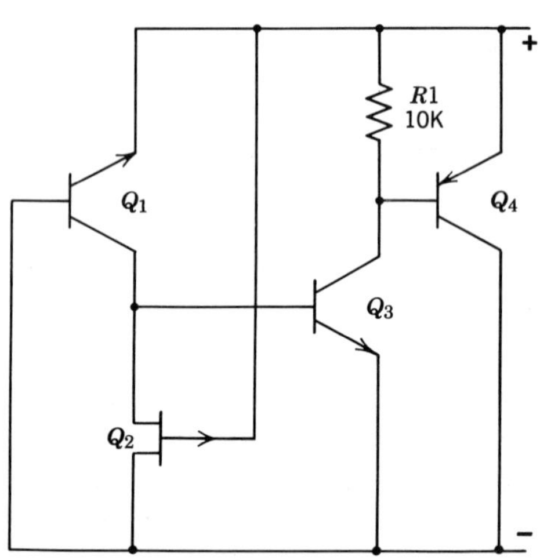

Figure 14.3. Schematic representation of the LM103.

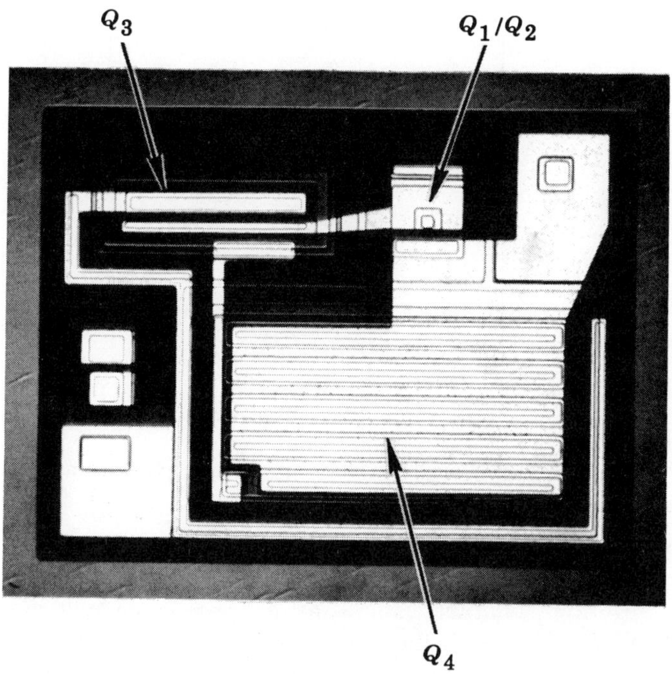

Figure 14.4. Photomicrograph of the integrated circuit.

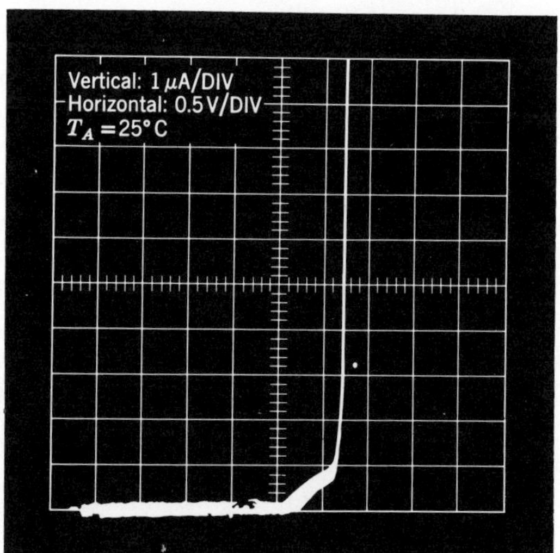

Figure 14.5. Low-current characteristics of the LM103.

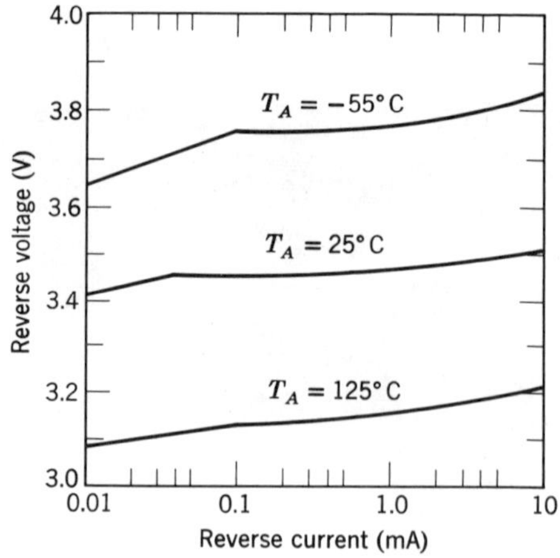

Figure 14.6. Temperature characteristics of the active-punchthrough diode.

continues until Q_3 turns on, and its added gain initiates breakdown and hardens the diode characteristic. The diode voltage will then vary by about 60 mV/decade until the current rises to 50 μA. At this point, Q_4 turns on, providing even more gain, and it reduces the voltage variation to the 15 mV/decade shown in Figure 14.6. At currents much above 1 mA, the base current of the output *pnp* becomes appreciable by comparison to the bleed current through R_1, so that the reverse characteristic softens somewhat.

The operation of the regulator diode over a −55°C to 125°C temperature range is also shown in Figure 14.6. It is evident that the diode characteristic becomes softer at both temperature extremes. At low temperatures regulation below 100 μA is worsened because the emitter-base voltage of Q_4 increases and the value of R_1 decreases, so that a larger current is required before Q_4 turns on. Above 100 μA, there is still some degradation because of reduced current gain in Q_3. At high temperatures, the softening is caused by lowered transconductance in Q_3.

An expanded plot of the reverse characteristics over a 10-nA to 10-mA current range is given in Figure 14.7. This demonstrates the low leakage of the device even at high temperatures.

The change of breakdown voltage with temperature is a nearly linear 3.3 mV/°C, independent of the nominal breakdown voltage. It has been verified experimentally that the temperature-induced voltage drift of the reverse punchthrough is not significantly affected by the initial breakdown voltage for the double-diffused silicon transistors used.

Because the LM103 is an active circuit with feedback, it is prone to frequency instability for some values of shunt capacitance and reverse current. However, the device will not oscillate at any current as long as the shunt capacitance is less than 100 pF or greater than 0.01 μF. It will also be stable if the shunt capacitance is isolated by at least 100 Ω.

14.4 APPLICATIONS

A zener diode is a fairly simple element with applications which are so numerous and so varied that elaborating on them in detail is difficult. However, the low voltage performance of the LM103 is enough improved over devices that have been available in the past that it permits the use of a single diode where much more exotic solutions had previously been required. Its low capacitance and hard characteristics down to 10 μA make it ideal as a low-level clipping or clamping diode. Furthermore, it is useful as a regulator diode in low voltage battery-powered systems, especially where power drain is important.

Nonetheless, one potential usage of the LM103 that does merit further discussion is its application in low-voltage regulators. Temperature compensated reference diodes have breakdown voltages above 5 or 6 V. Hence regulator designs using these diodes must have a bias supply that is greater than 6 V. This poses problems when it is necessary to operate with lower-voltage inputs.

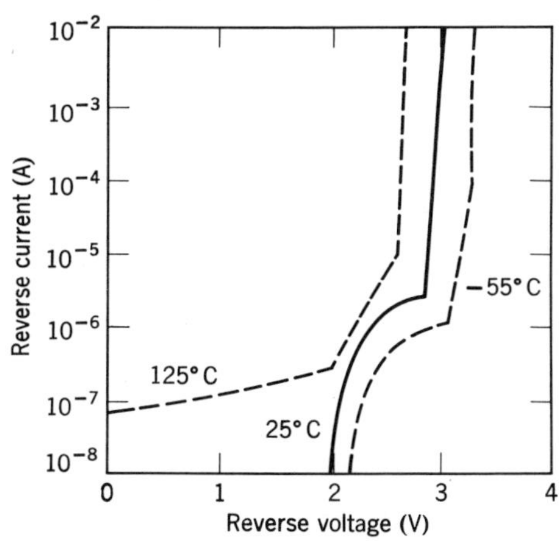

Figure 14.7. Reverse characteristics.

The LM103 will operate as a reference diode on input voltages below 4 V, providing 1% regulation. The device is not temperature compensated; however, there are ways of making a temperature-compensated regulator with them because of their predictable temperature characteristics.

Figure 14.8 shows a voltage regulator design using the LM103. The circuit will give regulated output voltages down to 3 V, and it works with input voltages as low as 4 V. Connected for a 5 V output voltage, its regulation is better than 1% at full load, for input voltages down to 5.5 V. The temperature stability is better than about 2% over a −55°C to 125°C temperature range.

D_1 is the reference diode for the regulator, and its operating current is supplied by R_7; R_7 is chosen to operate the LM103 in its best regulation range, between 100 μA and 1 mA. Q_1, operating common base, is the input stage of the error amplifier. This transistor drives the series pass element Q_3 through an emitter follower buffer Q_2. Q_3 feeds back to the input transistor through a resistive divider that determines the output voltage. Q_1 is connected so that its emitter-base voltage is subtracted from the reference voltage of D_1. Since the emitter-base of Q_1 has a negative temperature coefficient of 2.2 mV/°C, this subtracts from the 3.3 mV/°C negative coefficient of D_1 to produce a feedback sense voltage at the emitter of Q_1 with a temperature coefficient of 1.1 mV/°C.

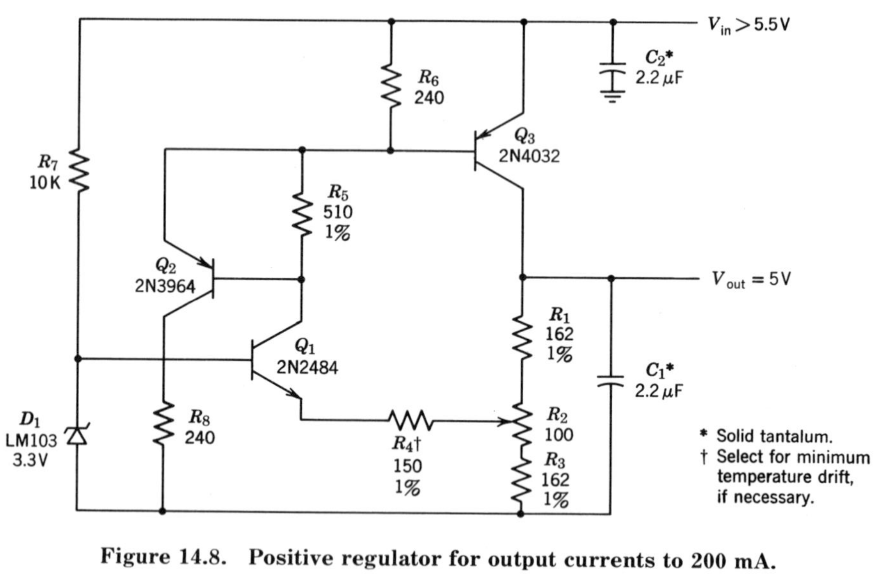

Figure 14.8. Positive regulator for output currents to 200 mA.

The collector current of Q_1 is proportional to the emitter-base voltage of Q_2, because R_5 is made small enough so that the current through it is much greater than the base current of Q_2. Since the current gain of Q_1 is high, its emitter current will also be proportional to the emitter-base voltage of Q_2 which has a temperature coefficient of 2.2 mV/°C. Therefore a resistance can be put in series with the emitter of Q_1 such that the negative temperature coefficient of the voltage across it (due to the emitter-base voltage change of Q_2) nearly compensates for the uncompensated temperature drift of the reference diode D_1. The resistance seen by the emitter of Q_1 is the sum of R_4 and the equivalent resistance seen by looking back into the divider consisting of R_1, R_2, and R_3. R_4 could be eliminated by making the divider resistance higher; it was kept separate, however, so that a single resistor could be selected to give minimum temperature drift, if desired.

The transistor types specified have very high current gain for optimum performance over a −55°C to 125°C temperature range with a minimum number of stages. Lower gain transistors can be used for limited temperature applications.

R_6 is chosen so that the current through it is about one-third the base current of Q_3 under worst-case conditions. This makes the emitter-base voltage of Q_2 pretty much independent of the load current. R_8 eliminates a potential latch-up condition that could occur during turn on, in which Q_2 tries to short out the unregulated supply.

The common-base, common-collector combination of Q_1 and Q_2 is roughly the equivalent of a common-emitter stage. This is combined with Q_5 to form a two-stage amplifier. This is frequency compensated with a single capacitor C_1 on the output of the regulator. A capacitor with good high-frequency characteristics, such as solid tantalum, should be used. It is also a good idea to make sure the input to the regulator is well bypassed, as with C_2.

It is a simple matter to build a negative low-voltage regulator by substituting complementary transistor types. This is shown in Figure 14.9.

A positive regulator that can deliver output currents up to 2 A is shown in Figure 14.10. It is similar to the circuit in Figure 14.8, except that an *npn* emitter follower has been added to the series pass transistor for higher output currents. Current limiting is provided by Q_5 and R_9. As the output current approaches the limit point, the voltage drop across R_9 becomes large enough to turn on Q_5. This starts to remove base drive from Q_3, resulting in a constant-current output characteristic. R_{10} is included to make sure that Q_5 is not burned out from excessive base current. C_1 is required to keep the regulator from oscillating in current limit.

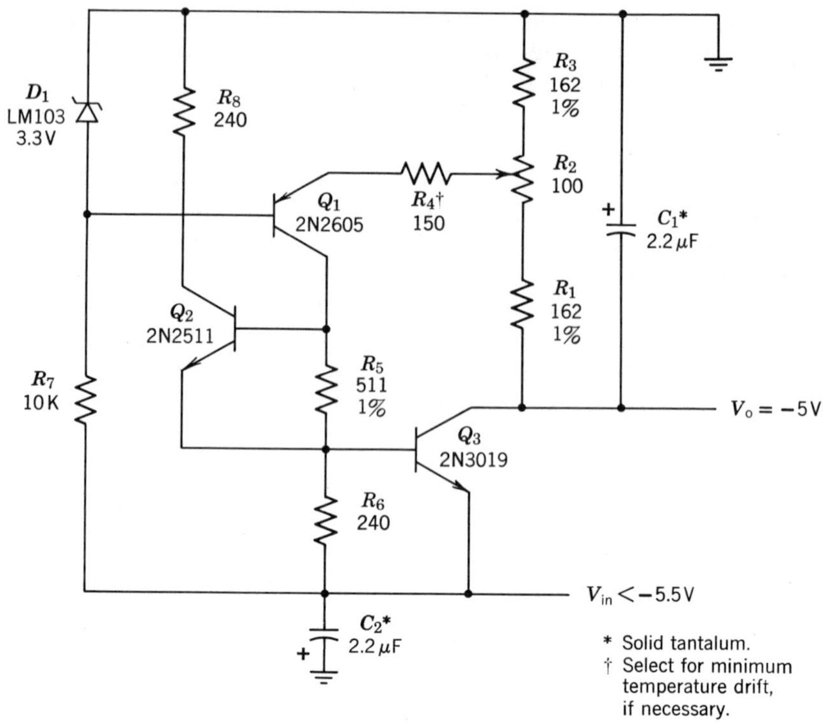

Figure 14.9. Negative regulator for output currents to 200 mA.

The regulation and current limiting characteristics of this regulator are shown in Figure 14.11. It should be noted that the addition of the emitter-follower output and current limiting increases the minimum input-output voltage differential to 1 V. A high-current negative regulator could, of course, be made by substituting complementary transistors. Because of the high cost of pnp power transistors, however, this is not an ideal solution.

An alternate circuit for a negative regulator is given in Figure 14.12. This is similar to the negative regulator in Figure 14.9, except that a modified Darlington connection is used for the series pass element. Current limiting is provided by Q_5 and R_{10}; when the output current becomes excessive, Q_5 turns on, shutting off the error amplifier Q_1. C_3 and C_4 are required to prevent oscillations in current limit. Furthermore, D_2 and D_3 must be added to pull down the reference as the output falls below 1.6 V, so that the collector base junction of Q_5 does not become forward biased.

There are methods of incorporating current limiting which are simpler than the one used here, but they will result in a higher power dissipation in Q_3 or an increased minimum input-output voltage differential. The performance of this regulator is given in Figure 14.13.

14.5 CONCLUSIONS

A two-terminal monolithic voltage regulator electrically equivalent to a breakdown diode has been described. At low voltages, it provides reverse characteristics which are ten times sharper than single-junction zener diodes. Devices have been manufactured in volume with breakdown voltages from 2.4 V to 5.6 V and specified for operation from 10 μA to 10 mA over the full military temperature range.

The regulator is fabricated on a 23 by 29 mil-square silicon die by using the six-mask, planar-epitaxial process. Because all junctions are passivated, reliable operation can be expected under extreme environmental conditions even at low operating currents. Tests underway point to

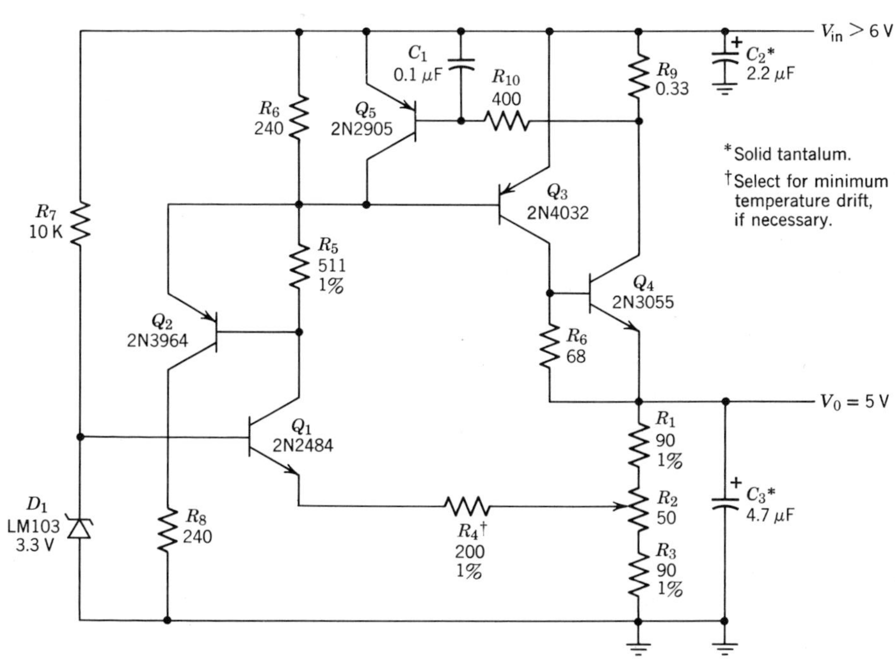

Figure 14.10. High-current positive regulator with current limiting.

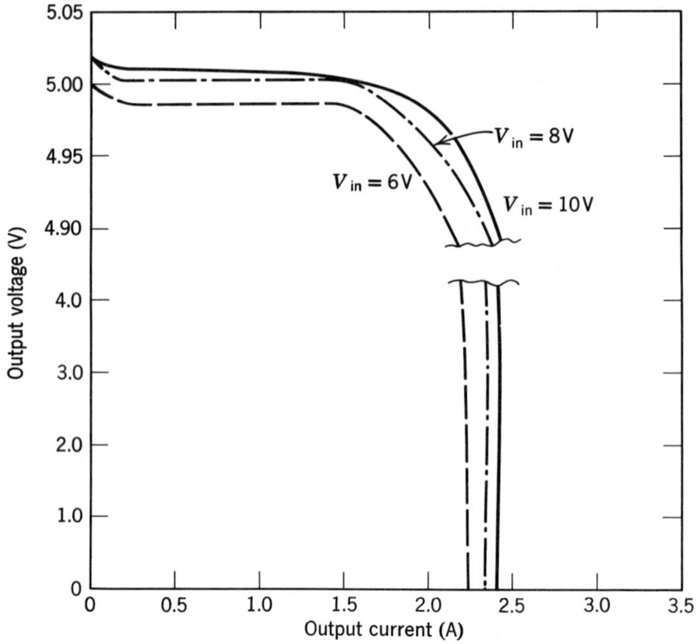

Figure 14.11. Regulation and current limiting characteristics of the 2-A positive regulator.

excellent long-term stability, which is to be expected since the breakdown mechanism is fairly quiet, giving an output noise of only 100 μV, peak-to-peak, over a 10-Hz to 10-kHz bandwidth.

Although the diode is not optimized for minimum capacitance, it is about an order of magnitude faster than alloy zeners in clipping and clamping applications. Also, because of its predictable temperature drift, it can be used in building fairly simple temperature compensated regulators for operation with input voltages down to 3 V and output voltages down to 1 V.

REFERENCES

[1] Dacey, G. C., "Space-Charge Limited Hole Current in Germanium," *Phys. Rev.*, **90**, No. 5 (June 1, 1953).

[2] Rediker, Robert H., "Narrow Base Planar Junction Punch-Thru Diode," *United States Patent Office*, No. 2,975,342, patented March 14, 1961.

[3] Widlar, R. J., International Solid State Circuits Conference, Philadelphia, February 1968.

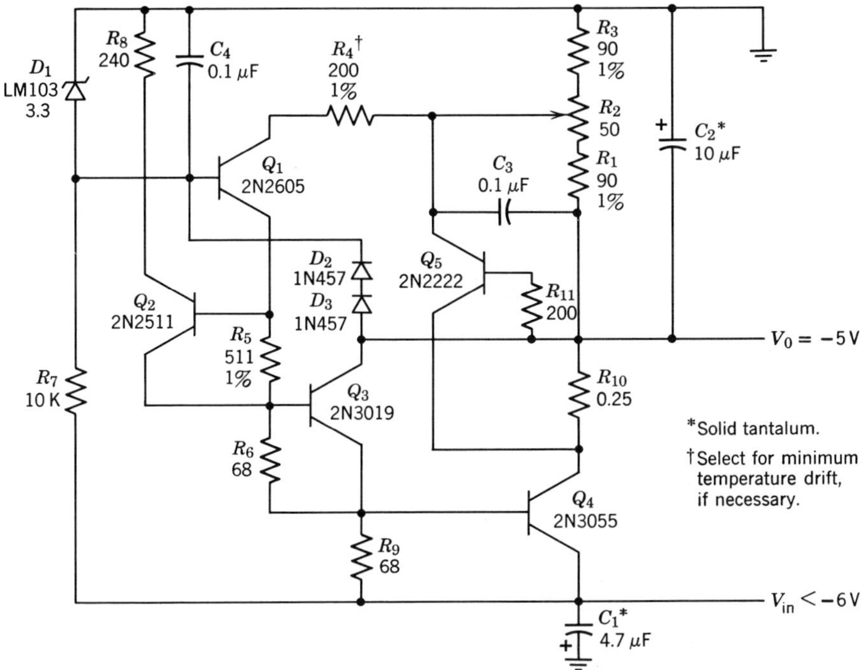

Figure 14.12. High-current negative regulator with current limiting.

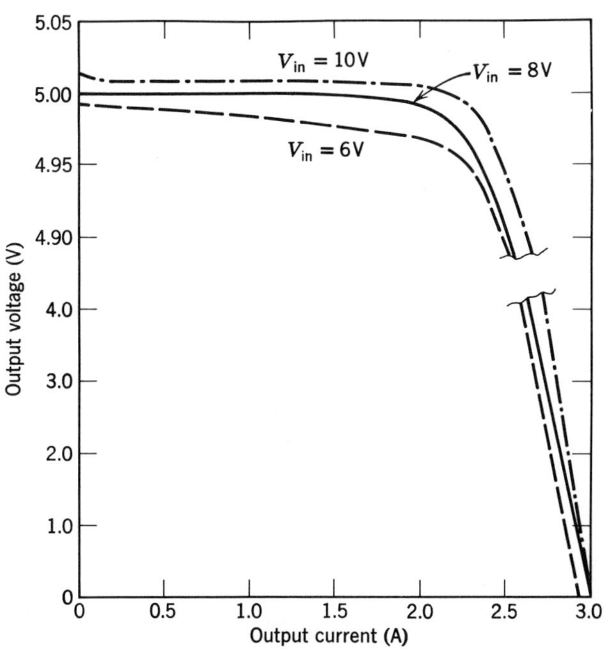

Figure 14.13. Regulation and current limiting characteristics of the 2-A negative regulator.

15

Functionalizing Linear ICs

by Ed Renschler
Motorola Semiconductor Products Inc.

15.1 INTRODUCTION

The following discussion is based on the use of a representative number of high-yield linear ICs in practical configurations. The approach is "cookbook" rather than theoretical and it is expected that the designer will apply data-sheet information to achieve performance requirements that differ somewhat from those of the given circuits. References are provided for more detailed information on design procedure.

15.2 HIGH-FREQUENCY AMPLIFIER

The MC1550 is a low-cost temperature-compensated differential amplifier (see Figure 15.1) capable of excellent high frequency response. It is useful, therefore, in such applications as rf and i-f amplifiers, video amplifiers, oscillators, amplitude modulators, etc. Temperature compensation is obtained through diode D_1 which closely tracks the V_{BE} characteristics of transistor Q_1, thereby holding power-gain variations to ± 1.5 dB over a temperature range from $-55°$C to 125°C.

Particularly well suited for purposes requiring agc, it minimizes changes in input impedance with the application of agc, thus preventing circuit detuning under varying signal conditions. Typically, input resistance and capacitance vary approximately 2% and 7% respectively for agc voltages causing a power-gain change of 55 dB.

15.3 TUNED AMPLIFIERS

In Figure 15.2 the MC1550 is shown in a narrow-band, tuned-amplifier configuration for 60-MHz operation. The typical measured performance of the circuit is shown in the accompanying graphs, while the table indicates the variance of gain and bandwidth with four different devices

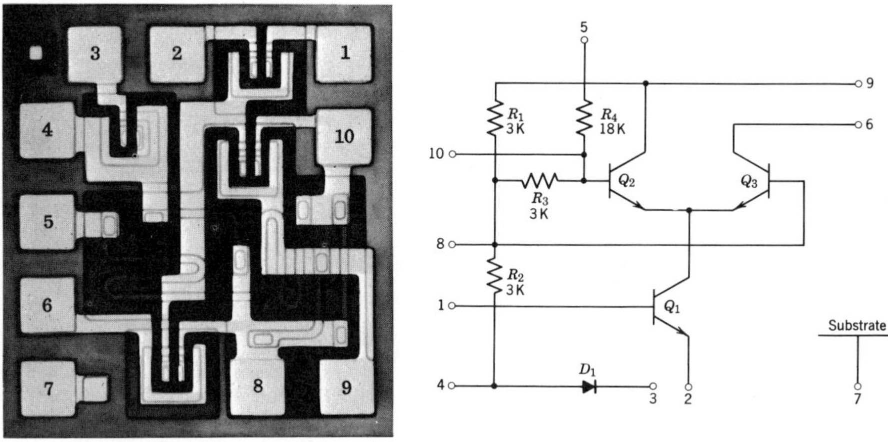

Figure 15.1. A basic integrated circuit rf/i-f amplifier (MC1550).

plugged into the test board. These variations were obtained without retuning after successive substitutions.

Similarly, broadbanding can be affected, as shown in the 45-MHz amplifier of Figure 15.3. Two stages of amplification are required to achieve the same amount of gain as before, and additional decoupling is used. Inductors with an unloaded Q of 30 or more must be used. Agc voltage is applied to both stages.

15.4 AMPLITUDE MODULATOR

The Agc characteristics of the MC1550 make it very useful as an amplitude modulator, as shown in Figure 15.4. Modulation is achieved by applying the rf signal to the base of transistor Q_1 and the modulating signal to the base of Q_2. Terminal 5 of the amplifier, the normal agc terminal, is fixed biased to 3.5 V, a condition that provides linear upward and downward modulation with a peak-to-peak modulation signal of 1.5 V. Nearly 90% modulation is achieved without distortion, as illustrated by the waveforms accompanying the diagram.

15.5 MULTI-STAGE VIDEO AMPLIFIERS

Utilizing the MC1550 in a video-amplifier configuration can provide a gain from better than 25 dB (with a 3-dB bandwidth of 22 MKz), for a single-stage circuit, to nearly 80 dB (with a 10-MHz bandwidth), for a 4-stage amplifier. Moreover, the single-stage circuit as well as the

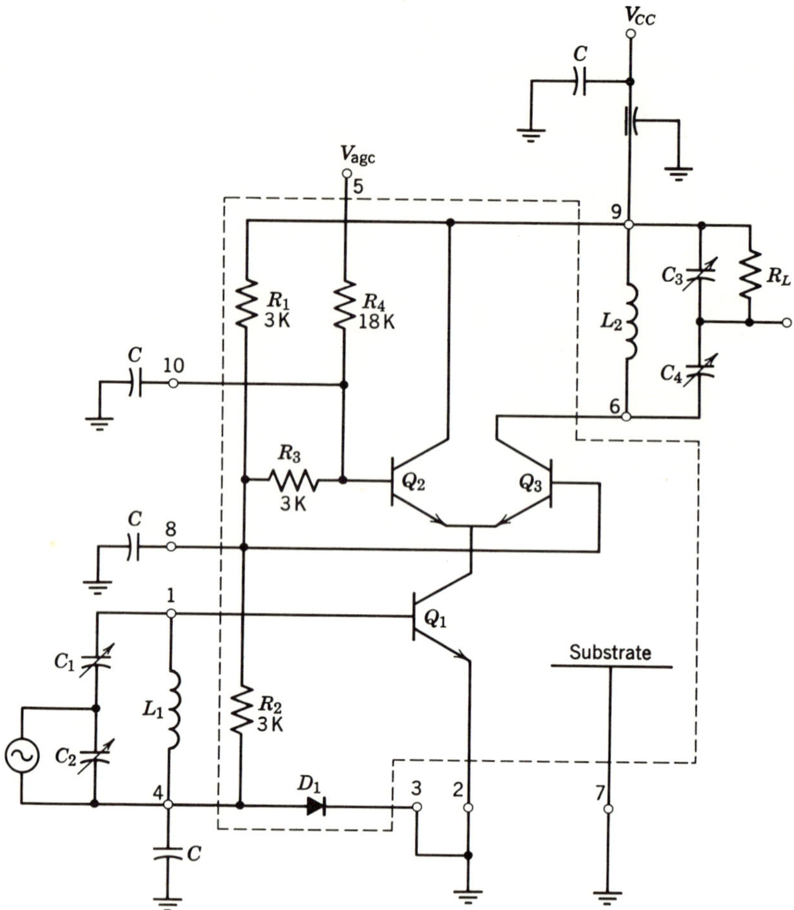

$L_1 = 0.23\,\mu\text{H}$
$C_1 = 48\,\text{pF}$
$C_2 = 88\,\text{pF}$
C (bypass) $= 1000\,\text{pF}$

$L_2 = 0.26\,\mu\text{H}$
$C_3 = 740\,\text{pF}$
$C_4 = 33\,\text{pF}$
$R_L = 50\,\Omega$

Schematic of 60 MHz tuned amplifier

Circuit No.	Center Frequency	Transducer Power Gain	Bandwidth
1	60 MHz	30.0 dB	0.6 MHz
2	60 MHz	30.0 dB	0.5 MHz
3	60 MHz	30.5 dB	0.6 MHz
4	60 MHz	30.4 dB	0.4 MHz

EFFECT OF CHANGING INTEGRATED CIRCUIT ON CIRCUIT PERFORMANCE

Figure 15.2. Performance of the MC1550 as a narrow-band 60-MHz tuned amplifier.

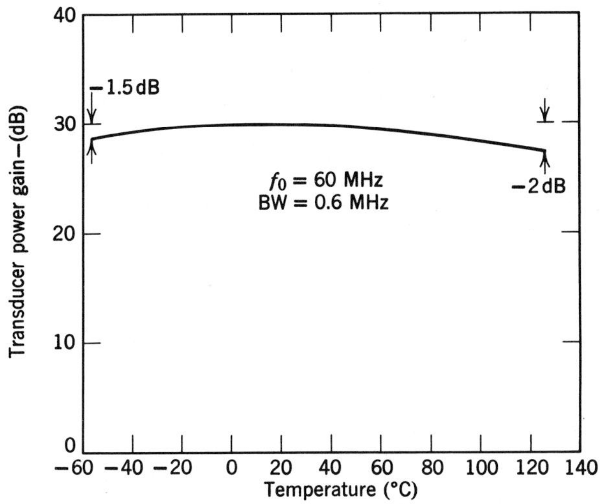

Gain variation with temperature change

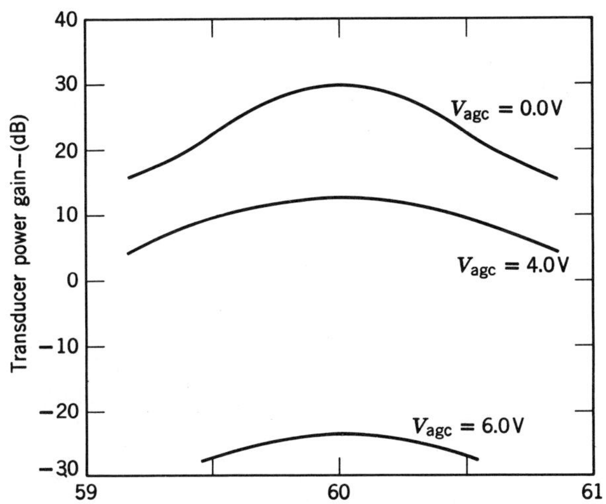

Effect of agc voltage on gain

(**Figure 15.2.** ***continued***)

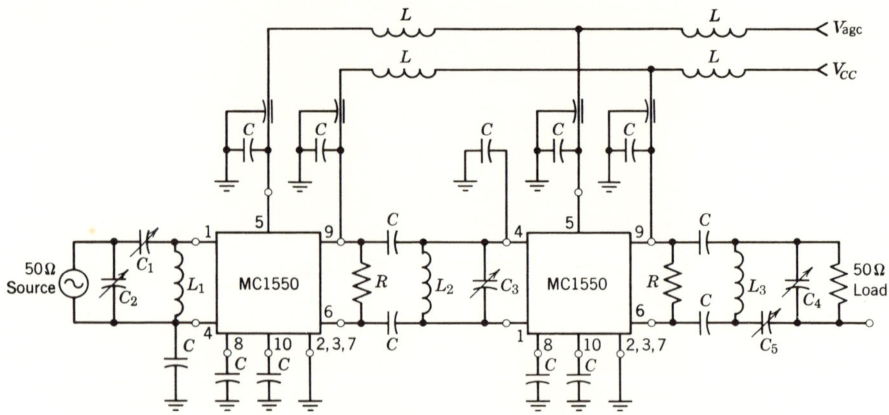

C = 2000 pF C_3 = 2–8 pF L_2 = 0.68 μH L = 1 μH

$C_1 = C_2 = C_4 = C_5$ = 9–35 pF L_1 = 0.42 μH L_3 = 0.55 μH R = 510 ohms

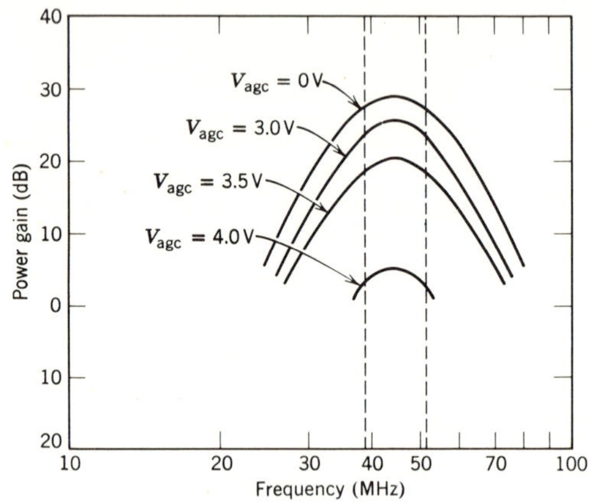

Unit	Center Frequency	Gain (dB)	Bandwidth
1	45 MHz	30.2	15.4 MHz
2	45 MHz	30.5	15.0 MHz
3	45 MHz	31.0	14.5 MHz
4	45 MHz	30.3	14.2 MHz

COMPARISON OF PERFORMANCE OF AMPLIFIER WITH CHANGE IN MC1550 AND NO RETUNING.

Figure 15.3. A 2-stage, 45-MHz broadband amplifier.

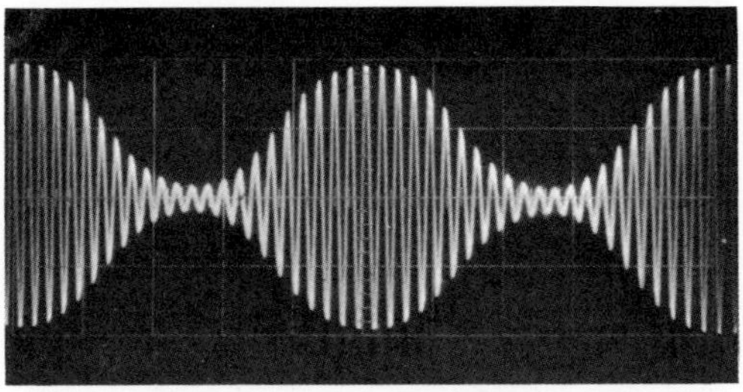

(a) **45-MHz rf, 2-MHz audio**

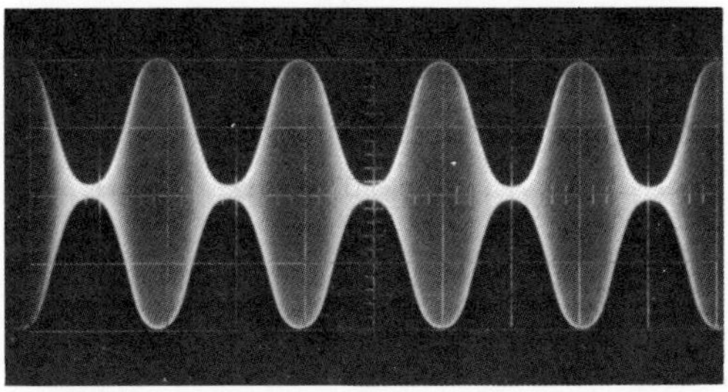

(b) **45-MHz rf, frequency, 5-kHz audio**

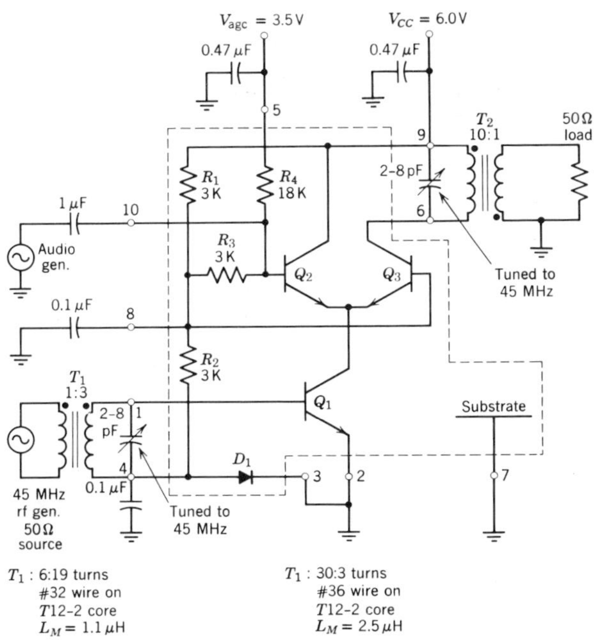

Figure 15.4. An rf modulator using the MC1550 and representative waveshapes.

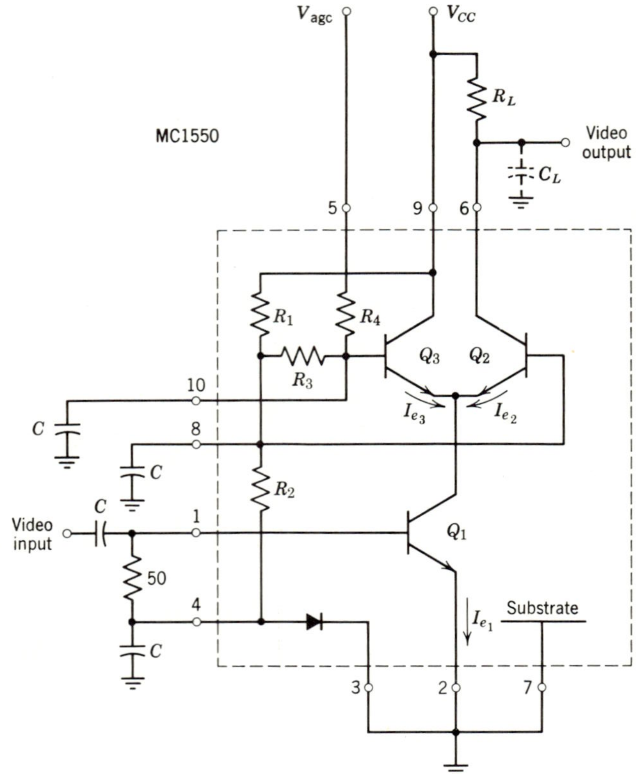

Left. Schematic of single–stage MC1550 video amplifier. Graph showing its performance is at right.

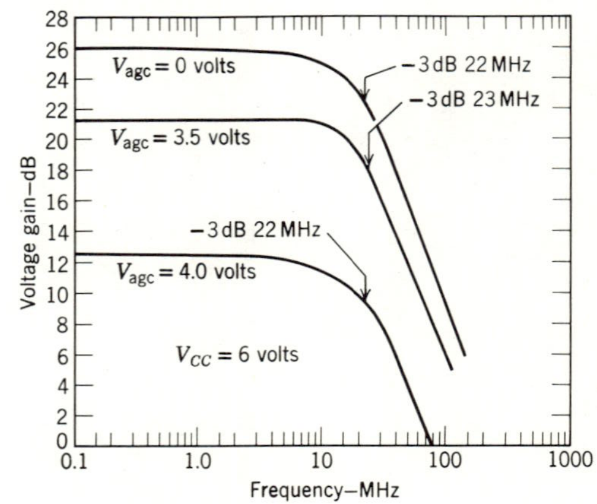

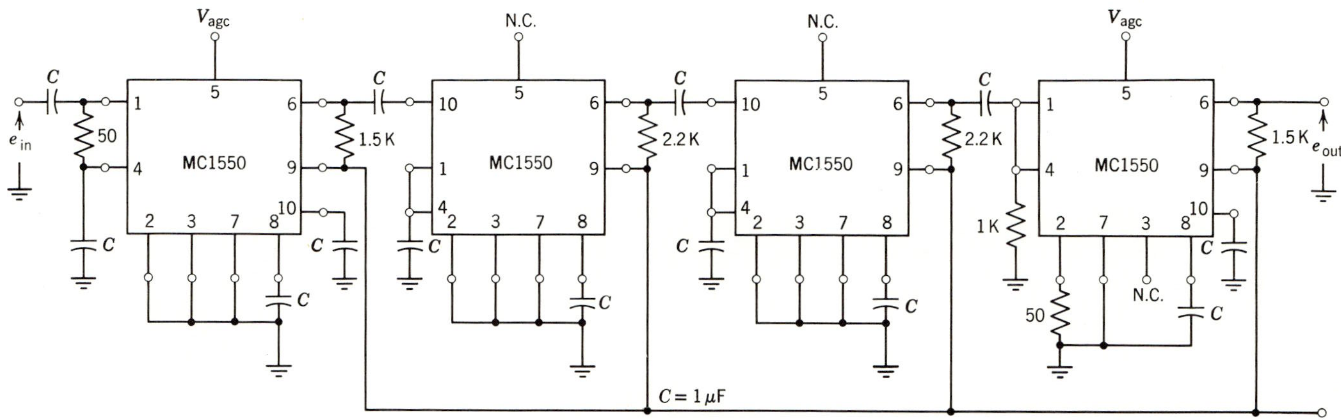

Above. Schematic of four-stage MC1550 video amplifier. Graph showing its performance is below.

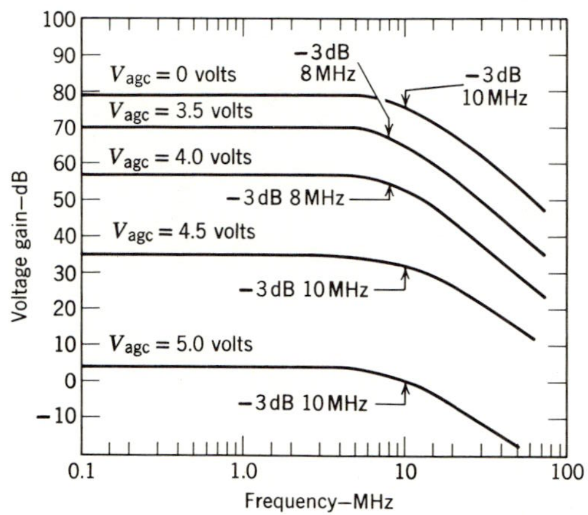

Figure 15.5. Single and multi-stage video amplifiers.

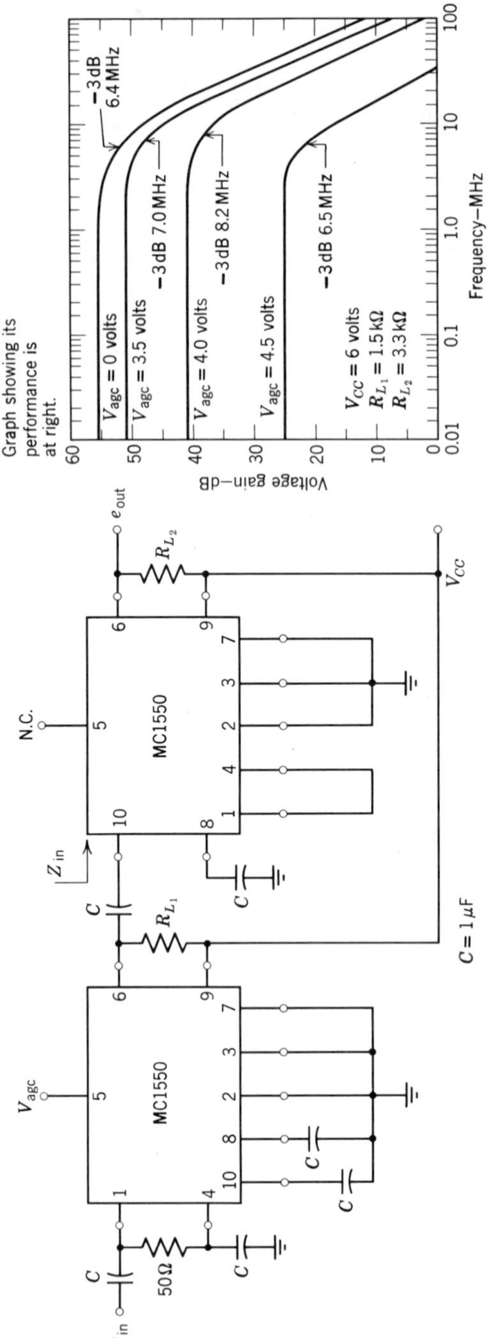

Left. Schematic of two-stage MC1550 video amplifier. Graph showing its performance is at right.

(Figure 15.5. *continued*)

multi-stage circuits exhibit excellent agc action and can be used in applications requiring symmetrical limiting.

Figure 15.5 shows the amplifier in single-stage, double-stage, and four-stage configurations, together with the resulting response curves for various agc voltages. Note that the second stage (and the third stage in the four-stage circuit) is operated as a common-collector, common-base amplifier, with transistor Q_1 operating as a dc current source. This permits the output from the first stage to be capacitively coupled directly into the second without the use of lossy matching networks or a buffer interstage.

Standard linear ICs, as a matter of form, are designed for a wide range of applications to give them the best possible chance of large-scale acceptance. While as a rule a circuit is synthesized with a specific application in mind, a good deal of engineering time is spent in researching and broadening its potential uses. Thus, while the MC1550 was primarily intended as an rf/i-f amplifier, its additional applications, some of which have been described, make it that much more versatile. Another example is the wideband amplifier, type MC1553, together with its performance characteristics, illustrated in Figure 15.6. Basically, this amplifier consists of three cascaded transistor stages (Q_1, Q_2, and Q_3), followed by an output emitter follower Q_4. Transistor Q_5 is incorporated in a dc feedback loop, used to establish and stabilize the output operating point. A current-source transistor Q_6 is used to establish symmetrical positive and negative load-current excursions, regardless of power supply voltage, temperature or load resistance.

A second feedback loop for ac operation consists of resistors R_6, R_4, and R_5. This provides for a wide variety of gain options obtained by varying the amount of feedback through external connections and by taking the output at various points of the circuit. In those instances where gain is increased at the expense of reduced feedback, bandwidth is proportionally reduced. Inverting and noninverting outputs may be obtained to provide flexibility when the unit is used to process complex video pulses or pulse trains. Automatic gain control may be achieved by replacing the external gain control resistor with the variable resistance of a small-signal diode.

15.6 OSCILLATORS

Both the MC1550 and the MC1553 amplifiers make excellent oscillators within their operating frequency range. The schematic diagram of a tuned oscillator utilizing the MC1550 rf/i-f amplifier is shown in

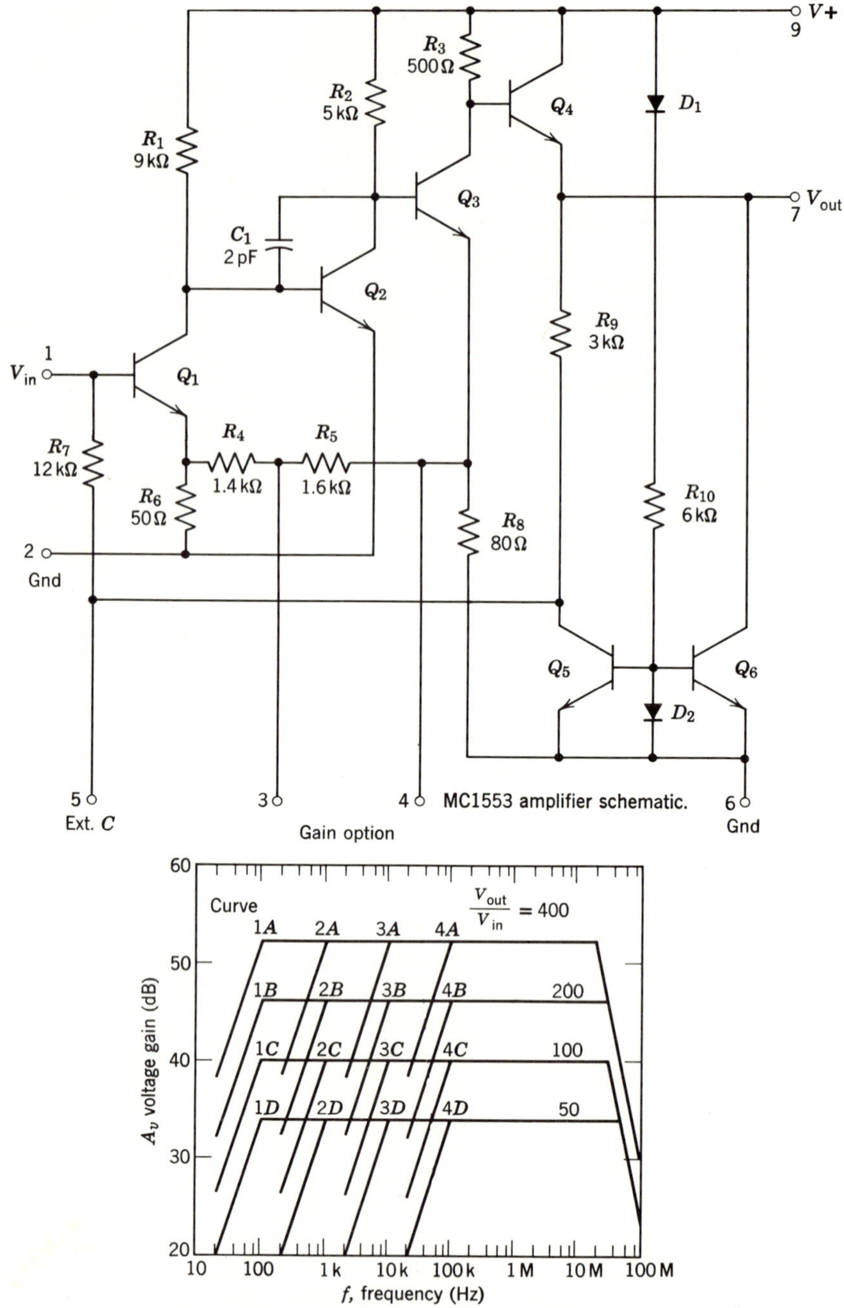

Low frequency responses curves for various circuit configurations and component values.

Figure 15.6. The MC1553 as a wideband amplifier.

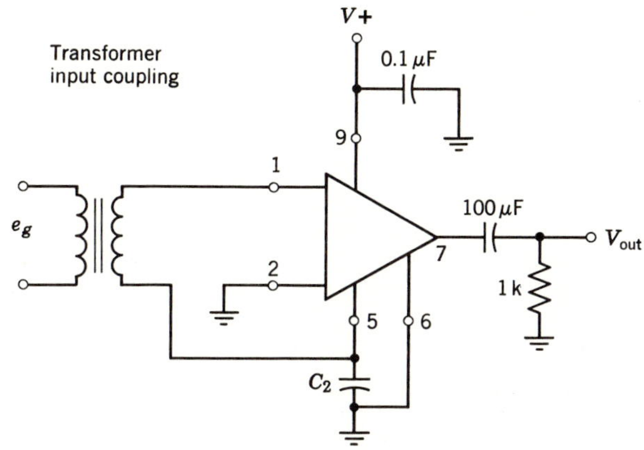

CURVE NO.	$C_2(\mu F)$
1A	200
1B	100
1C	70
1D	30
2A	20
2B	10
2C	7.0
2D	3.0
3A	20
3B	10
3C	7.0
3D	0.3
4A	0.2
4B	0.1
4C	0.07
4D	0.03

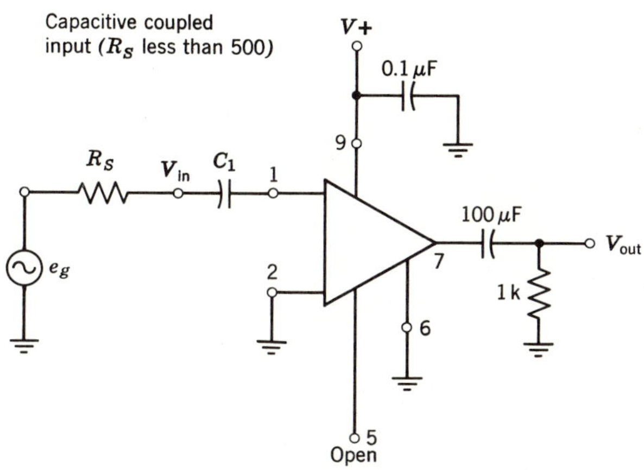

CURVE NO.	$C_1(\mu F)$
1A	20
1B	10
1C	7.0
1D	3.0
2A	3.0
2B	1.0
2C	0.8
2D	0.5
3A	0.4
3B	0.2
3C	0.1
3D	0.06
4A	0.04
4B	0.02
4C	0.01
4D	0.007

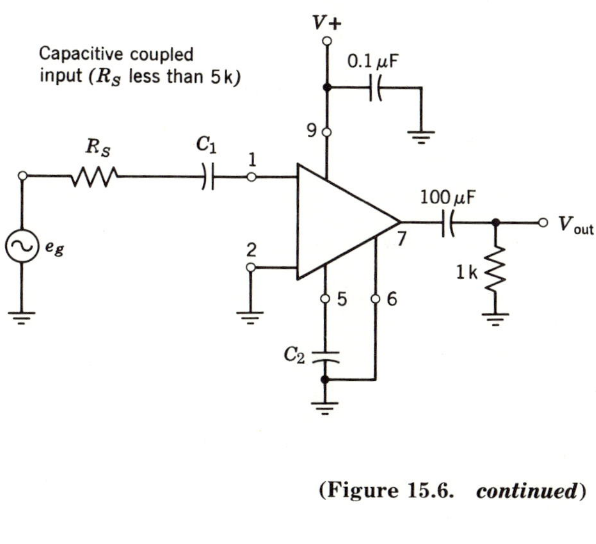

CURVE NO.	$C_1(\mu F)$	$C_2(\mu F)$
1A	0.1	250
1B	0.1	150
1C	0.1	70
1D	0.1	40
2A	0.01	30
2B	0.01	18
2C	0.01	8.0
2D	0.01	4.0
	(pF)	
3A	1000	3.0
3B	1000	1.8
3C	1000	0.8
3D	1000	0.4
4A	100	0.3
4B	100	0.18
4C	100	0.08
4D	100	0.04

(Figure 15.6. *continued*)

Voltage Gain Volts/Volt	Comments
400	Output @ Pin 7
61	Non-inverting. Output @ Pin 4
29	Non-inverting. Output @ Pin 3
200	Connect Pin 3 to Pin 4. Output @ Pin 7
30	Non-inverting. Connect Pin 3 to Pin 4. Output @ Pin 3/4
400 to 3000	Connect external resistance from Pin 3 to gnd. Output @ Pin 7
61 to 480	Non-inverting. Connect external resistance from Pin 3 to gnd. Output @ Pin 4
29	Non-inverting. Connect external resistance from Pin 3 to gnd. Output @ Pin 3

VOLTAGE GAIN OBTAINABLE.

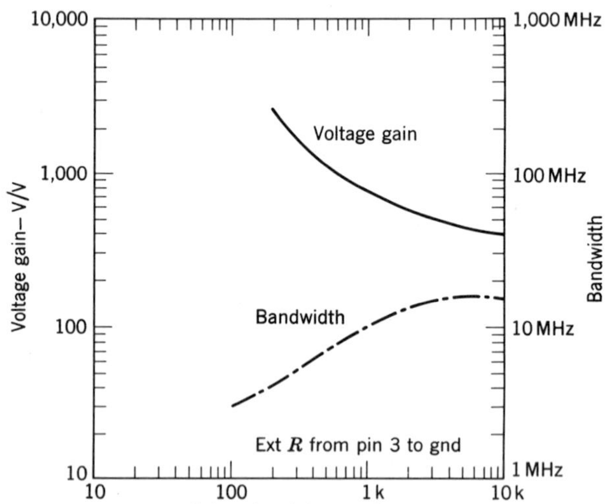

Voltage gain and bandwidth as a function of external resistance

(Figure 15.6. *continued*)

Figure 15.7 together with a tuning-range graph and an oscillograph showing a typical output waveform. Figure 15.8 illustrates the external connections to an MC1553 wideband amplifier for crystal-controlled oscillator applications.

The output of the oscillator is taken from pin 7 which is buffered from the oscillator circuit proper by a stage of gain and an emitter follower. As shown in the diagram, no provisions were made in the design to control the loop gain or the amplitude of oscillation to reduce harmonic content.

Instead, a "brute-force" π filter was inserted at the output (pin 7) to extract the fundamental frequency. The oscillograph shows the resultant output waveform.

For more sophisticated designs or for higher frequencies, a tuned feedback network can be used in conjunction with the crystal to provide the necessary phase shift adjustment and feedback control to ensure oscillation and reduce harmonic content.

15.7 AUDIO APPLICATIONS

Of all the various linear functions, the audio circuit is probably the most difficult to integrate in monolithic form. There are two reasons for this. First, conventional audio circuits have required large-value capacitors, and second, integrated circuits capable of dissipating large amounts of power, as required for high-power amplifiers, are still beyond standard practice. Nevertheless, a considerable audio capability does exist even now, so that a good portion of a low-frequency system can be integrated.

15.8 A POWER AMPLIFIER AND ITS APPLICATIONS

A case in point is the MC1554 monolithic power amplifier capable of an "undistorted" output of 1 W. The amplifier schematic and some typical performance curves are illustrated in Figure 15.9.

At first glance, one may be surprised to find as many as 16 transistors and diodes when the goal is modest gain and output power; however, the many low level devices contribute only about 25% of the total die area and produce a circuit which has (a) reasonable independence of device h_{fe}, (b) independence of absolute resistor values, (c) stable performance with temperature variations, (d) simplified processing through the requirement of only *npn* transistors, and (e) the linearity of an overall feedback amplifier.

The diagrams and tables in Figure 15.10 show typical connections and performance for a variety of applications.

15.9 AUDIO PREAMPLIFIER

While power type amplifiers still leave something to be desired in the maximum amount of available power, the requirements for voltage amplifiers or preamplifiers can be quite adequately satisfied. This is due to a whole host of differential and operational monolithic amplifiers that can, with the proper external compensation, provide about every dc or low-frequency function a system designer might want, in just about any gain range. One of the latest is the MC1302 dual preamplifier shown in Figure 15.11. A high voltage version, the MC1303, will be introduced shortly.

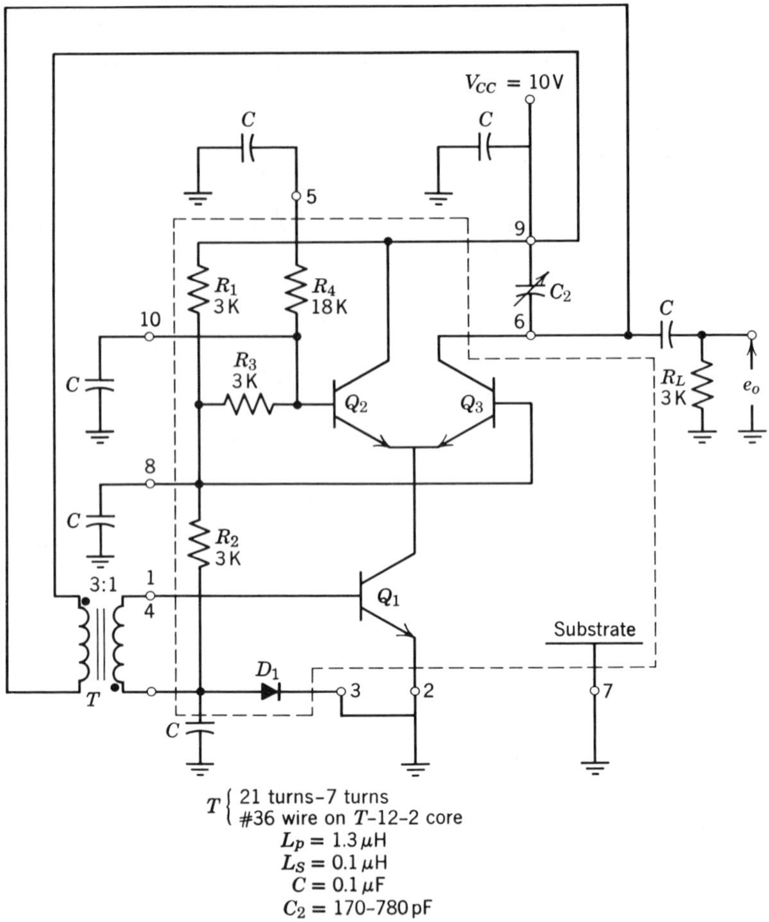

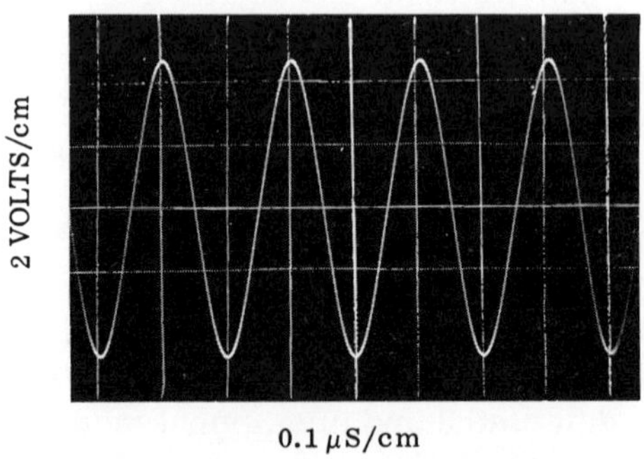

Figure 15.7. The MC1550 rf/i-f amplifier connected as a 5-10 MHz oscillator. The curves show oscillator frequency as a function of tuning capacitance, and output voltage as a function of power supply voltage and load resistance.

The obvious application for such a device is a dual preamplifier for stereo sound systems and, indeed, it has been tested for just such a function. A single package containing one dual operational amplifier can provide the necessary record equalization and the addition of a package will give up to 50 dB of treble and base tone control. The necessary equalization and filter component must, of course, be supplied externally. Typical circuits and their measured performance graphs are given in Figure 15.11.

15.10 LOW-DRIFT AMPLIFIER

The uses of a properly designed dual operational amplifier go far beyond those of stereophonic preamplification. In fact, for certain purposes, the dual amplifier on a single chip performs functions that could not be adequately served with two separate circuits. One of these is a dc amplifier with extremely low drift characteristics. This function, normally accomplished through the use of complex chopper circuits, can be achieved with a dual operational amplifier, provided that the drift characteristics of the two amplifiers are very nearly identical. When the two amplifiers are built on a single chip, the change in characteristics of the two (as a function of temperature) are almost the same. By proper selection of feedback elements, a feedback configuration can be generated which will effectively cancel the drift. The circuit configuration, utilizing a selected version of MC1437, is shown in Figure 15.12. For the resistor values indicated, a gain of 100 is obtained.

15.11 ANALOG MULTIPLIER

Probably the classic application for operational amplifiers is analog computation. An example of this use is the "electronic slide rule" shown in Figure 15.13. By taking advantage of the functional characteristics of operational amplifiers, the multiplication $Z = XY$ can be obtained for a wide range of inputs. The circuit consists of four parts. Two parts are logarithmic function generators that provide log X and log Y. The next stage is a summing amplifier to combine the outputs of the first two stages into log X + log Y. The fourth stage is the complement of the first stages, an antilogarithmic generator to supply the output function $Z = XY$.

The first two (logarithmic amplifier) stages utilize the logarithmic characteristics of *npn* transistors in the common-base configuration. The fourth (antilog) stages uses a *pnp* device, also in a common-base configuration. In each case the transistors are in the external feedback loop of the operational amplifiers.

The table in Figure 15.13 shows measured results over a three-decade operating range. For accuracy over a wide range of input and outputs, it is particularly important that input offset voltage of the amplifier and base current of the logarithmic transistor in the final stage be adjusted carefully.

15.12 VOLTAGE REGULATOR

A high-gain operational amplifier can be used very effectively as an error-sensing amplifier and series-element driver. The high gain, low drift, and high power-supply rejection ratio make the MC1533 well suited for regulator applications. It can be used in a fixed output-voltage

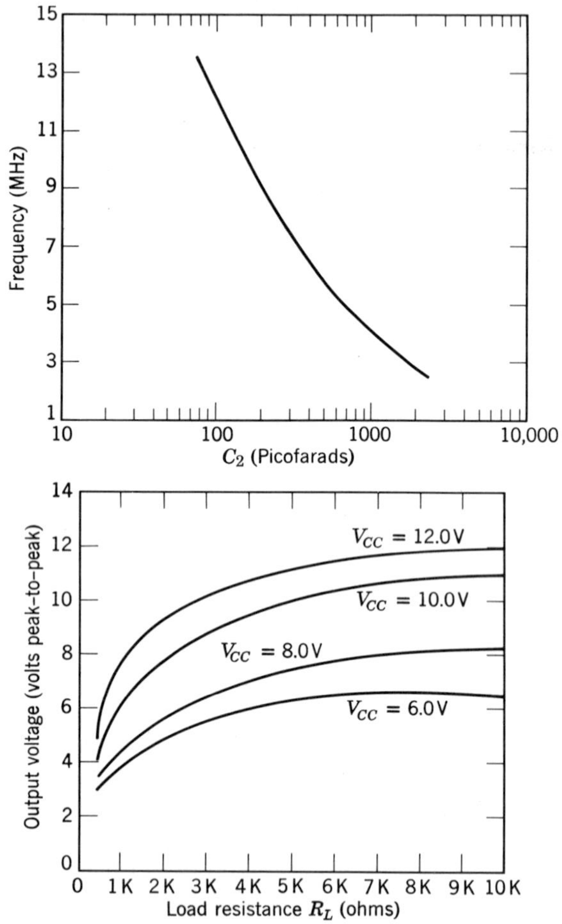

(**Figure 15.7.** ***continued***)

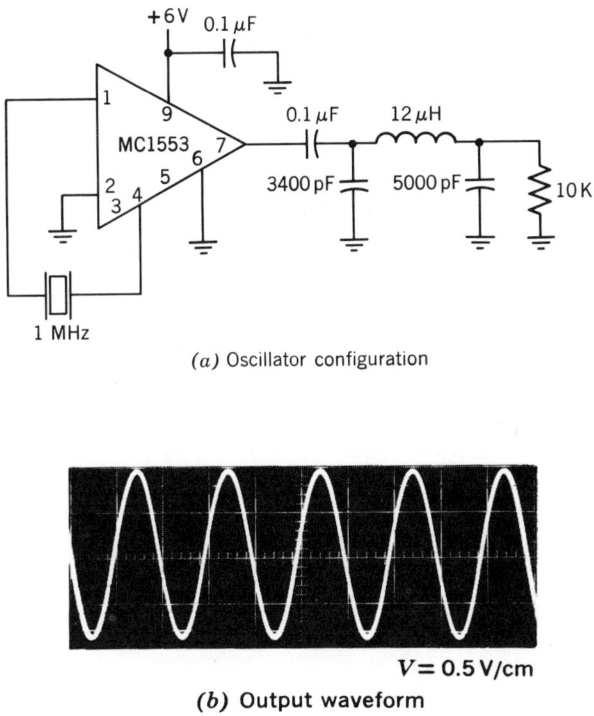

Figure 15.8. MC1553G connected as a 1-MHz crystal-controlled oscillator: (*a*) oscillator configuration; (*b*) output waveform.

configuration using a single voltage source, or a variable output voltage can be obtained through the use of two input voltage supplies.

Figure 15.14 shows a regulator using dual input supplies. The output voltage can be varied from 0 to 25 V. In addition, current limiting is provided at the output. The current limiting is set by the value of resistor R_1. The selection of the resistor value can be accomplished by using the graph shown. First one determines the desired load current and then picks the resistor value that will provide that amount of current.

The V^- voltage should be ripple-free because any ripple that appears on the reference zener will be amplified by the ratio of $|(R_3 + R_6)/R_7|$. The ripple that occurs on the V^+ supply will be greatly attenuated by the error-sensing amplifier. In general, the input V^+ ripple is less than the noise on the output voltage which is approximately 0.5 mV rms. The voltage regulation is better than 1 mV from no load to full load.

The regulator performs well with resistive loads and with capacitive loads up to 150 μF. Care must be taken, however, to eliminate ground loops. Otherwise regulation may suffer and oscillation occur.

REFERENCES

[1] "An Integrated Circuit Rf/I-f Amplifier," Motorola AN-247.
[2] "The MC1554 One-Watt Monolithic Power Amplifier," Motorola AN-401.
[3] "A High Voltage Monolithic Operational Amplifier," Motorola AN-248.
[4] "A Monolithic Dual Operational Amplifier," Motorola AN-411.
[5] "Transistor Logarithmic Conversion Using an Integrated Operational Amplifier," Motorola AN-261.

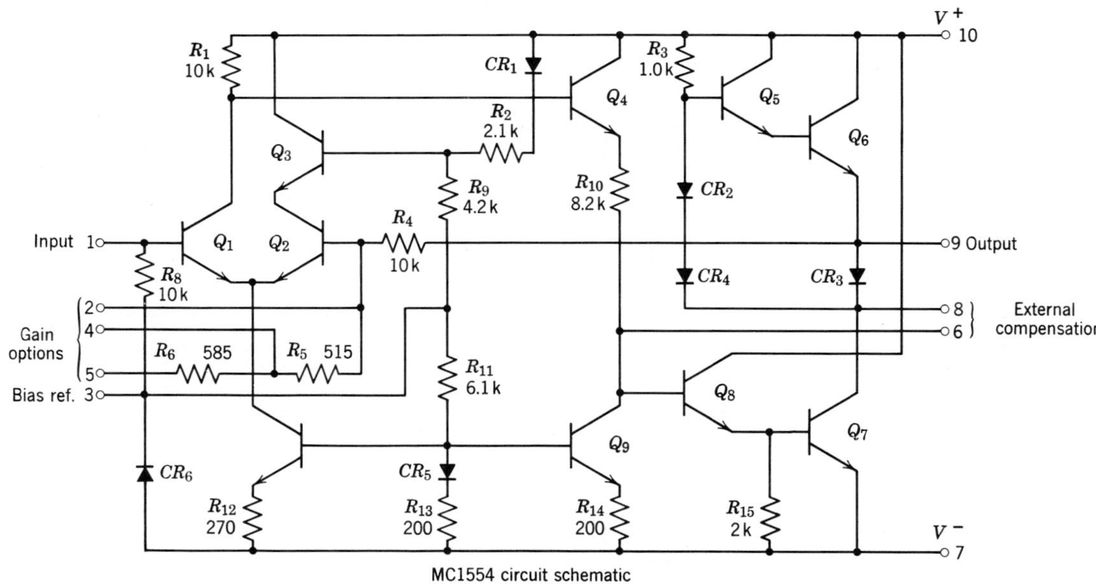

Figure 15.9. Integrated circuit audio-power amplifier (MC1554).

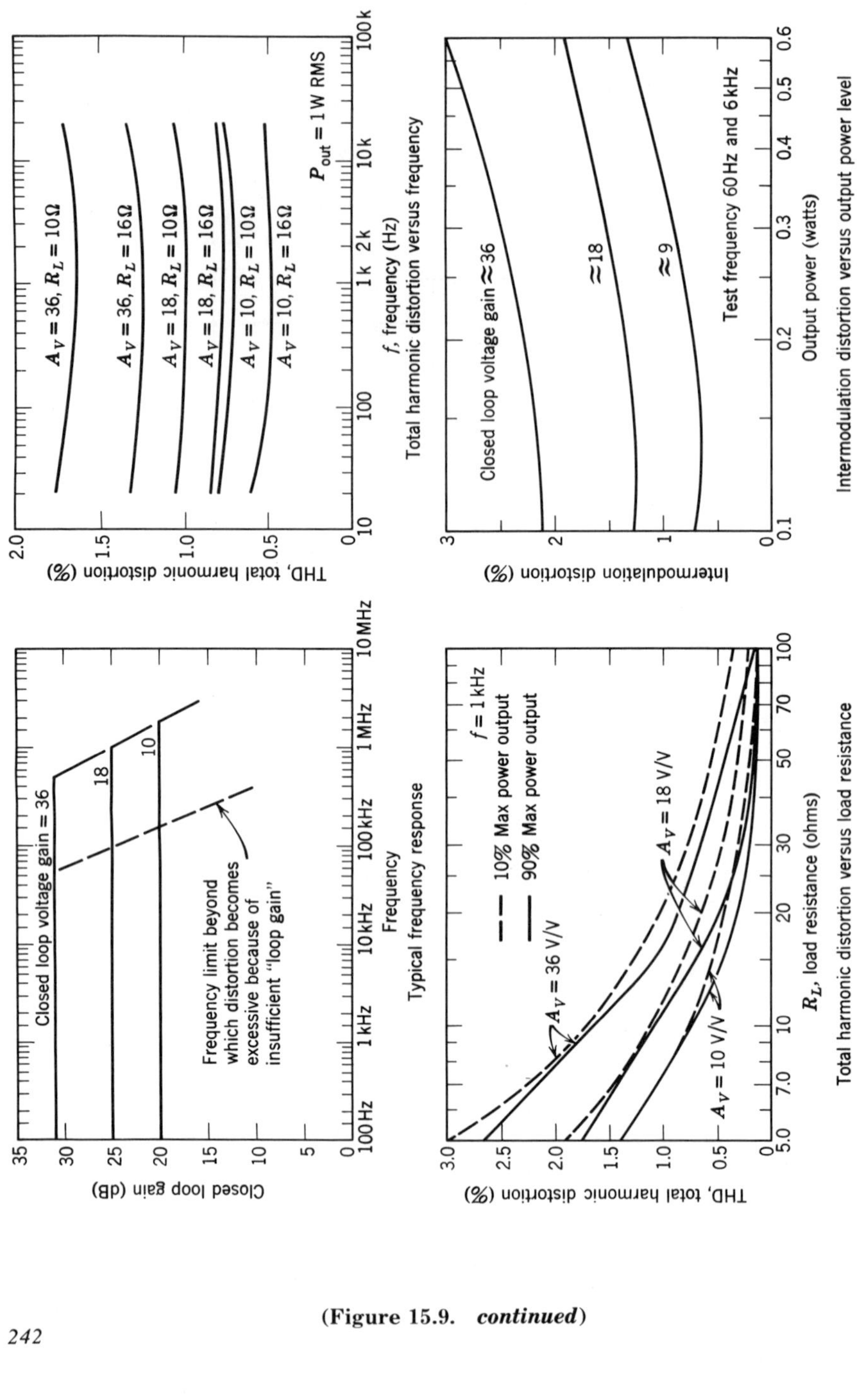

(**Figure 15.9.** ***continued***)

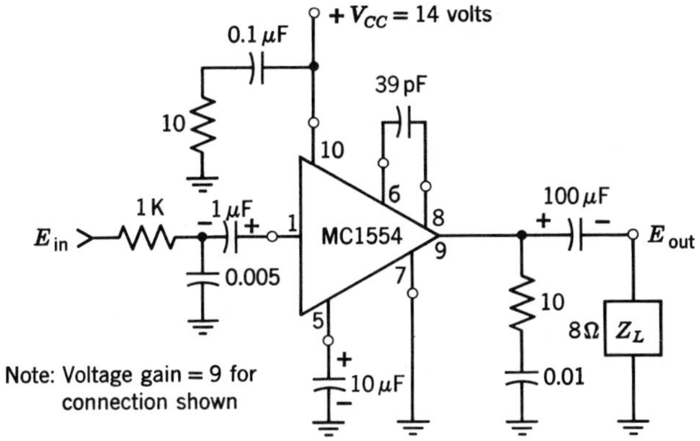

MC1554 typical connection single supply operation

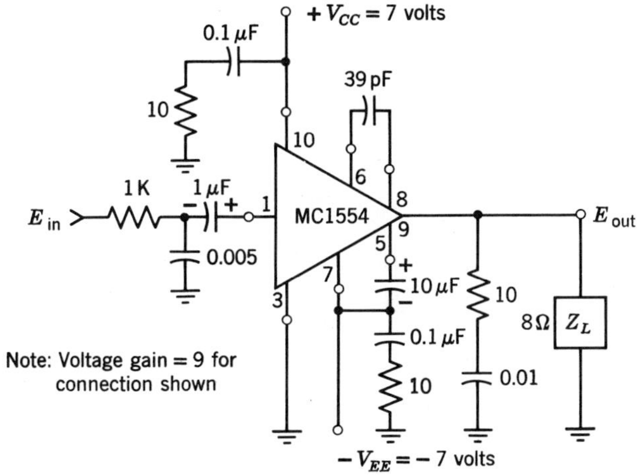

MC1554 typical connection split supply operation

AC Voltage Gain $\frac{E_{out}}{E_{in}}$:	9
Max Available Power Output:	1 Watt (Sine Wave)
Approximate Power Dissipated Inside Amplifier Can:	1.4 Watt (External Heat Sink Required)
Total Harmonic Distortion: 20 → 20 kHz	<0.75%
Frequency Response:	200 Hz → 22 kHz (−3 dB) Single Supply 40 Hz → 22 kHz (−3 dB) Split Supplies
Input Impedance:	10 k

ELECTRICAL CHARACTERISTICS OF THE ONE-WATT AMPLIFIER SHOWN ABOVE

Figure 15.10. Uses of the MC1554 IC power amplifier.

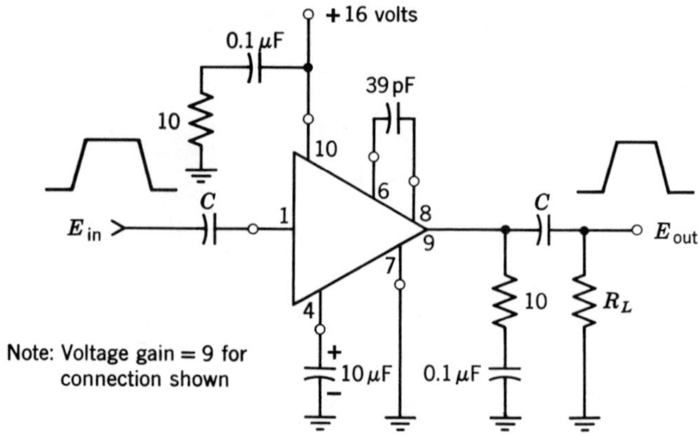

Pulse power amplifier

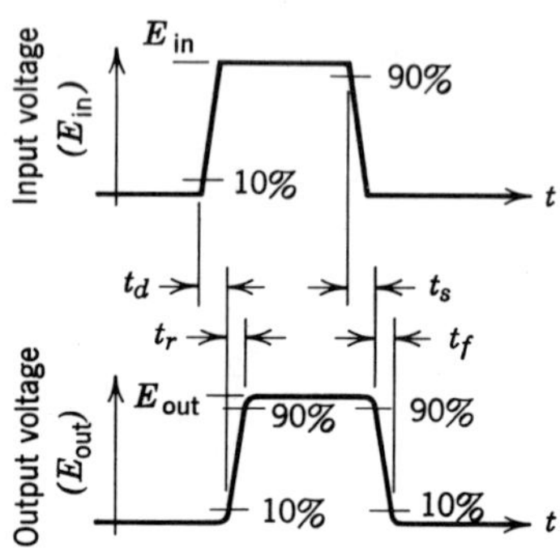

Pulse response of MC1554

Input Conditions:

Input rise time is such that

$$\frac{d\,E_{in}}{dt} > 200 \text{ volt}/\mu\text{s}$$

Pulse Amplifier Gain:

$$\frac{E_{out}}{E_{in}} = 20$$

Output Response:

t_d (delay time): 180 ns typ

t_s (storage time): 195 ns typ

t_r and t_f are a function of the total output voltage swing (E_{out}).

Slew Rate or Rate of Output Voltage Change per unit of time remain constant: $\frac{dE_{out}}{dt} = \frac{0.8\,E_{out}}{t_r}$ or $\frac{0.8\,E_{out}}{t_f}$

$$\text{Positive Going Slew Rate} = \frac{0.8\,E_{out}}{t_r} = 10 \text{ volts}/\mu\text{s typ}$$

$$\text{Negative Going Slew Rate} = \frac{0.8\,E_{out}}{t_f} = 13 \text{ volts}/\mu\text{s typ}$$

(Figure 15.10. *continued*)

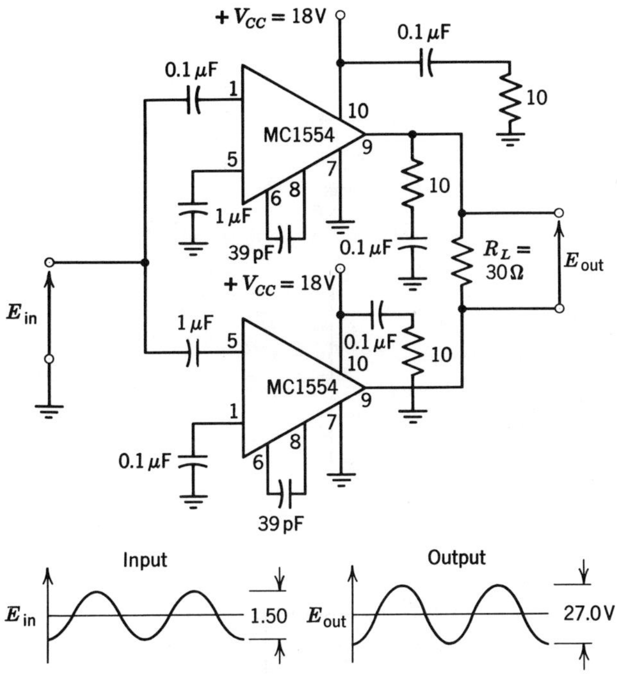

Individual amplifier voltage gain = 9
Effective voltage gain of differential output = 18
Note: Supply voltage = 18 volts
Yet peak–to–peak output swing = 27.0 volts
Power out ≈ 3 watts

A 3–watt differential power amplifier

(Figure 5.10. *continued*)

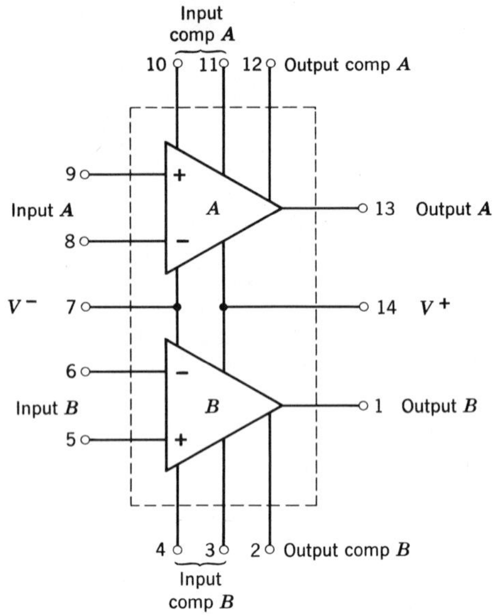

Schematic of the MC1302 IC dual preamplifier

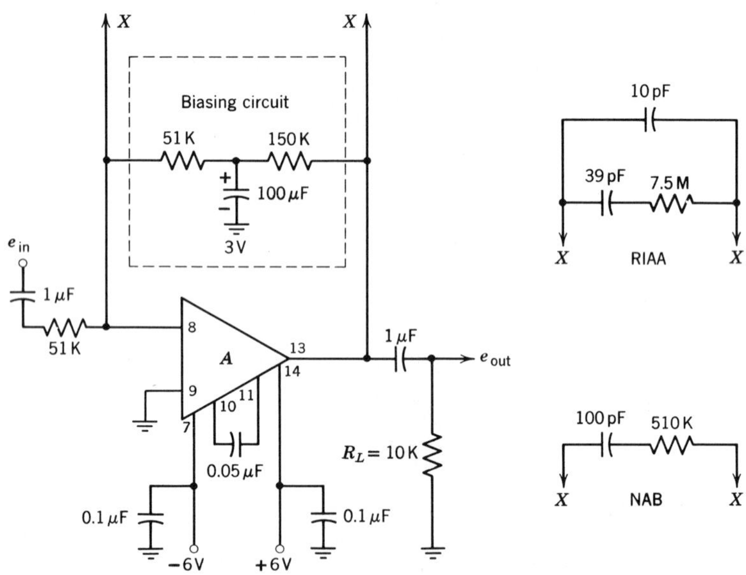

Figure 15.11. Applications of the MC1302 dual preamplifier.

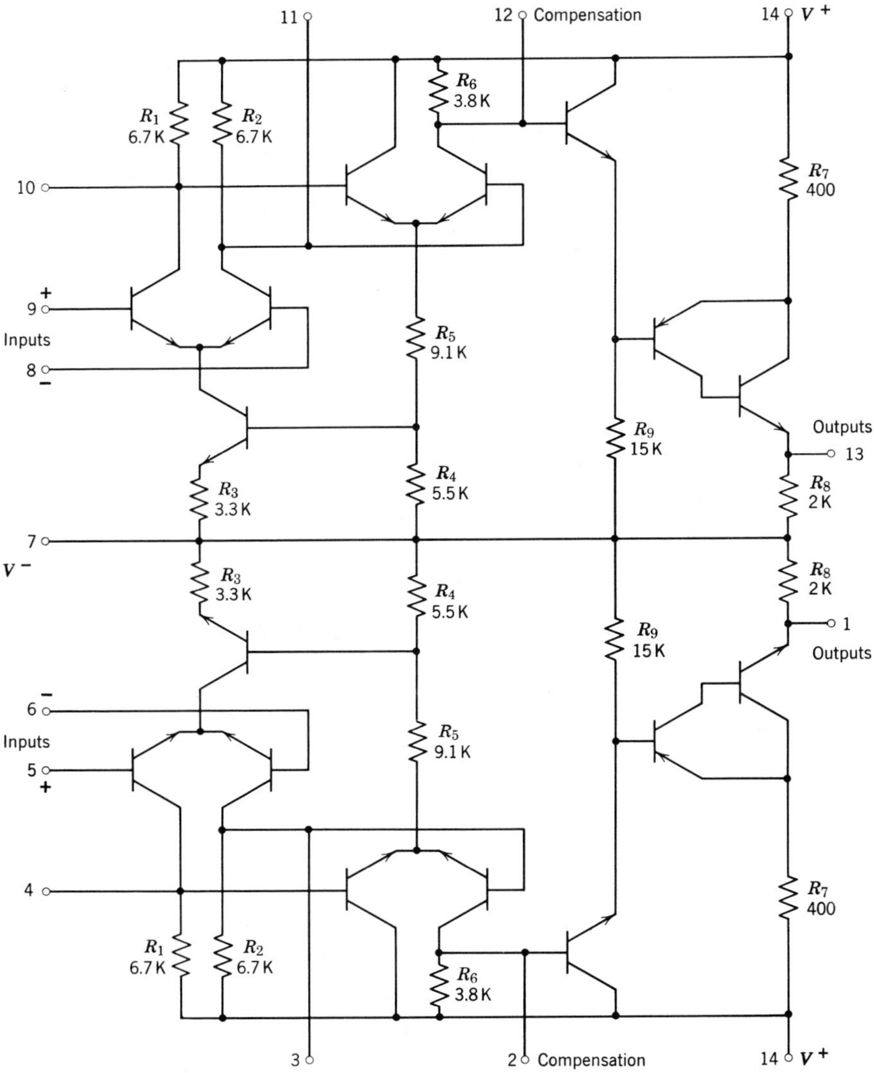

Schematic of the MC1302 IC dual preamplifier

(Figure 15.11 *continued*)

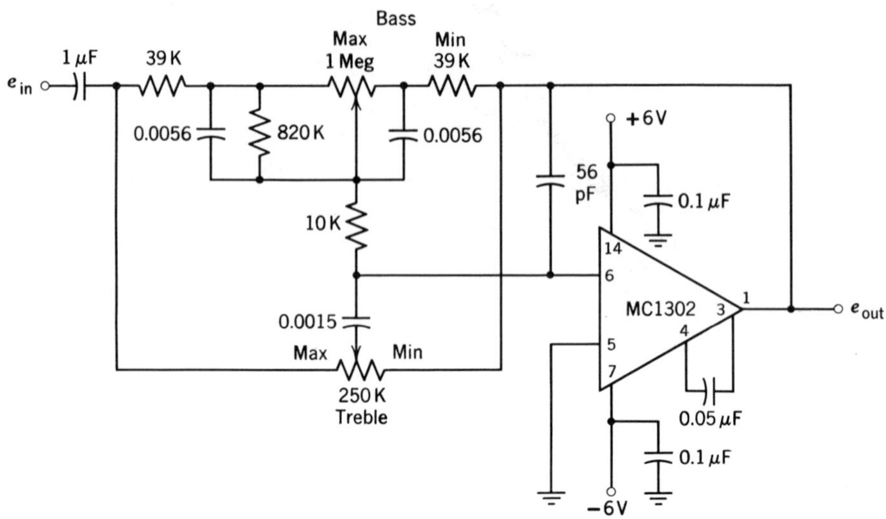

Active tone control

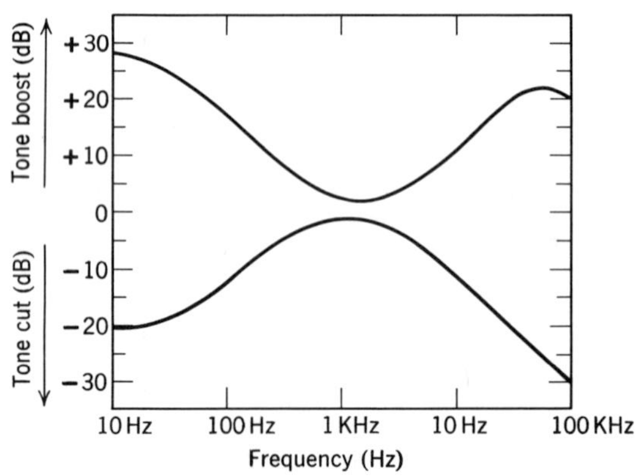

Tone control circuit response

(**Figure 15.11** ***continued***)

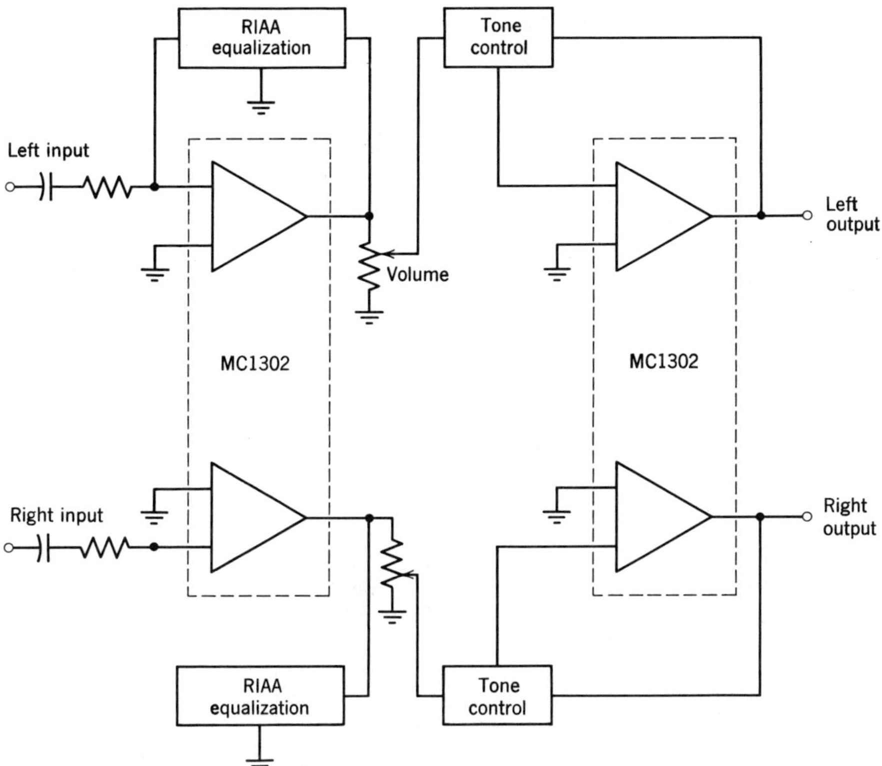

Stereophonic preamplifier

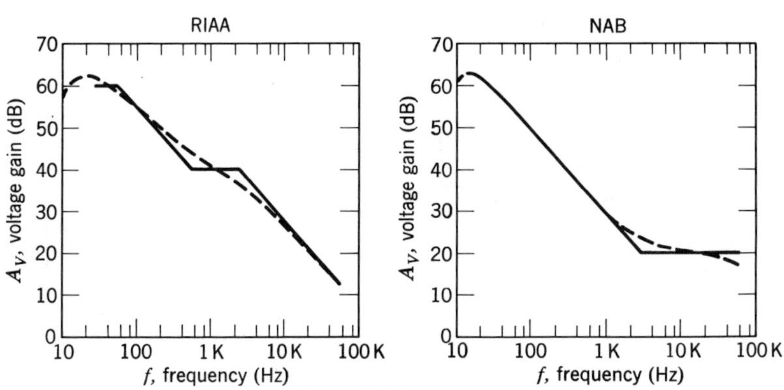

RIAA and NAB equalization

(Figure 15.11 *continued*)

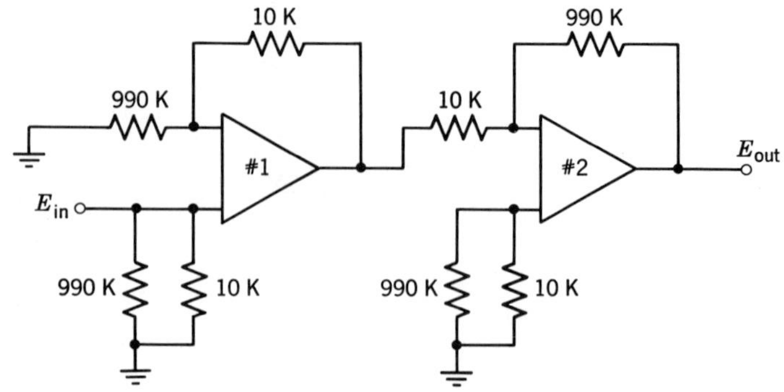

Figure 15.12. Zero-drift amplifier with voltage gain of 100.

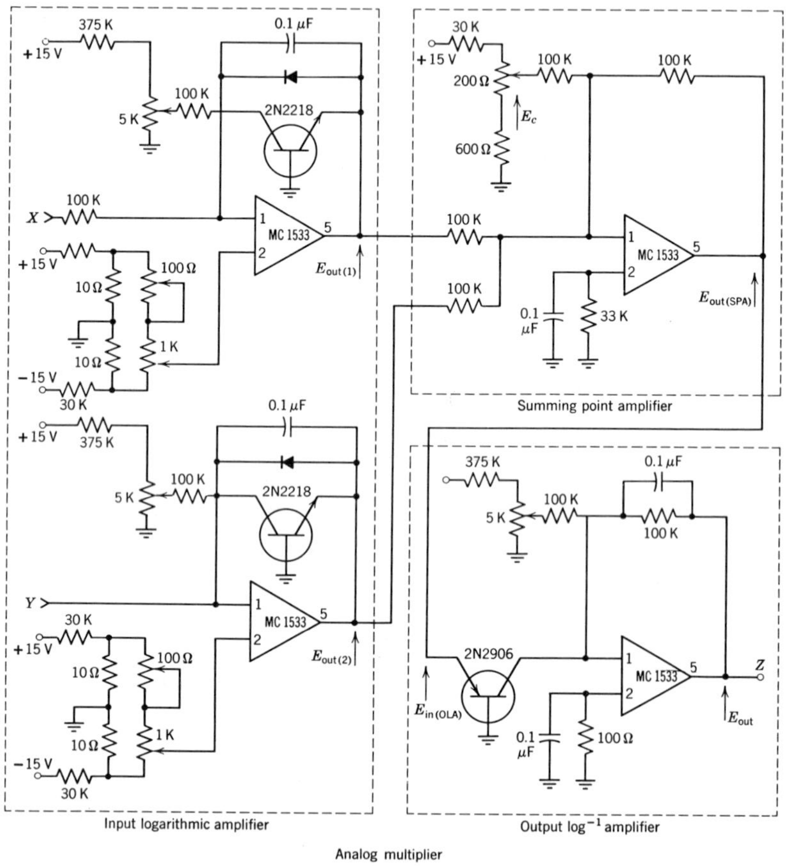

Figure 15.13. The MC1533 operational amplifier used as an analog multiplier.

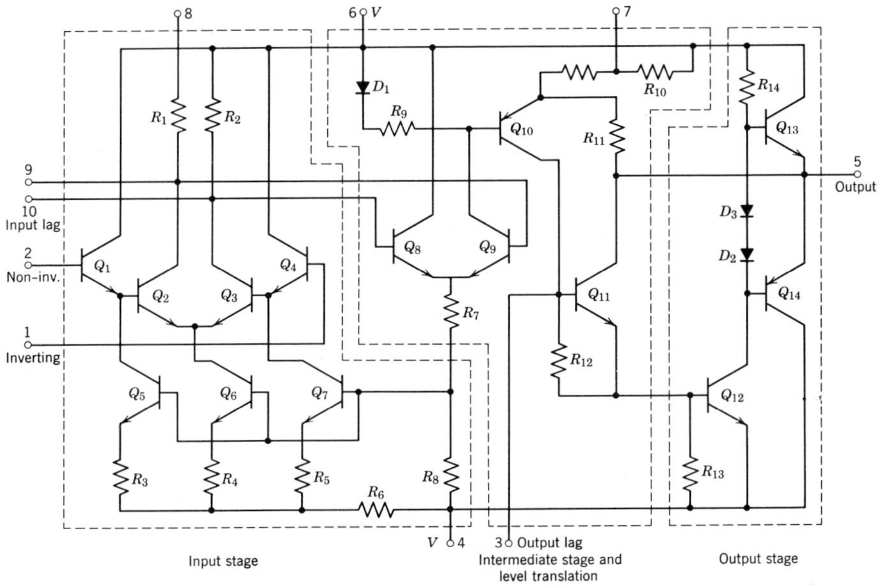

Schematic diagram of high-performance MC1533 operational amplifier

X Input #1	Y Input #2	Z Output	Calculated $Z=XY$	Difference	% Error
9.95 mV	13.8 mV	14.0 mV	13.72 mV	0.28	2.04
	81 mV	82.4 mV	18.6 mV	1.8	2.2
	297 mV	308 mV	295.5 mV	12.5	4.2
	605 mV	695 mV	647 mV	48	7.4
	1.042 V	1.110 V	1.038 V	0.072	6.9
	1.346 V	1.44 V	1.338 V	0.102	7.7
	2.98 V	2.99 V	2.965 V	0.025	0.8
	4.99 V	5.04 V	4.97 V	0.07	1.4
	10.11 V	10.34 V	10.06 V	0.28	2.8
	12.97 V	12.5 V	12.4 V	0.1	0.8
40.0 mV	14.44 mV	57.0 mV	57.8 mV	0.8	1.4
	40.9 mV	154.9 mV	163.8 mV	8.9	5.4
	102.3 mV	397 mV	410 mV	13	3.2
	250 mV	1.02 V	1.00 V	0.02	2.0
	500 mV	2.05 V	2.00 V	0.05	2.5
	730 mV	3.08 V	2.95 V	0.13	4.4
	1.003 V	4.28 V	4.02 V	0.26	6.5
98.0 mV	7.45 mV	72.9 mV	73 mV	0.1	0.14
	14.5 mV	143 mV	142 mV	1	0.7
	40.9 mV	407 mV	401 mV	6	1.5
	85 mV	837 mV	833 mV	4	0.48
	199 mV	2.04 V	1.95 V	0.09	4.6
	250 mV	2.54 V	2.45 V	0.09	3.7
500 mV	14 mV	754 mV	700 mV	54	7.7
	50 mV	2.56 V	2.5 V	0.06	2.4
	102 mV	5.31 V	5.10 V	0.21	4.1
	155 mV	7.94 V	7.75 V	0.19	2.5
	207 mV	10.38 V	10.35 V	0.03	0.29
	250 mV	12.54 V	12.50 V	0.04	0.32
1.003 V	5.6 mV	598 mV	562 mV	36	6.4
	14.5 mV	1.44 V	1.455 V	0.015	1.0
	50 mV	4.91 V	5.02 V	0.11	2.2
	102 mV	10.2 V	10.23 V	0.03	0.29
	120 mV	11.87 V	12.04 V	0.17	1.4
4.99 V	5.07 mV	2.48 V	2.53 V	0.05	2.0
	10.08 mV	5.02 V	5.04 V	0.02	0.4
	20.15 mV	9.30 V	10.07 V	0.77	7.7

TABULATED RESULTS FROM MULTIPLIER CIRCUIT $Z=XY$

(Figure 15.13. *continued*)

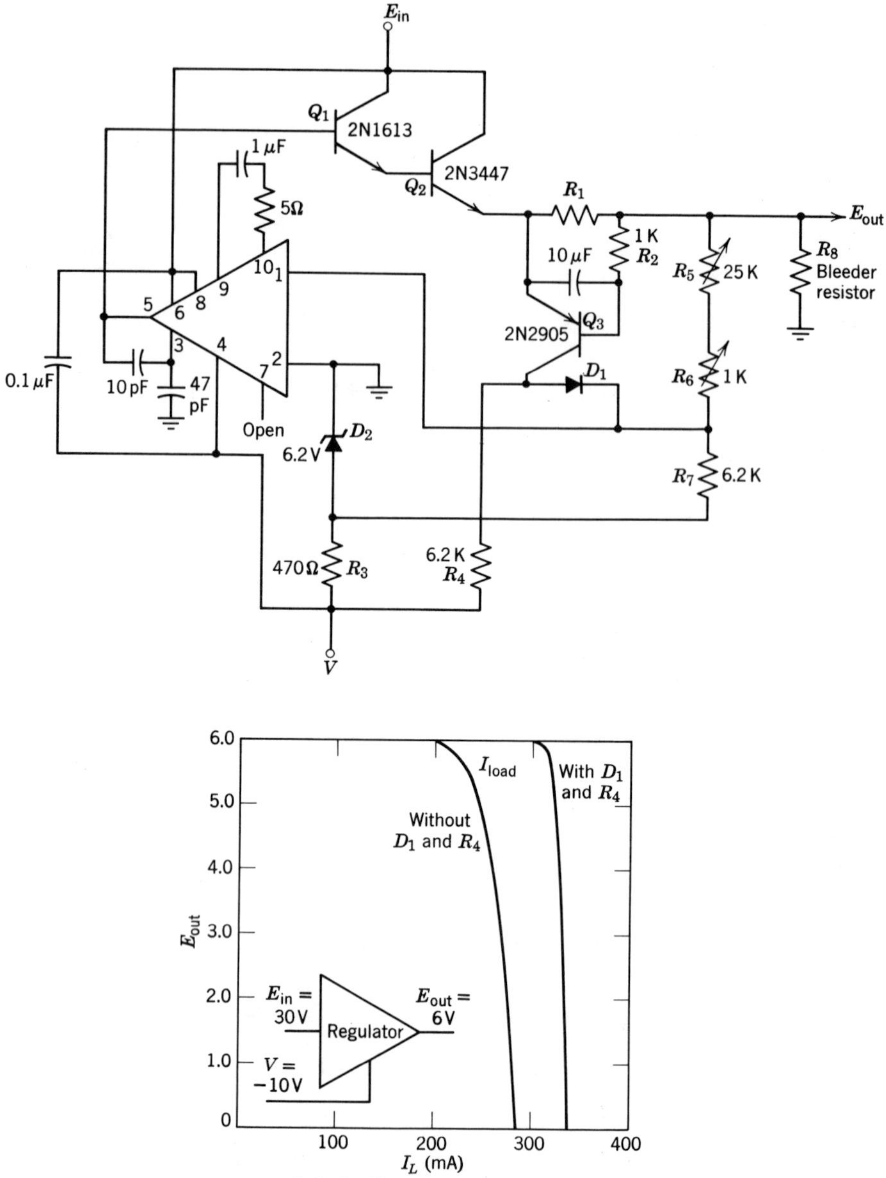

Output voltage versus load current

Figure 15.14. Voltage regulator.

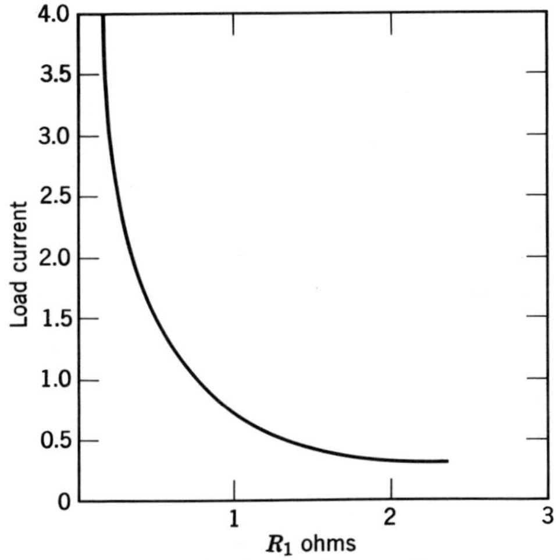

Load current versus R_1

(Figure 15.14. ***continued*****)**

16

Switching Regulators

by R. J. Widlar
National Semiconductor

16.1 INTRODUCTION

In a conventional series regulator, an element, usually a transistor, is used as a variable resistor to drop an unregulated input voltage down to a fixed output voltage. This element must be able to dissipate the voltage difference between the input and output at the load current. The power generated can become excessive, particularly if the input voltage is not well regulated and if the difference between the input and output voltages is large.

On the other hand, switching regulators are capable of high efficiency operation even if there are large differences between the input and output voltages. The efficiency is negligibly affected by the voltage difference, since this type of regulator acts as a continuously variable power converter.

Switching regulators are, therefore, useful in battery-powered equipment in which the required output voltage is considerably lower than the battery voltage. An example of this is a missile with a 30-V battery as its only power source, containing a large number of integrated logic circuits which require a 5-V supply. Switching regulators are also useful in space vehicles in which conservation of power is extremely important. In addition they are frequently the most economical solution in commercial and industrial applications in which the increased efficiency reduces the cost of the series-pass transistors and simplifies heat sinking.

Switching regulators are more complex than linear regulators, but this is often a substitution of electrical complexity for the thermal and mechanical complexity of high-power linear regulators. They also have higher output ripple. However, this can be held to a minimum (about 10 mV), and it is at a high enough frequency so that it can be easily filtered out. Another limitation is that the response to load transients is not always as fast as with linear regulators, but this can be largely overcome by proper design. The rejection of line transients, however, is every bit as good if not better than linear regulators. Finally, switching regulators throw current transients back into the unregulated supply that are somewhat larger than the maximum load current. These, in some cases, can be troublesome unless adequate filtering is used.

This chapter will cover the use of a monolithic voltage regulator in a number of switching regulator applications. These include both self-oscillating and synchronously driven regulators in the 0.1 A to 5 A range. Circuits are shown for both positive and negative regulators with output voltages in the 2 V to 30 V range. Methods of isolating the integrated circuit from the input voltage are given, permitting input voltages in excess of 100 V. Further, current limiting schemes which keep peak currents and dissipation well within safe limits for both overload and short-circuit conditions are presented. Finally, component selection details peculiar to switching regulators are covered.

16.2 SWITCHING REGULATOR OPERATION

The method by which a switching regulator produces a voltage conversion with high efficiency can be explained with the aid of Figure 16.1. Q_1 is a switch transistor that is turned on and off by a pulse waveform with a given duty cycle, and D_1 is a catch diode that provides a continuous path for the inductor current when Q_1 turns off. The voltage waveform at the collector of Q_1 is shown in the figure. The output of the LC filter will be the average value of the switched waveform V_1. If the voltage drops across the transistor and diode are neglected, the output voltage will be

$$V_{out} = V_{in} \frac{t_{on}}{T}. \tag{16.1}$$

V_{out} is independent of the load current. It is obvious from the equation that changes in input voltage can be compensated for by varying the duty cycle of the switched waveform. This is how a switching regulator works.

Figure 16.2 shows a self-oscillating switching regulator which produces this duty-cycle control. A reference voltage V_{ref}, equal to the desired output voltage, is supplied to one input of an operational amplifier A_1. The operational amplifier in turn drives the switch transistor. The resistive divider, arranged such that $R_1 \gg R_2$, provides a slight amount of positive feedback at high frequencies to make the circuit oscillate. At lower frequencies at which the attenuation of the LC filter is less than the attenuation of the resistive divider, there is net negative feedback to the inverting input of the operational amplifier.

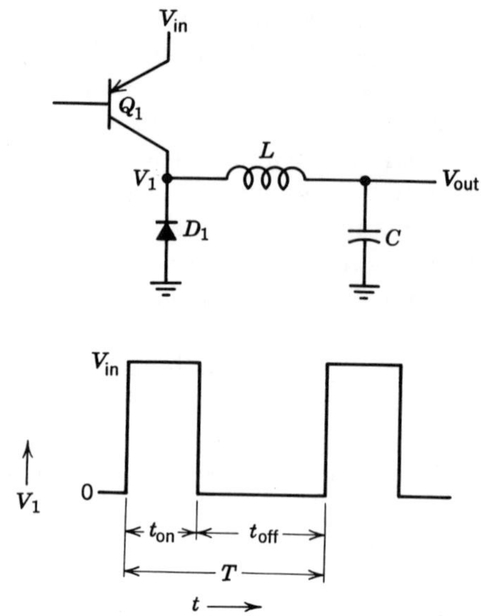

Figure 16.1. Switching circuit for voltage conversion.

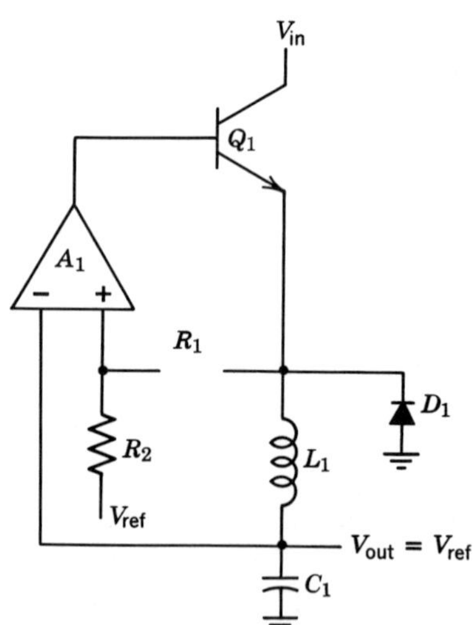

Figure 16.2. Self-oscillating switching regulator.

In operation, when the circuit is first turned on, the output voltage is less than the reference voltage so that the switch transistor is turned on. When this happens, current flow through R_1 raises the voltage on the noninverting input of the operational amplifier slightly above the reference voltage. The circuit will remain switched on until the output rises to this voltage. The amplifier now goes into the active region, causing the switch to turn off. At this point, the reference voltage seen by the amplifier is lowered by feedback through R_1, and the circuit will stay off until the output voltage drops to this lower voltage; hence the output voltage oscillates about the reference voltage. The amplitude of this oscillation (or the output ripple) is nearly equal to the voltage fed back through R_1 to R_2 and can be made quite small.

16.3 VOLTAGE REGULATOR IC

The switching regulator circuits described here use the LM100 integrated voltage regulator as the control element. This device contains, on a single silicon chip, the voltage reference, the operational amplifier, and the circuitry for driving a *pnp* switch transistor. Discrete switch transistors, catch diodes, and reactive elements are employed since these components are not easily integrated.

Figure 16.3 shows a schematic diagram of the LM100. The voltage reference portion of the circuit starts with a breakdown diode D_1 that is supplied by a current source from the unregulated input (one of the collectors of Q_2). The output of the reference diode, which has a positive temperature coefficient of 2.4 mV/°C, is buffered by an emitter follower Q_4 which increases the temperature coefficient to +4.7 mV/°C. This is further increased to 7 mV/°C by the diode-connected transistor Q_6. A resistor divider reduces this voltage as well as the temperature coefficient to exactly compensate for the negative temperature coefficient of Q_7, producing a temperature-compensated output of 1.8 V.

The transistor pair Q_8 and Q_9 form the input stage of the operational amplifier. The gain of the stage is made high by the use of a current source, one of the collectors of Q_2, as a collector load. The output of this stage drives a compound emitter follower, Q_{11} and Q_{12}. The output of Q_{12} is taken across R_8 to drive the *pnp* switch transistor. An additional transistor Q_{10} is used to limit the output current of Q_{12} to the value required for driving the *pnp* transistor. This current is determined by a resistor placed between the current limit and regulated output terminals. The value of the drive current can be determined from Figure 16.4 which plots the output current as a function of temperature for various current limit resistors.

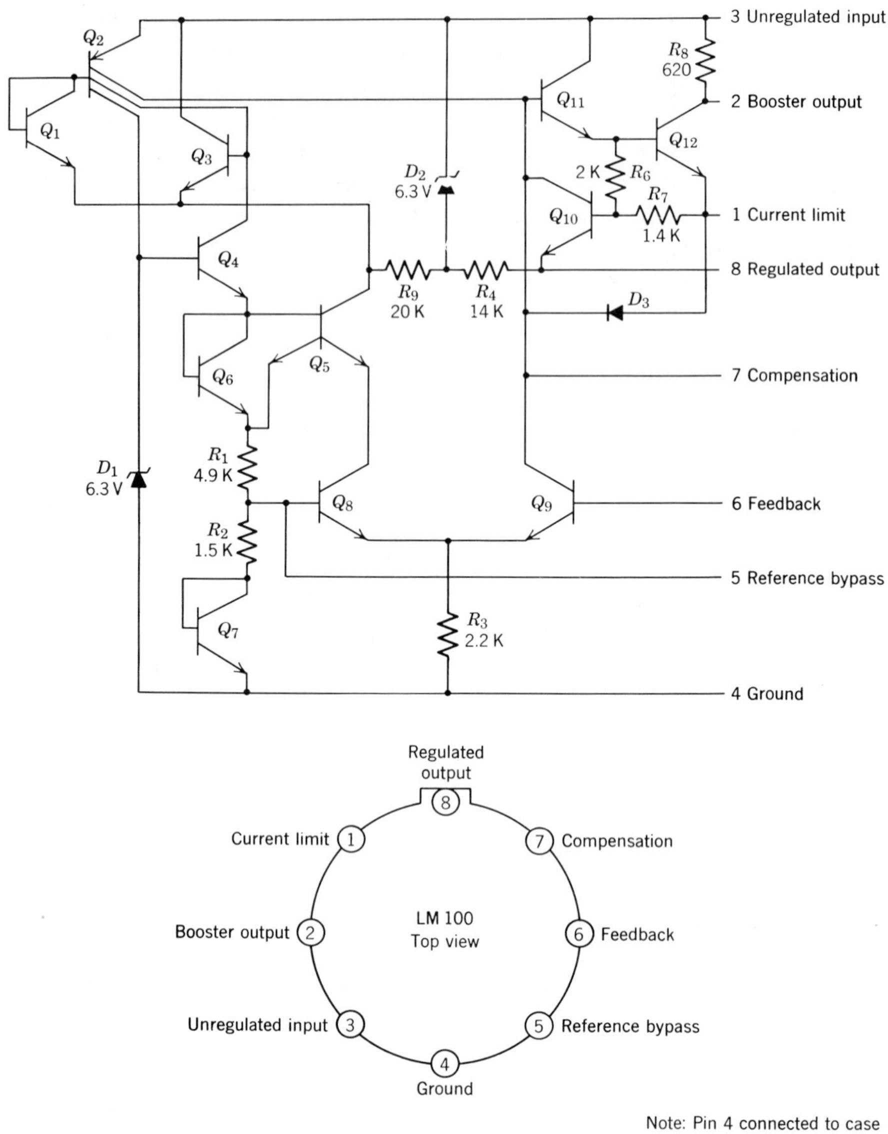

Figure 16.3. Schematic and connection diagrams of the LM100 voltage regulator.

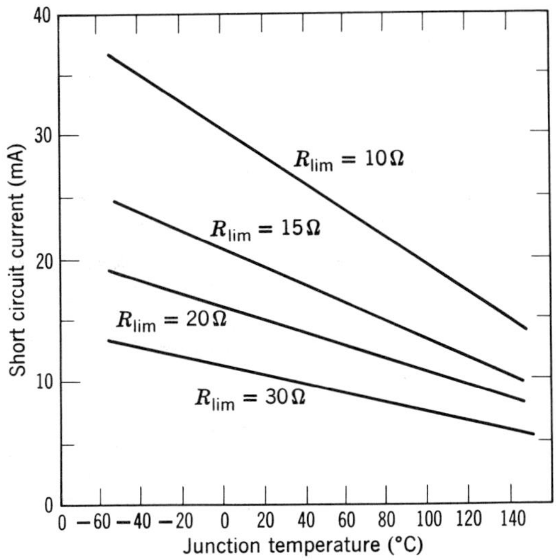

Figure 16.4. Switched output current as a function of temperature for various values of current limit resistors.

As for the remaining details of the circuit, Q_5, Q_3, and Q_1 are part of a bias stabilization circuit for Q_2 to set its collector current at the desired value. R_9, R_4, and D_2 serve the sole function of starting the regulator. Finally, D_3 is a clamp diode which keeps Q_9 from saturating when it is switching.

16.4 SWITCHING REGULATOR CIRCUITS

Figure 16.5 demonstrates the use of the LM100 as a switching regulator. Feedback to the inverting input of the operational amplifier (pin 6 of the LM100) is obtained through a resistive divider that can be used to set the output voltage anywhere in the 2-30 V range. R_3 determines the base drive for the switch transistor Q_1, providing enough drive to saturate it with maximum load current. R_4 works into the 1-kΩ impedance at the reference terminal, producing the positive feedback. C_2 serves to minimize output ripple by causing the full ripple to appear on the feedback terminal. The remaining capacitor C_3 removes the fast-risetime transients which would otherwise be coupled into pin 5 through the shunt capacitance of R_4. It must be made small enough so that it does not seriously integrate the waveform at this point.

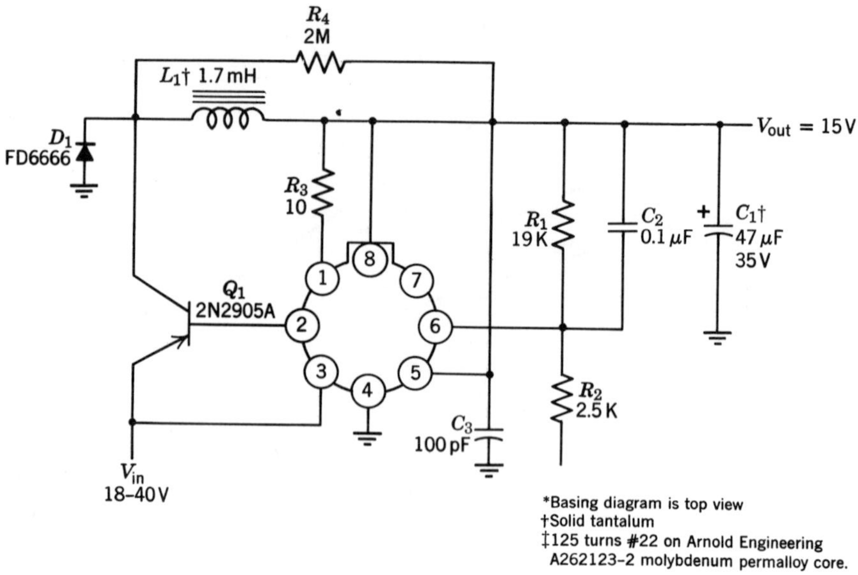

Figure 16.5. Switching regulator using the LM100.

The circuit shown in Figure 16.5 is suitable for output currents as high as 500 mA. This limit is set by the output current available from the LM100 to saturate the switch transistor Q_1. For lower currents, the value of R_3 should be increased so that the base of Q_1 is not driven unnecessarily hard.

The optimum switching frequency for these regulators has been determined to be between 20 kHz and 100 kHz. At lower frequencies, the core becomes unnecessarily large; and at higher frequencies, switching losses in Q_1 and D_1 become excessive. It is important, in this respect, that both Q_1 and D_1 be fast-switching devices to minimize switching losses.

The output ripple of the regulator at the switching frequency is mainly determined by R_4. It should be evident from the description of circuit operation that the peak-to-peak output ripple will be nearly equal to the peak-to-peak voltage fed back to pin 5 of the LM100. Since the resistance looking into pin 5 is approximately 1000 Ω, this voltage will be

$$\Delta V_{\text{ref}} \simeq \frac{1000\, V_{\text{in}}}{R_4}. \tag{16.2}$$

In practice, the ripple will be somewhat larger than this. When the switch resistor shuts off, the current in the inductor will be greater than the load current so that the output voltage will continue to rise above the value required to shut off the regulator. An important consideration in choosing the value of the inductor is that it be large enough so that the current through it does not change drastically during the switching cycle. If it does, the switch transistor and catch diode must be able to handle peak currents which are significantly larger than the load current. The change in inductor current can be written as

$$\Delta I_L \simeq \frac{V_{out} t_{off}}{L}. \qquad (16.3)$$

In order for the peak current to be about 1.2 times the maximum load current, it is necessary that

$$L_1 = \frac{2.5\, V_{out} t_{off}}{I_{out} \max}. \qquad (16.4)$$

A value for t_{off} can be estimated from

$$t_{off} = \frac{1}{f}\left(1 - \frac{V_{out}}{V_{in}}\right), \qquad (16.5)$$

where f is the desired switching frequency and V_{in} is the nominal input voltage.

The size of the output capacitor can now be determined from

$$C_1 = \left(\frac{V_{in} - V_{out}}{2L_1 \Delta V_{out}}\right)\left(\frac{V_{out}}{f V_{in}}\right)^2, \qquad (16.6)$$

where ΔV_{out} is the peak-to-peak output ripple and V_{in} is the nominal input voltage.

It now remains to determine if the component values obtained above give satisfactory load-transient response. The overshoot of the regulator can be determined from

$$\Delta V_{out} = \frac{L_1 (\Delta I_L)^2}{C_1 (V_{in} - V_{out})} \qquad (16.7)$$

for increasing loads, and

$$\Delta V_{out} = \frac{L_1 (\Delta I_L)^2}{C_1 V_{out}} \qquad (16.8)$$

for decreasing loads, where ΔI_L is the load-current transient. The recovery time is

$$t_r = \frac{2L_1 \Delta I_L}{V_{in} - V_{out}}, \tag{16.9}$$

and

$$t_r = \frac{2L_1 \Delta I_L}{V_{out}}, \tag{16.10}$$

for increasing and decreasing loads, respectively.

In order to improve the load transient response, it is necessary to allow larger peak to average current ratios in the switch transistor and catch diode. Reducing the value of inductance given by (16.4) by a factor of 2 will reduce the overshoot by 4 times and double the response time. This, of course, assumes that the output capacitance is doubled to maintain a constant switching frequency.

The above equations outline a design procedure for determining the values for R_4, L_1 and C_1, given the switching frequency and the output ripple. These equations are not exact, but they do provide a starting point for designing a regulator to fit a given application.

As an example, this design method will be applied to a regulator which must deliver 15 V at a maximum current of 300 mA from a 28 V supply. To start, a 40 kHz switching frequency will be selected along with an output ripple of 14 mV, peak-to-peak. From (16.2), R_4 is calculated to be 2 MΩ. In determining L_1, t_{off} is found to be 11.6 μs from (16.5). Inserting this into (16.4) gives a value of 1.45 μH for L_1; the value of C_1 obtained from (16.6) is then 57.5 μF.

In the actual circuit of Figure 16.5, a standard value of 47 μF is used for C_1; and L_1 is adjusted to 1.7 μH. The switching frequency obtained experimentally for this circuit is 60 kHz and the peak-to-peak output ripple is 20 mV. The fairly-large disagreement between the calculated and experimental values is not alarming because many simplifying assumptions were made in the derivation of the equations. They do, however, provide a convenient method of handling a large number of mutually dependent variables to arrive at a working circuit. More exact expressions would involve a design procedure that is too cumbersome to be of practical value.

The variations of switching frequency with input voltage and load current are shown in Figures 16.6 and 16.7. The sharp rise in frequency at low output currents occurs because the output transistor of the LM100 (Q_{12}) begins to supply an appreciable portion of the load current directly.

The efficiency of the regulator over a wide range of input voltages and output currents is given in Figures 16.8 and 16.9.

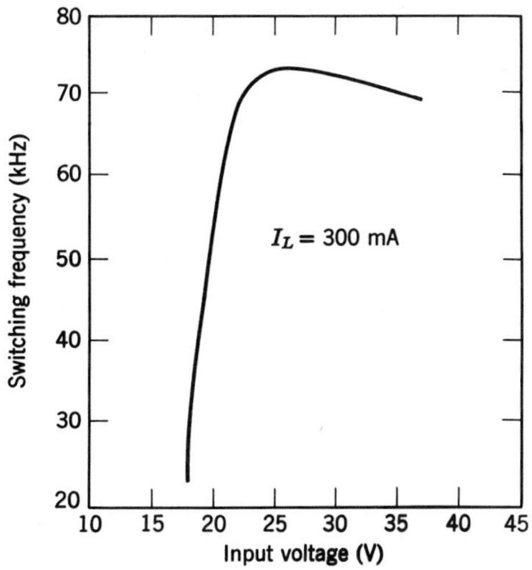

Figure 16.6. Switching frequency as a function of input voltage.

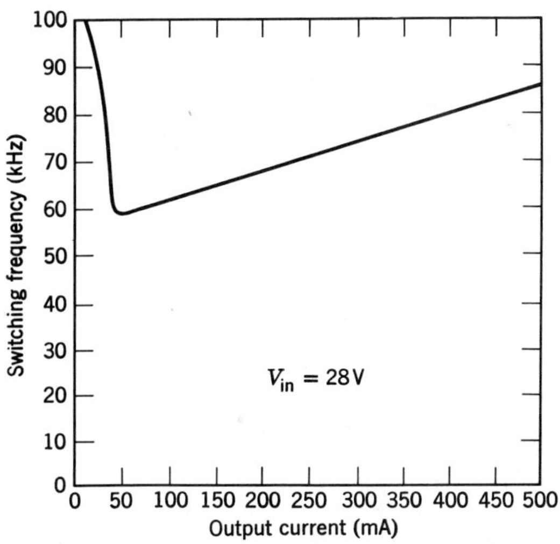

Figure 16.7. Switching frequency as a function of output current.

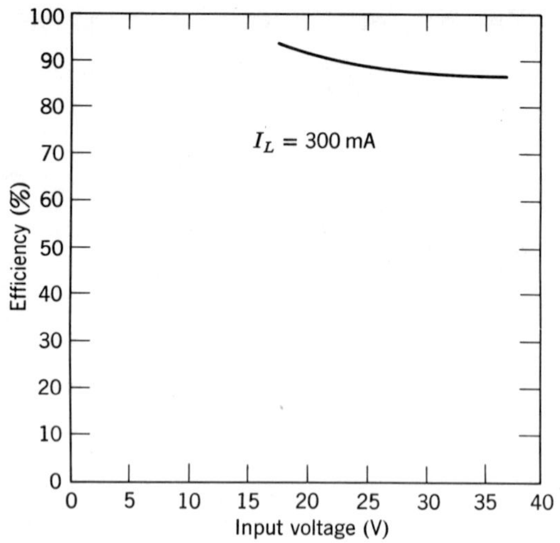

Figure 16.8. Efficiency as a function of input voltage.

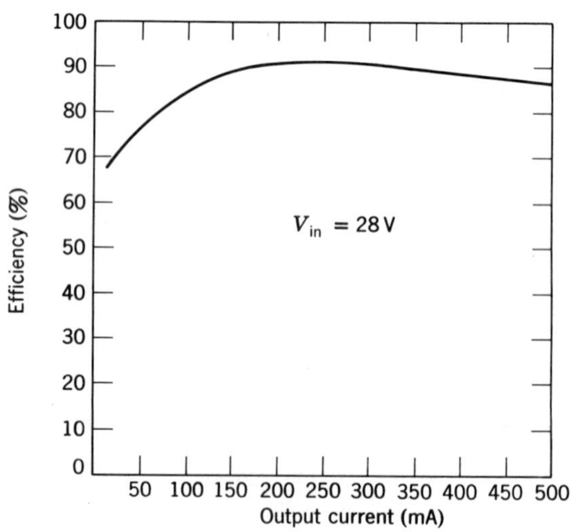

Figure 16.9. Efficiency as a function of output current.

16.5 HIGHER CURRENT REGULATORS

If output currents greater than about 500 mA are required, another switch transistor should be added to obtain more current gain. This is illustrated in Figure 16.10. With the exception of the added *npn* power switch Q_2, this circuit is the same as that described previously.

A photograph of a high-current regulator is shown in Figure 16.11. It is capable of delivering output currents of 3 A continuously with only a small heat sink. Figure 16.12 shows that the efficiency is better than 80% at this level. Output currents to 5 A can be obtained at reduced efficiency. However, the case temperature of the power switch and catch diode approach 100°C under this condition, so that continuous operation is not recommended unless more heat sink is provided.

Figure 16.13 shows that the efficiency is not significantly affected by input voltage. In Figure 16.14 it can be seen that the switching frequency is fairly constant over a wide range of input voltages. Figure 16.15 shows that the switching frequency increases with increasing load current. The higher dc current through the inductor reduces the incremental inductance causing the frequency to go up. The last graph, Figure 16.16, illustrates the line regulation of the device. This can be improved by putting a small capacitor (0.01 μF) in series with the positive feedback resistor R_3 to isolate the reference terminal from the dc input voltage changes.

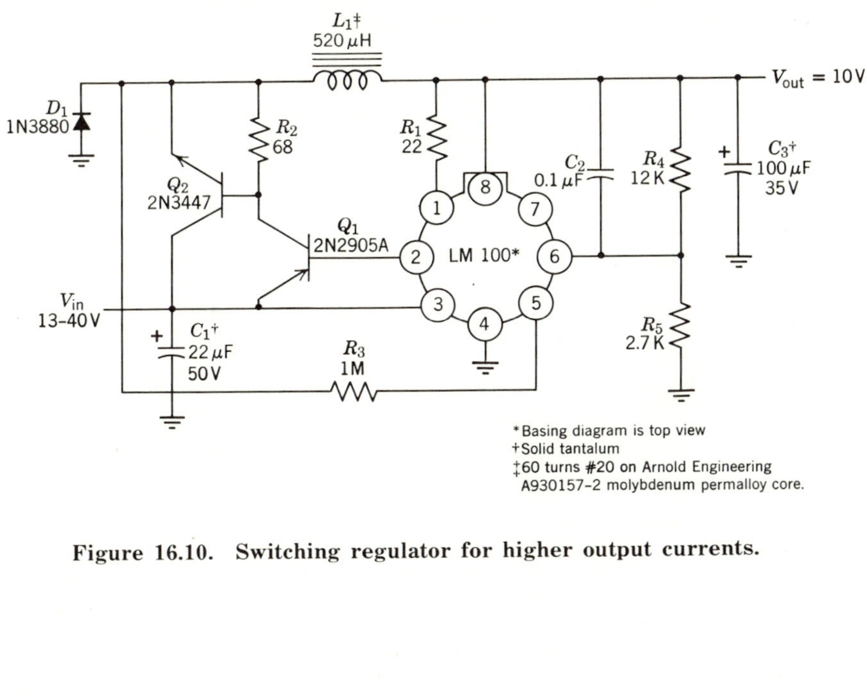

Figure 16.10. Switching regulator for higher output currents.

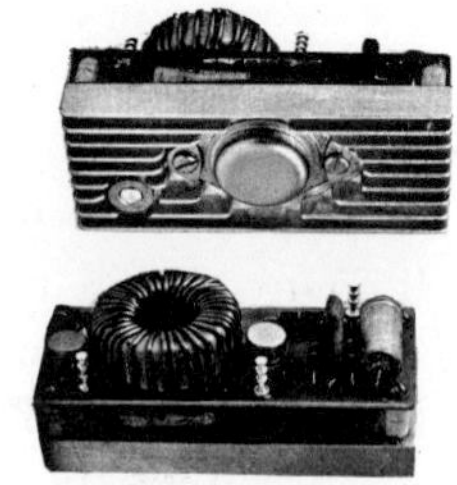

Figure 16.11. High current switching regulator.

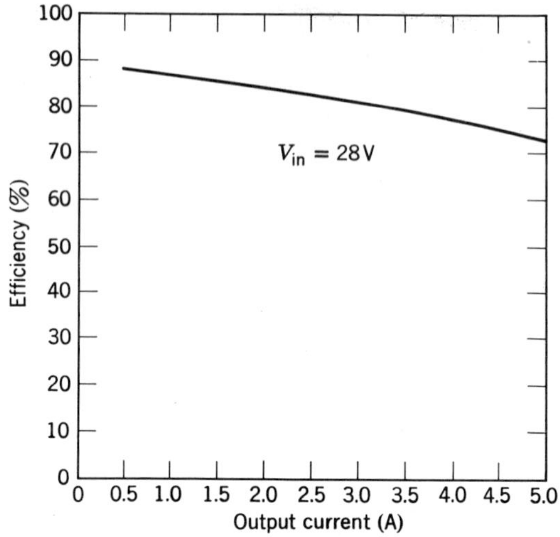

Figure 16.12. Efficiency as a function of output current.

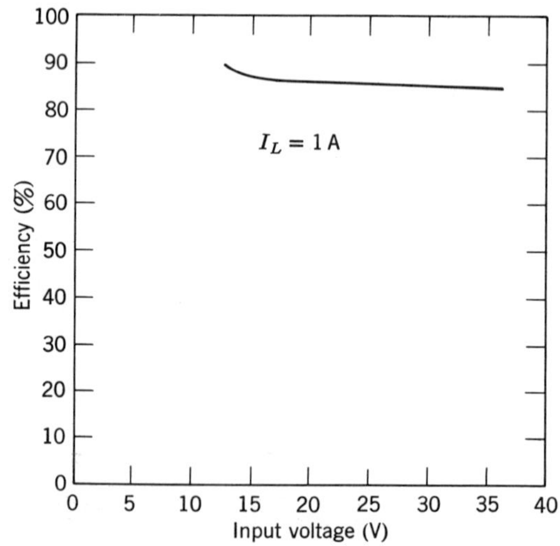

Figure 16.13. Efficiency as a function of input voltage.

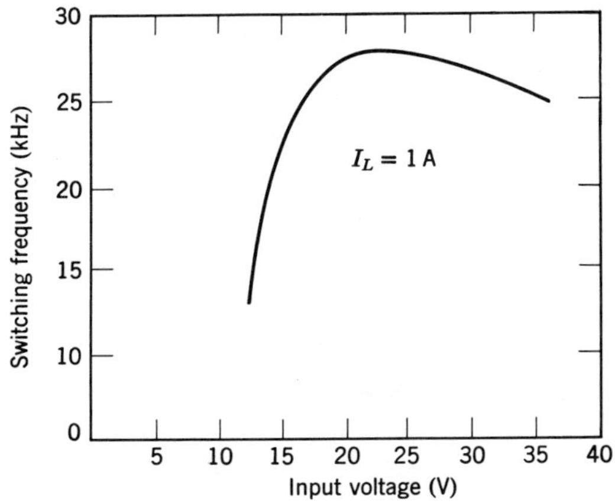

Figure 16.14. Variation of switching frequency with input voltage.

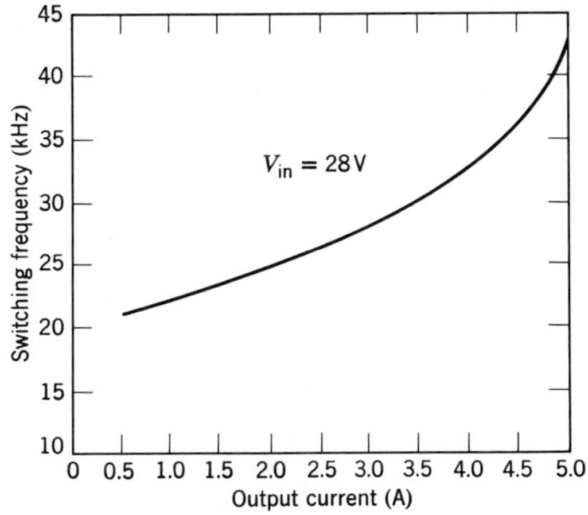

Figure 16.15. Variation of switching frequency with output current.

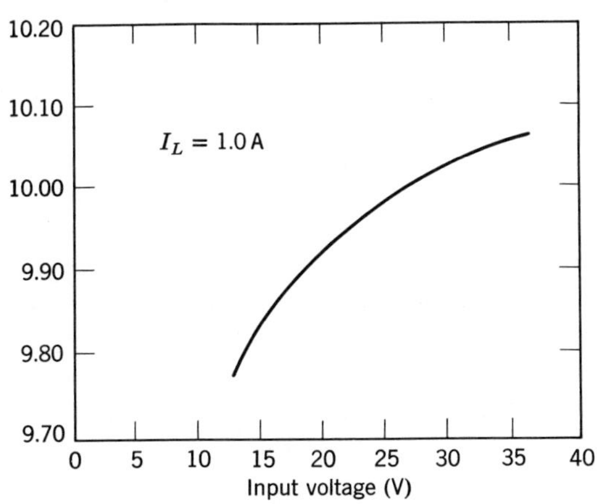

Figure 16.16. Line regulation.

At low output currents the inductor current can drop to zero at some time after the switch transistor turns off. When this happens, ringing occurs on the switching waveform. This is perfectly normal and causes no ill effects.

The use of solid tantalum capacitors for C_1 and C_3 is recommended when the regulator is expected to perform over the full military temperature range. The reason for using 35 V capacitors on the output, even though the output voltage is only 10 V, is that the 40 mV peak-to-peak ripple on the output would, for example, exceed the ratings of a 100 μF, 15 V capacitor.

Aluminum electrolytic capacitors have been used successfully over a limited temperature range. And there is basically no reason why wet foil or wet slug tantalums could not be used as long as their equivalent series resistance is low enough so that they behave like capacitors with the high frequency switched-current waveform. It is also important that manufacturer's data be consulted to ensure that they can withstand the high frequency ripple.

As was mentioned with respect to the low current regulator, it is necessary to use fast-switching diodes and transistors in these circuits. Ordinary silicon rectifiers or low-frequency power transistors will operate at drastically reduced efficiencies and will quickly overheat in these circuits.

16.6 DRIVEN SWITCHING REGULATOR

When a number of switching regulators are used together in a system, it is sometimes desirable to synchronize their operation to more uniformly distribute the switched current waveforms on the input line. Synchronous operation is also desirable when a switching regulator is operated in conjunction with a power converter.

A circuit for synchronizing the switching regulator with a square wave drive signal is shown in Figure 16.17. In this circuit positive feedback is not used. Instead, the square wave drive signal is integrated; and the resulting triangular wave (about 40 mV peak-to-peak) is applied to the reference bypass terminal of the LM100. This triangular wave will cause the regulator to switch because its gain is so high that the waveform overdrives it. The duty cycle of the switched waveform is controlled by the voltage on the feedback terminal, pin 6. If this voltage goes up, the duty cycle will decrease because it is picking off a smaller portion of the triangular wave on pin 5. By the same token, the duty cycle will decrease if the voltage on pin 6 drops. This action produces the desired regulation: if the output voltage starts to go up, it will raise the voltage on pin 6 such that a smaller portion of the triangular wave is picked off.

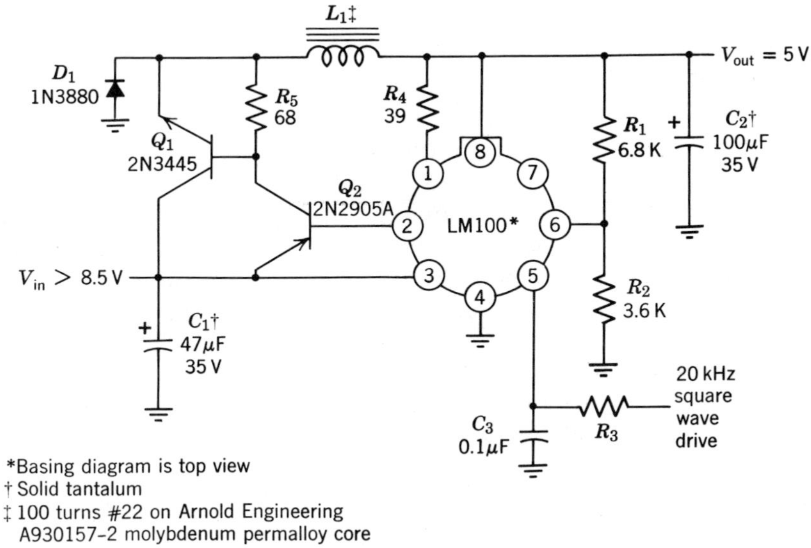

Figure 16.17. Driven switching regulator.

This reduces the duty cycle, counteracting the output voltage increase.

In order for this circuit to work properly, the ripple voltage on pin 6 should be less than a quarter of the peak-to-peak amplitude of the triangular wave. If this condition is not satisfied, the regulator will try to oscillate at its own frequency. Furthermore, since the resistance looking into pin 5 is about 1 kΩ, the integrating capacitor C_3 should have a capacitive reactance of less than 100 Ω at the drive frequency. The value of R_3 is determined so that the amplitude of the triangular wave on pin 5 is about 40 mV.

Driven regulators also have other advantages. For one, it is possible to design the LC filter independent of switching frequency considerations; hence lower output ripple and better transient response can be realized. A second advantage is the frequency stability. In a self-oscillating regulator the switching frequency is controlled by a relatively large number of factors; as a result, it is not well determined when normal tolerances are taken into account. With low and medium power regulators, this is not usually a problem since the efficiency does not vary greatly with frequency. High power regulators, however, tend to be more frequency sensitive, and it is desirable to operate them at constant frequency.

16.7 CURRENT LIMITING

In the circuits described previously the regulator is not protected from overloads or short-circuited output. Providing short-circuit protection is no simple problem, since it is necessary to keep the regulator switching when the output is shorted. Otherwise, the dissipation will become excessive even though the current is limited.

A circuit that does this is shown in Figure 16.18. The peak current through the switch transistor is sensed by R_8. When the voltage drop across this resistor becomes large enough to turn on Q_3, the output voltage begins to fall, since current is being supplied to the feedback terminal of the regulator from the collector of Q_3 so that less has to be supplied from the output through R_1. Furthermore, the circuit will continue to oscillate, even with a shorted output, because of positive feedback through R_6 and the relatively long discharge time constant of C_2.

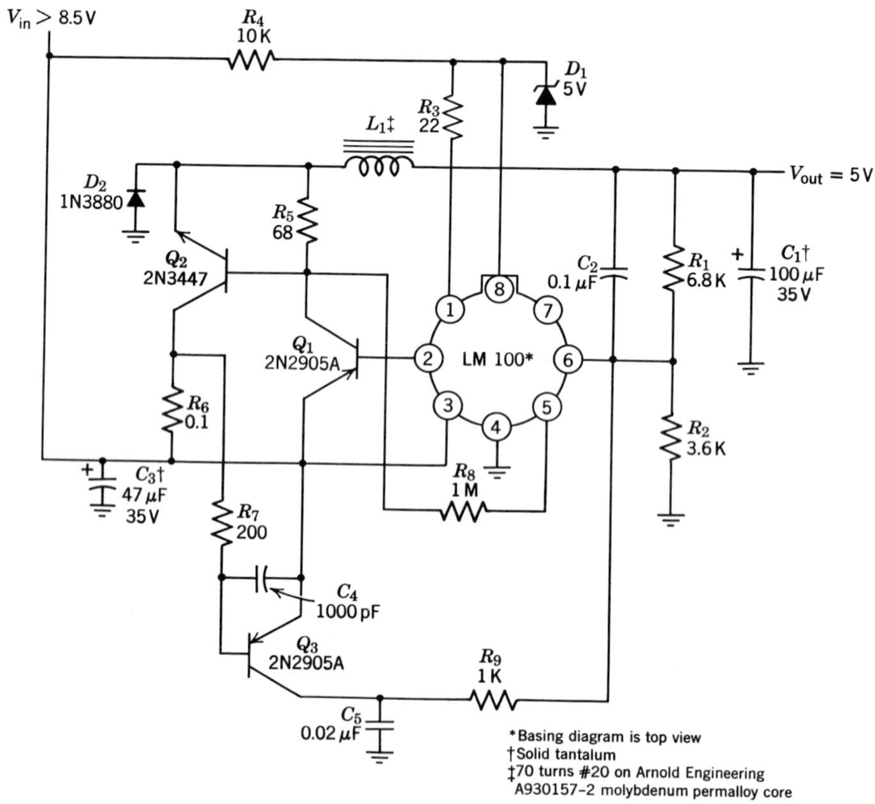

Figure 16.18. Switching regulator with current limiting.

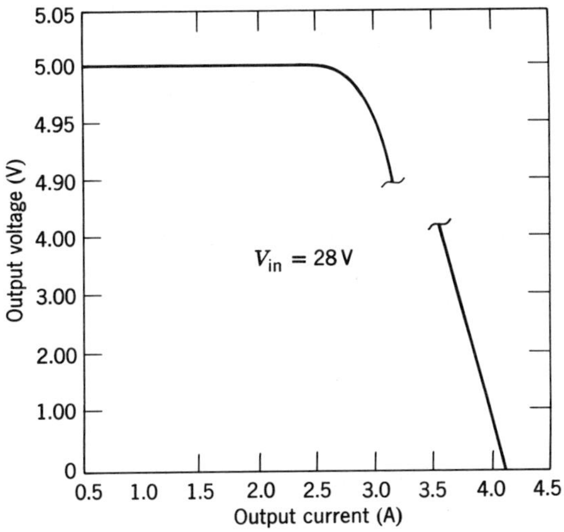

Figure 16.19. Current limiting characteristics.

It is necessary to put a resistor R_7 in series with the base of Q_3 to insure that excessive current will not be driven into the base. In addition a capacitor C_4 must be added across the input of Q_3 so that it does not turn on prematurely on the large current spike (about twice the load current) through the switch transistor caused by pulling the stored charge out of the catch diode. A zener diode bias supply must also be used on the output of the LM100 since the current limiting will not work if the voltage on this point drops below about 1 V.

The current limiting characteristics of this circuit are shown in Figure 16.19. Figure 16.20 shows how the average input current actually drops off as the circuit goes into current limiting. This current limiting scheme protects the switching transistors from overload or short-circuited output. However, the drop-out current and short-circuit current are not well controlled, so that it is difficult to prove that the circuit will sustain continuous short circuit under worst-case conditions. This is particularly true with high current regulators where the required amount of overdesign can become quite expensive.

Figure 16.21 shows a circuit that is more easily designed for continuous short-circuit protection under worst-case conditions. In this circuit the current-sensing resistor is located in series with the inductor. Therefore, the peak-limiting current can be more precisely determined since the current spike generated by pulling the stored charge out of the catch diode does not flow through the sense resistor.

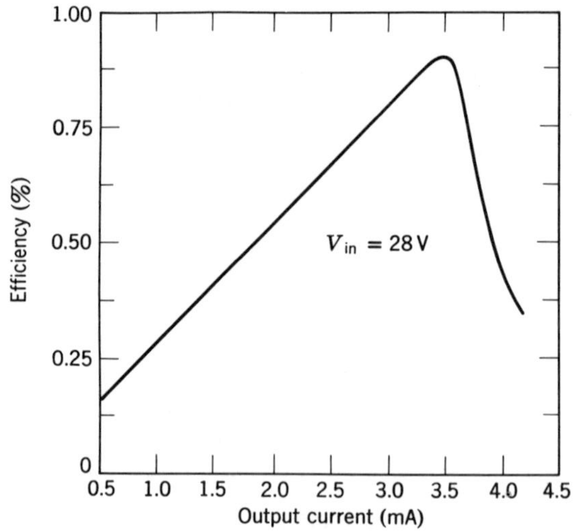

Figure 16.20. Illustrating drop in input current as regulator goes into limiting.

Operation of this circuit is essentially the same as the previous one in that an *npn* transistor, Q_4, senses the overcurrent condition and turns on Q_3 which supplies the current-limit signal to the feedback terminal. The zener diode D_3 is required on the feedback terminal to guarantee that this terminal cannot go more than 0.5 V higher than pin 1. If this does happen, the circuit can latch up and burn out. The performance of this current-limit scheme is illustrated in Figures 16.22 and 16.23.

With this circuit it is not only possible to determine more accurately the limiting current, but, as can be seen from Figures 16.22 and 16.23, the limiting characteristic is considerably sharper. One disadvantage of this circuit is that the load current flows continuously through the current sense resistor, reducing efficiency. As an example, with a 5 V regulated output the efficiency will be reduced by 10% at full load.

16.8 NEGATIVE REGULATORS

The circuits discussed so far were intended for regulators with positive outputs. Although negative regulators can be obtained by floating the unregulated supply and grounding the output, this is not always convenient.

Figure 16.24 shows a circuit for a negative switching regulator in which the unregulated input and regulated output have a common ground. The only limitation of the circuit is that there must be a positive voltage greater than 3 V available in order to properly bias the negative regulator.

In this circuit, the normal output terminal of the LM100 (pin 8) is grounded and the ground terminal (pin 4) is connected to the regulated negative output. Hence, as before, it regulates the voltage between the output and ground terminals. The unregulated input terminal (pin 6) is run from a positive voltage for proper biasing. A *pnp* booster transistor Q_3 is connected in the normal manner, and it in turn drives a Darlington-connected *npn* switch. Positive feedback is developed by the resistive divider, R_8 and R_{12}.

It is necessary to use a Darlington switch even though the current gain is not needed. The power switch transistor Q_4 cannot be operated with a fixed base drive: if the base drive is made large enough to ensure saturation at maximum load current, it will overstore so badly at lower currents that the output ripple will increase radically. With the extra transistor, however, it is kept out of saturation at lower output currents, eliminating the problem.

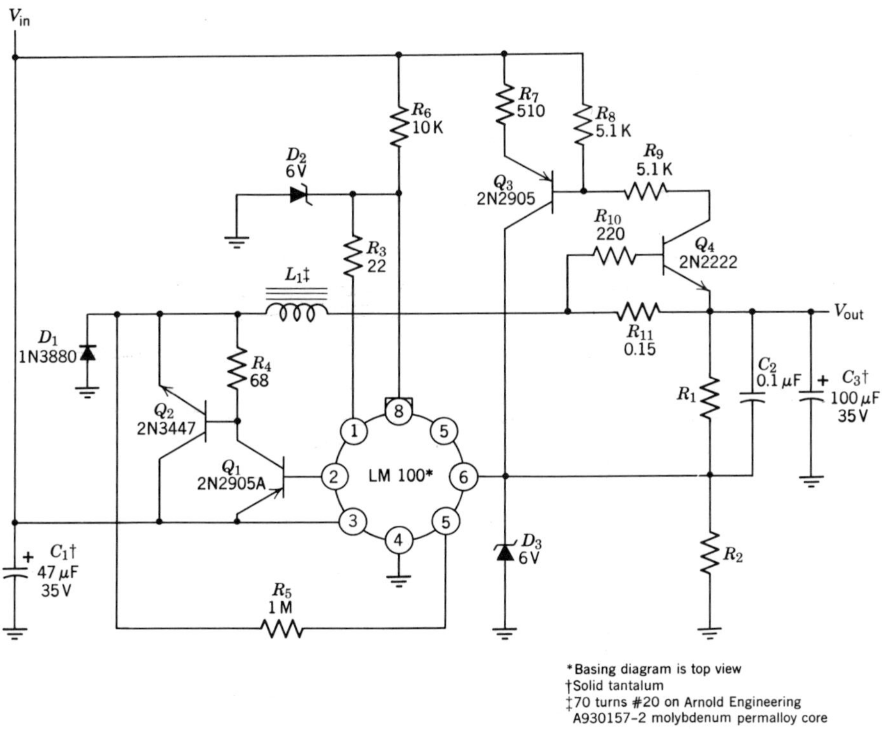

Figure 16.21. Switching regulator with continuous short-circuit protection.

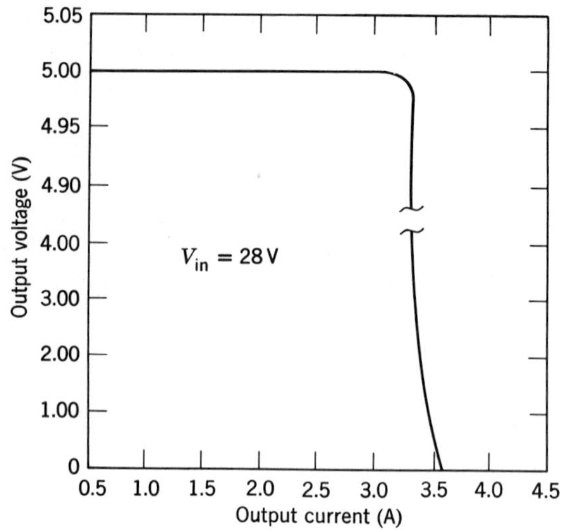

Figure 16.22. Current limiting characteristics.

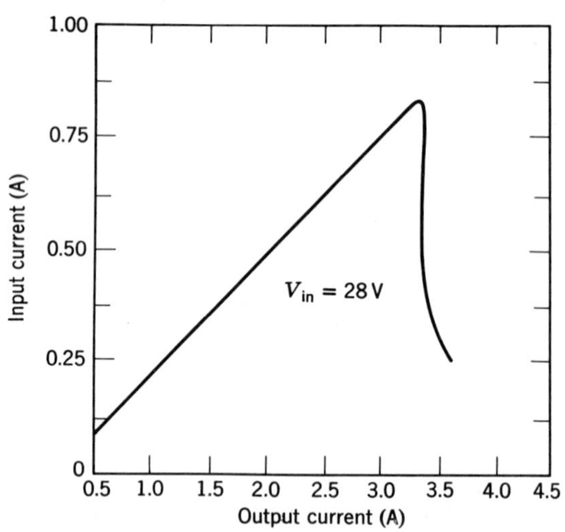

Figure 16.23. Plot of input current as regulator goes into limiting.

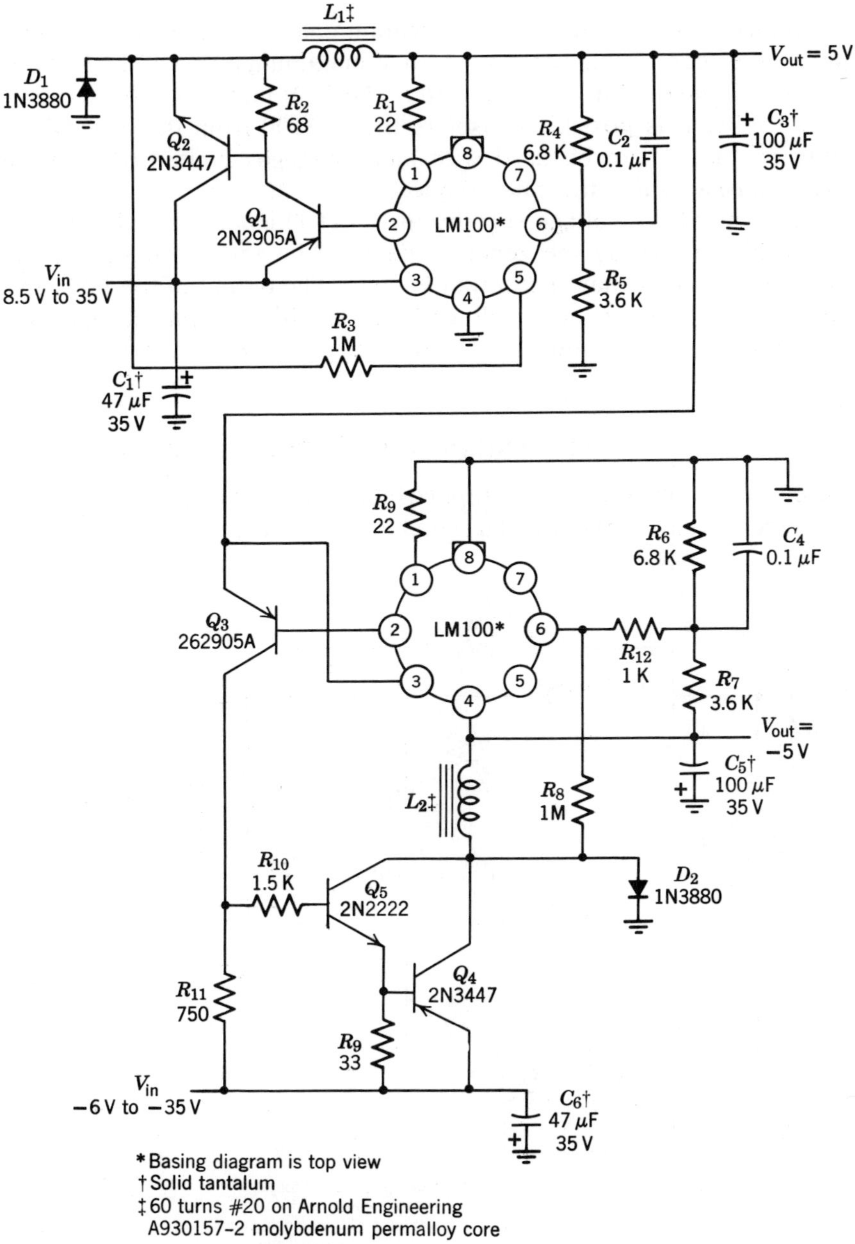

Figure 16.24. Positive and negative switching regulators.

16.9 HIGH VOLTAGE REGULATORS

With switching regulators, an application can easily arise in which the input voltage can be higher than the 40 V maximum rating of the LM100, even though the output voltage is within the 30 V maximum. As shown in Figure 16.25, it is possible to isolate the LM100 from the unregulated supply so that it can be used with input voltages limited only by the switch transistors and the catch diode.

In this circuit the voltage seen by the LM100 is maintained at a fixed level within ratings by the zener diode D_2. The zener voltage must be at least 3 V greater than the output voltage. The output of the LM100 is level-shifted up to the input voltage by an additional *npn* transistor Q_3, which is operated common base. This drives the *pnp* switch driver in the normal manner.

16.10 SWITCHING AND LINEAR REGULATOR COMBINATION

In certain applications the output ripple and load transient response requirements rule out the use of a switching regulator, yet the input-output

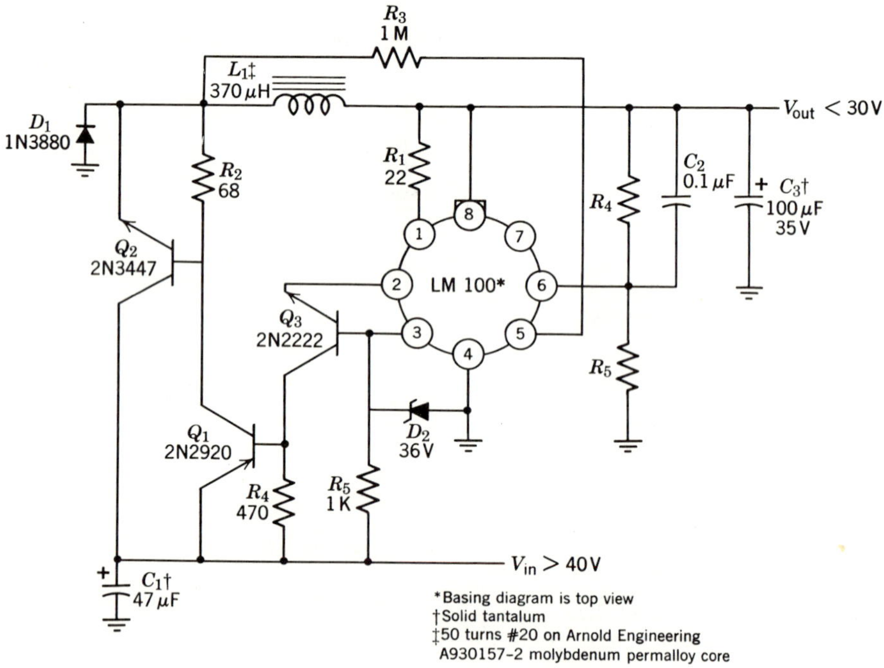

Figure 16.25. Switching regulator for high-voltage inputs.

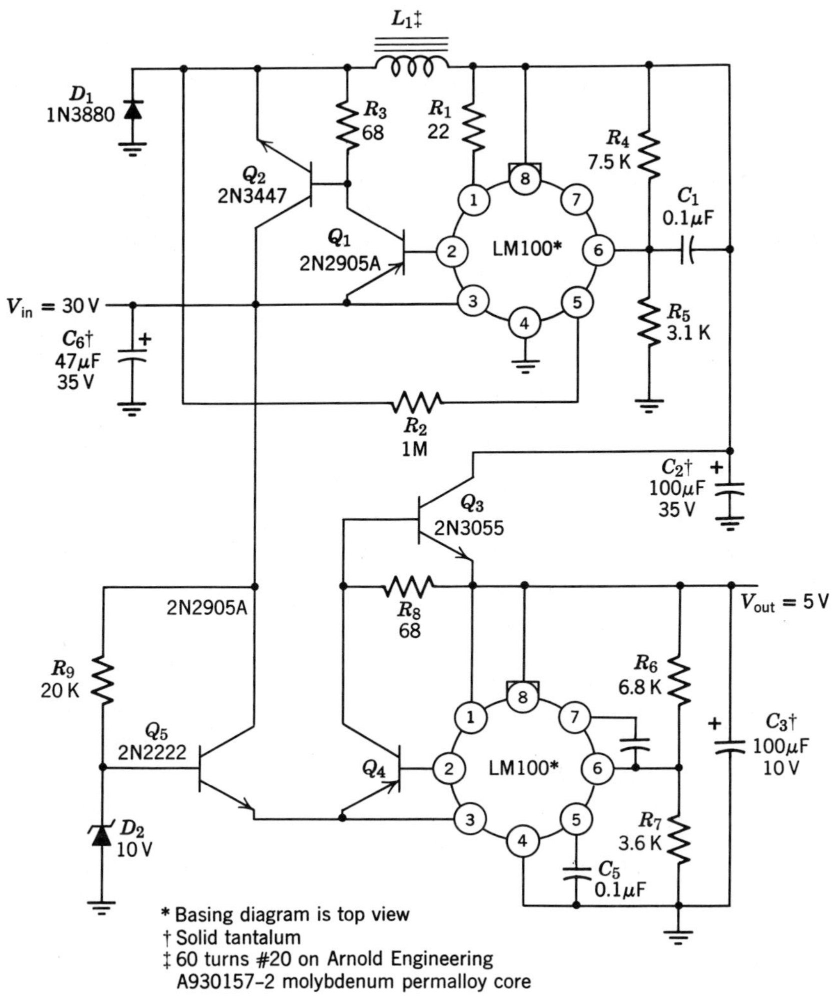

Figure 16.26. Switching and linear regulator combination for obtaining very low ripple and fast transient response.

voltage differential is still high. In this case a power converter might be used to reduce the input voltage, and this reduced voltage would be regulated by a linear regulator. This arrangement, however, is not nearly as efficient as the switching and linear regulator combination shown in Figure 16.26. The switching regulator not only reduces the input voltage with high efficiency, but it also regulates it. Therefore, the linear regulator operates with a fixed input-output voltage differential that holds dissipation to a minimum.

In this circuit, the linear regulator is biased by a zener pre-regulator (R_9, D_2, and Q_5) to isolate it from noise on the unregulated supply. This separate bias supply permits the linear pass transistor Q_3 to operate right down into saturation. The collector of Q_3 is supplied by the output of a switching regulator which is made sufficiently higher than the linear regulator output to allow for the maximum overshoot of the switching regulator plus the saturation of Q_3.

16.11 SUMMARY

A number of switching regulator circuits that use a readily available monolithic voltage regulator as the voltage reference and control circuitry have been described. These regulators are useful over a 2 V to 30 V range for either positive or negative supplies. Although the discussion was limited to circuits providing maximum output currents from 100 mA to 5 A, it is possible to obtain even high output currents. The output current is, in fact, limited by the discrete components—not by the basic design or the integrated circuit.

The majority of the circuits shown were self-oscillating regulators; however, a method of driving the regulator in synchronism with an external clock signal was given. In addition circuits which provide overload protection, limiting both the output current as well as the power dissipation, were presented. The performance of the regulator circuits was described in detail, and a design procedure was outlined. Suggestions were also made on the selection of components for switching regulators.

The circuits which have been described here for the LM100 work equally well with the LM200. The LM200 is identical, except that it is specified for operation over a 0°C to 70°C temperature range instead of −55°C to 125°C.

17

Applications for Audio-Power-Amplifier ICs

by J. E. Byerly
General Electric

17.1 INTRODUCTION

Versatility can be designed into audio-amplifier ICs to make them useful in a wide variety of applications. Some of these applications include phonograph, tape player, a-m-fm-television audio, intercom, servo-amplifiers and industrial uses.

The broad range of applications that an amplifier such as the PA237 can fit into exists because of the IC's ability to operate over a wide range of supply voltage and bias conditions. Various forms of feedback, applied to the amplifier, allow for adjustment of stability, input and output impedance, and amplifier sensitivity. Both ac and dc feedback is employed to provide stability with frequency and temperature.

The circuit for the PA237 is shown in Figure 17.1. The output is a quasi-complementary push-pull circuit consisting of transistors Q_4, Q_5, Q_6, Q_7, and Q_8. The Darlington configuration enables the driving circuit to operate at low-current levels and therefore provides good sensitivity and noise characteristics. The driving circuit consists of transistors Q_1 and Q_2 in a differential-amplifier configuration. This provides good bias stability against variations of h_{FE} and temperature.

The biasing network is external to the chip to provide overall user economy and allow maximum versatility of application. The IC can be biased into class A, class A-B, or class-B operation. Many biasing arrangements may be employed, including current-source biasing, various degrees of voltage source biasing, common-mode biasing, etc. The standby idle current is established by the current in the diode-chain regulator. Increasing the diode current moves the amplifier toward

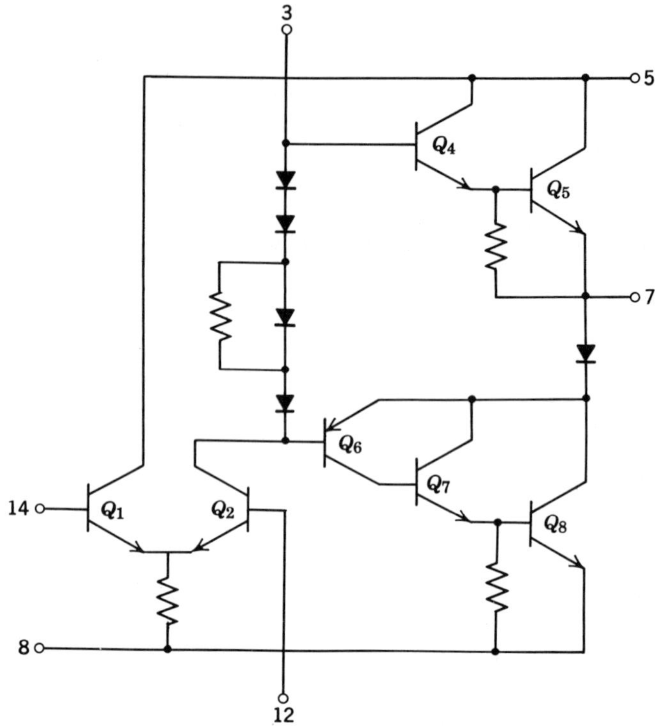

Figure 17.1. Circuit diagram for the PA237.

class-A operation. The center point voltage (pin 7) is established by the voltage drop across the load (R_3) in series with the diode string (pin 3) as shown in Figure 17.2. Thus, by properly choosing the value of R_3 and diode current, the desired center point voltage and output idle current can be obtained.

The operating voltage for a standard PA237 is 9 to 27 volts. The output has a peak capability in excess of one ampere which enables the amplifier to drive low impedance loads.

17.2 TEST CIRCUIT

The amplifier test circuit is shown in Figure 17.2. The circuit has voltage-source biasing, high and low frequency rolloff, bootstrapping, negative dc feedback and negative ac feedback loop closure. The amplifier is biased into class A-B operation. The sensitivity is 120 mV for a 2 W output or a voltage gain > 45. The input impedance is 40 kΩ and the output impedance is 0.85 Ω. The noise output is 75 dB below 2 W.

The ratio of R_4 to R_f determines the amount of ac feedback closure. Setting the 6.8-kΩ feedback resistor (R_f) to other values enables the sensitivity, input impedance, and distortion to vary as shown in Table 17.1. Making this feedback resistor the volume control may be advantageous in that as the output level is decreased by increasing the negative ac feedback, the distortion decreases. This is particularly desirable for lowering crossover distortion at low levels, because at low levels crossover becomes a significant portion of the total distortion.

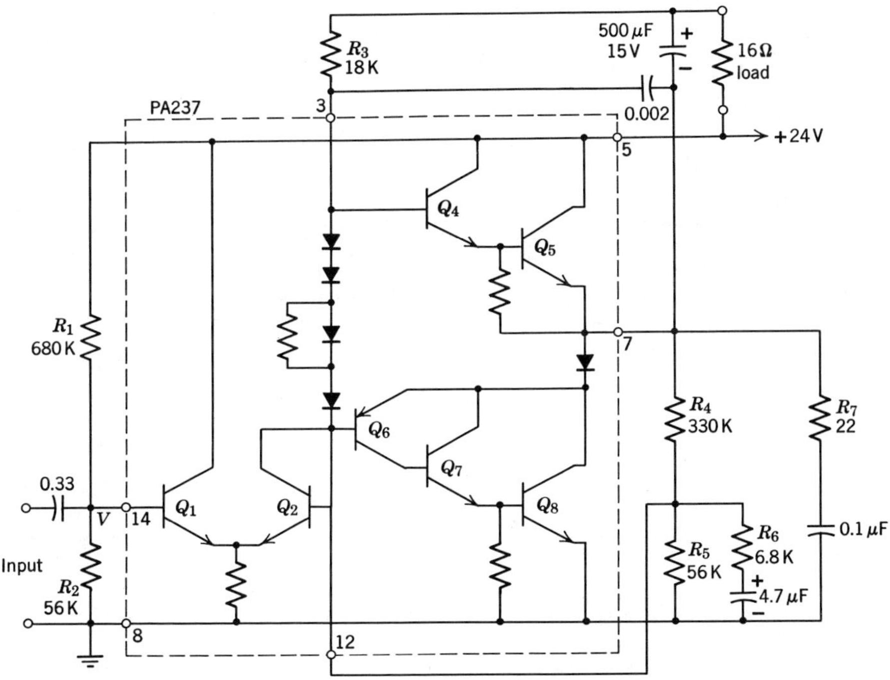

Figure 17.2. Test circuit for the PA237.

TABLE 17.1 Performance Data in Test Circuit of Figure 17.2

R_f	0	1 kΩ	5 kΩ	10 kΩ
Sensitivity at 2 W	8.0 mV	22 mV	86 mV	150 mV
Total harmonic distortion at 2 W	5.2%	2.9%	1.7%	1.5%
Z_{in}	20 kΩ	35 kΩ	40 kΩ	40 kΩ
Frequency response −3 dB	80−24k Hz	35−50k Hz	18−100k Hz	17−110k Hz

17.3 POWER IC APPLICATIONS

Figure 17.3 shows the IC in a 24-V, 16-Ω open-loop application. It has a sensitivity of 5 mV for 2-W output or a voltage gain of > 1100. The input impedance is > 15 kΩ. This sensitivity enables the amplifier to be driven directly from a diode detector or magnetic phono cartridge. The same configuration can be used with other combinations of voltage and load impedance simply by changing the values of the biasing resistors. Figure 17.4 shows such a change for 12-V and 8-Ω operation. 7 mV drives the amplifier to ¾ W output for a voltage gain of 350 at 3.5% total harmonic distortion. Maximum power out is 1 W at 10% total harmonic distortion. The input impedance is 15 kΩ.

The number of external components can be reduced when the application does not require all of the protective elements that were included in Figure 17.2. The configuration of Figure 17.5 employs current-source biasing. Adjustable loop closure and low frequency rolloff call for a total external component count of four resistors and two capacitors. The current gain (h_{FE}) of the differential-amplifier transistors Q_1 and Q_2 must be known in order to choose the current-source resistor values. The configuration for 24-V and 16-Ω class A-B operation is shown in Figure 17.5. Table 17.2 shows the amplifier performance for different degrees of loop closure.

TABLE 17.2 Performance Data for Current-Source Bias Configuration of Figure 17.5

R_f	0	1 kΩ	5 kΩ	10 kΩ
Sensitivity at 2 W	5 mV	11 mV	26 mV	40 mV
Total harmonic distortion at 2 W	5.0%	4.5%	3.5%	1.8%
Z_{in}	24 kΩ	50 kΩ	80 kΩ	100 kΩ
Frequency response −3 dB	60−19k Hz	40−29k Hz	20−41k Hz	10−50k Hz

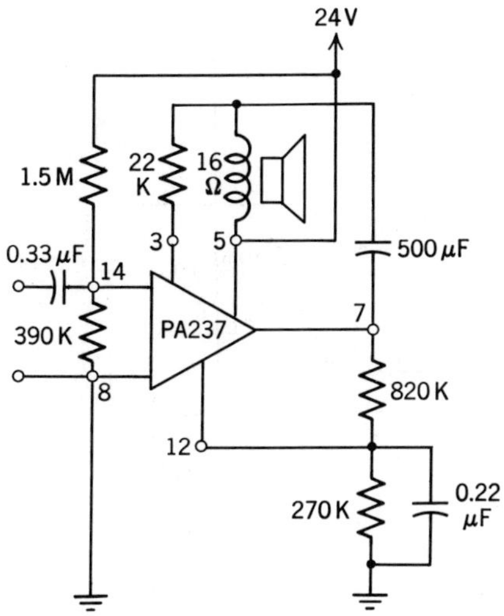

Figure 17.3. 24-V open-loop configuration.

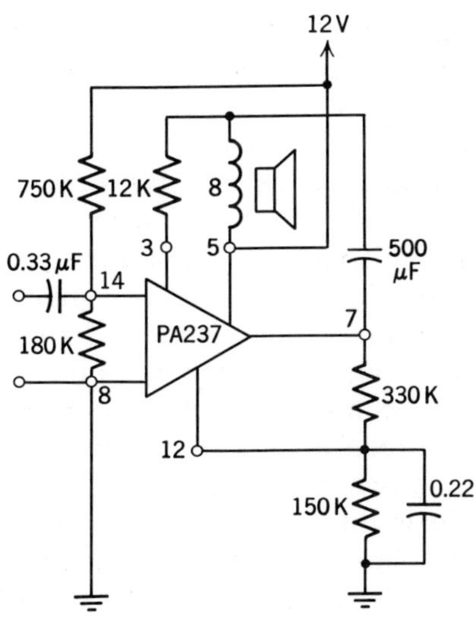

Figure 17.4. 12-V open-loop configuration.

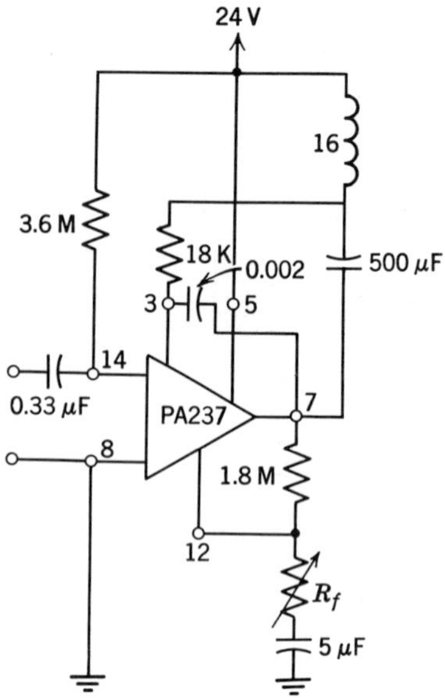

Figure 17.5. Current-source bias configuration.

The large output coupling capacitor can be eliminated by using two supplies, if available. Figure 17.6*a* shows such a configuration employing voltage-source biasing. Table 17.3 gives performance data for the circuit. The same circuit but with bootstrapping added is shown in Figure 17.6*b*; this circuit's performance is charted in Table 17.4.

The circuit in Figure 17.7 employs a minimum number of external components. The performance, as a result, is not optimum and not up to that of some of the other circuits described.

For operation from high-source impedance transducers such as ceramic and crystal phono cartridges, the amplifier input impedance may be increased by use of the configuration shown in Figure 17.8. The PA237 is used with additional input bootstrapping to increase input impedance. The ratio of the parallel combination of resistors R_2 and R_5 to R_4 determines the amount of ac feedback closure. The loop gain can be adjusted by choosing the biasing resistors so that the parallel combination of R_2 and R_5 result in the desired loop closure. Another method is to place an ac shunt consisting of a capacitor and the desired resistor value across R_5.

For operation with crystal phono cartridges or other high-impedance sources, the circuit of Figure 17.9 may be considered using voltage ratio feedback. The circuit performance for different degrees of feedback closure is listed in Table 17.5. The circuit of Figure 17.10 may be used in place of the one in Figure 7.9, saving a resistor and capacitor if the degree of negative dc feedback required is not as great.

TABLE 17.3 Performance Data for Dual-Supply Configuration in Figure 17.6*a*

R_f	0	1 kΩ	5 kΩ	10 kΩ
Sensitivity at 2 W	40 mV	80 mV	150 mV	220 mV
Total harmonic distortion at 2 W	4.5%	3.2%	2.5%	2.2%
Z_{in}	15 kΩ	20 kΩ	30 kΩ	35 kΩ
Frequency response −3 dB	10−75k Hz	12−90k Hz	9−100k Hz	8−130k Hz

TABLE 17.4 Performance Data for Dual-Supply Configuration with Bootstrapping in Figure 17.6*b*

R_f	0	1 kΩ	5 kΩ	10 kΩ
Sensitivity at 2 W	10 mV	23 mV	85 mV	140 mV
Total harmonic distortion at 2 W	7.5%	4.5%	2.0%	1.5%
Z_{in}	24 kΩ	20 kΩ	35 kΩ	40 kΩ
Frequency response −3 dB	50−20k Hz	25−40k Hz	12−80k Hz	10−95k Hz

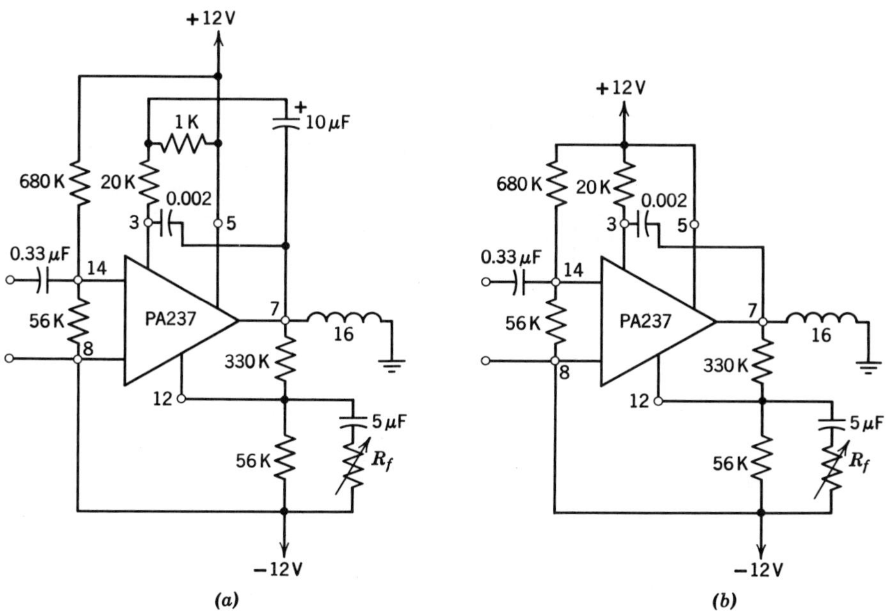

Figure 17.6*a*. Dual-supply configuration.

Figure 17.6*b*. Dual-supply configuration with bootstrapping.

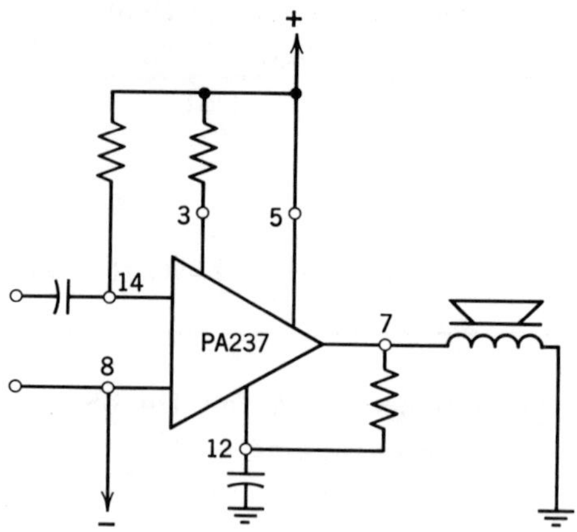

Figure 17.7. Minimum external component configuration.

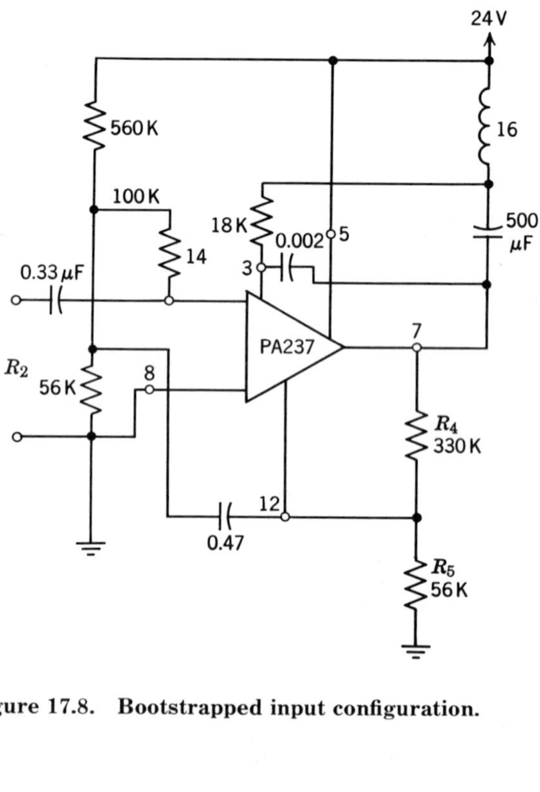

Figure 17.8. Bootstrapped input configuration.

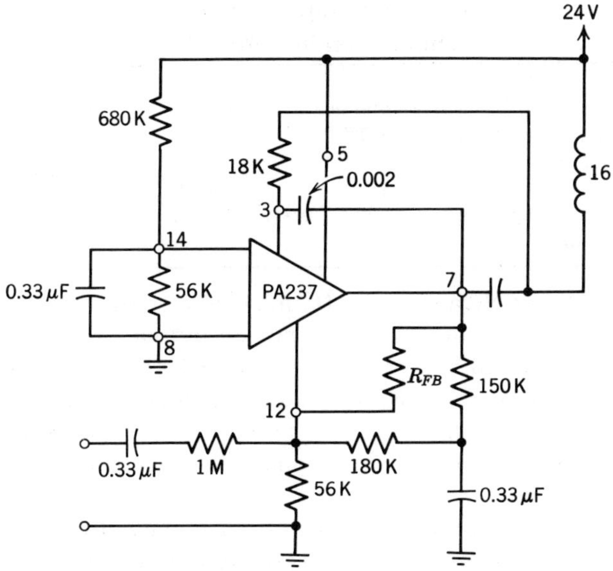

Figure 17.9. Voltage-ratio feedback configuration.

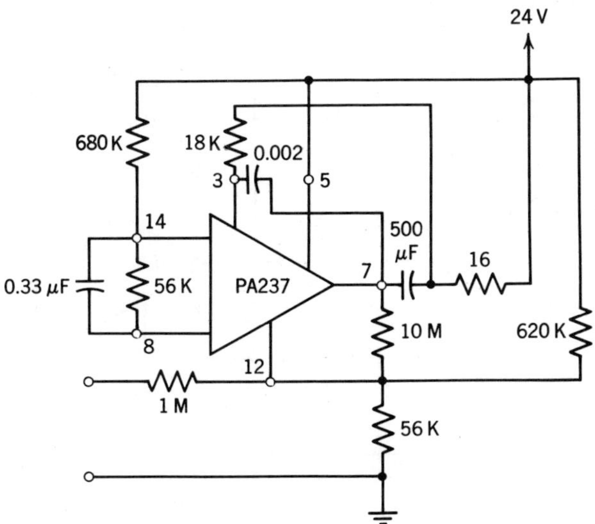

Figure 17.10. Simplified version of Figure 17.9.

TABLE 17.5 Performance Data for Voltage-Ratio Feedback Configuration in Figure 17.9

R_{FB}	∞	10 MΩ	5 MΩ
Sensitivity at 2 W	650 mV	1.1 V	1.9 V
Total harmonic distortion at 2 W	3.0%	1.8%	1.25%
Z_{in}	1 MΩ	1 MΩ	1 MΩ
Frequency response −3 dB	150–10k Hz	80–21k Hz	50–34k Hz

For high-power applications the IC may be used in a bridge configuration, thereby doubling the available power output. Figures 17.11 and 17.12 show two typical bridge circuits, one with bootstrapping and one without. Resistors R_1 and R_2 are adjusted for the desired amplifier sensitivity and input impedance. The various biasing configurations previously discussed for the single amplifier also apply to the bridge circuits.

The PA237 may also be used as a voltage regulator as is shown in Figure 17.13. The current capability is limited by the difference between input voltage and regulated voltage so that the voltage-current product

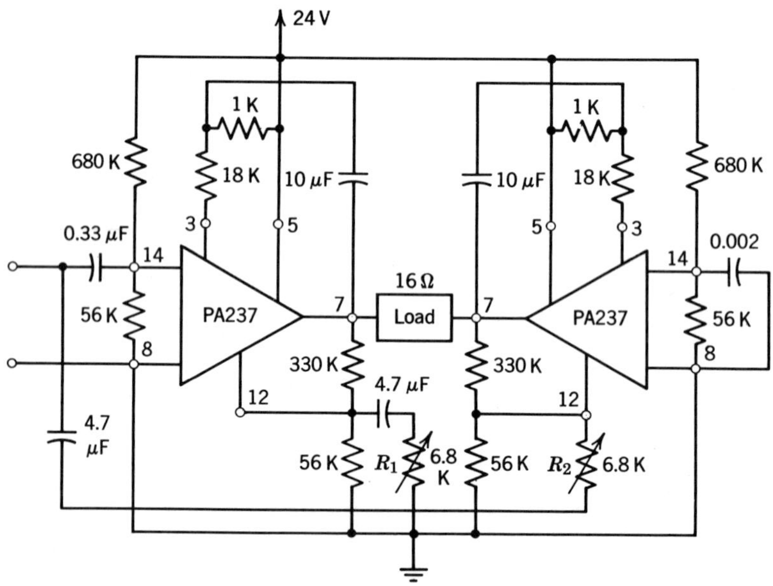

Figure 17.11. Bootstrapped bridge configuration.

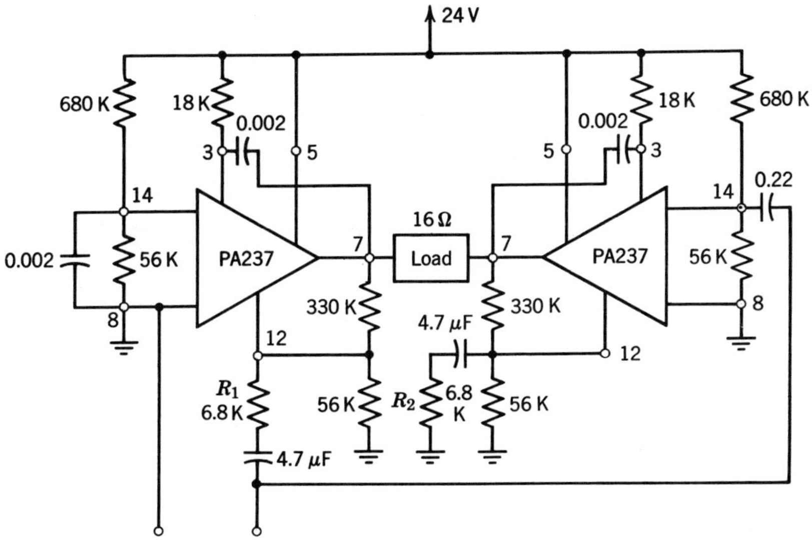

Figure 17.12. Bridge configuration without bootstrapping.

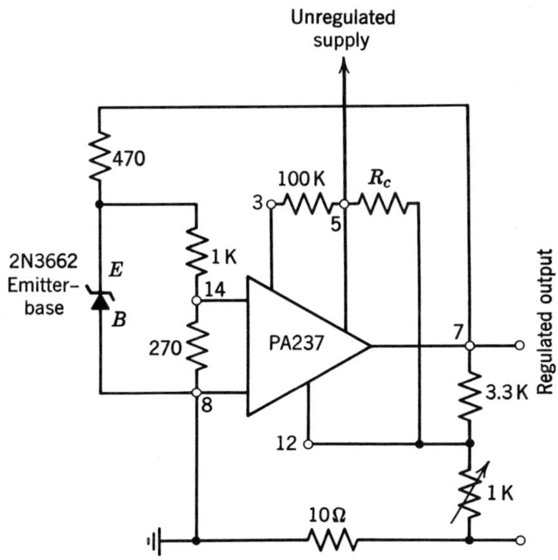

Figure 17.13. Voltage regulator.

does not exceed the power rating of the IC. Short-circuit protection is provided by the resistor in the ground return path. An alternate method of current limiting that provides a sharper cutoff is to use a transistor to shunt the reference diode at the desired current. Table 17.6 reveals the performance of the circuit in Figure 17.13. Various regulator output voltages are shown for unregulated input voltages.

TABLE 17.6 Performance Data for Voltage-Regulator Circuit in Figure 17.13

	Unregulated Voltage Input (V)					
Regulated voltage output and compensating resistor R_C	15	20	25	30	35	40
$R_C = 100$ kΩ	9.9	10.0	10.0	10.0	10.0	10.05
$R_C = 70$ kΩ	11.66	12.01	12.01	12.00	11.97	11.99
$R_C = 50$ kΩ		14.8	15.0	15.0	15.0	15.0
$R_C = 40$ kΩ		16.5	17.0	17.0	17.0	17.0
$R_C = 30$ kΩ			19.9	20.1	20.1	20.1
$R_C = 20$ kΩ			22.4	22.6	22.5	22.5

18

Fully Compensated Operational Amplifier

by William S. Routh
Radiation, Inc.

18.1 INTRODUCTION

Circuitry is simplified by the use of an operational-amplifier integrated circuit that requires no external frequency compensation. The Radiation RA909 is an example of this type of IC; introduced in March 1968, it was the first unconditionally stable op-amp to be marketed.

The RA909 uses dielectric isolation, making possible the diffusion of vertical *pnp* as well as vertical *npn* transistors. Figure 18.1 compares the structure of an integrated circuit using dielectric isolation with that in common use for integrated circuits built with junction isolation. The RA909 consists of six *pnp* transistors; it also has 13 *npn* transistors and seven resistors (a total of 103.2 kΩ).

Although it has blank leads where 709-type ICs have compensation leads, it can be custom tailored thanks to separate leads for offset (offset can be adjusted to zero without loading inputs), bias, and bandwidth control. The circuit diagram for the RA909 is given in Figure 18.2; the chip is shown in Figure 18.3.

18.2 DC SERVO PREAMPLIFIER

The use of the RA909 as a dc servo preamplifier is shown in Figure 18.4; in this configuration, accurate determination of lag and lead frequencies is possible. The transfer relationship is

$$\frac{E_{\text{out}}}{E_{\text{in}}} = -\left(\frac{R_2+R_3}{R_1}\right)\left(\frac{1+j\omega\tau_1}{1+j\omega\tau_2}\right),$$

where $\tau_1 = R_2C$ and $\tau_2 = R_3C$. The frequency of lag and lead are defined by

$$f_{\text{lag}} = 1/2\pi\tau_2;\, f_{\text{lead}} = 1/2\pi\tau_1.$$

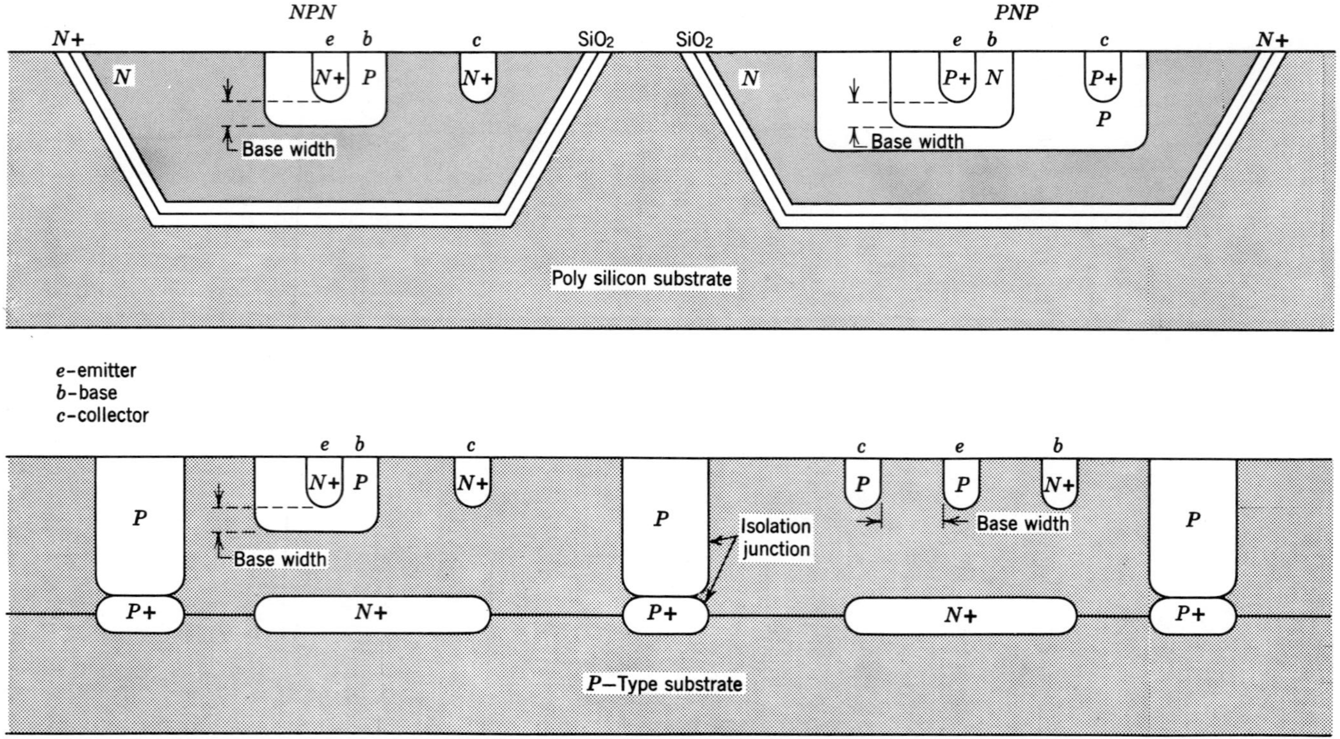

Figure 18.1. Dielectric isolation used in RA909 (top) compared with junction isolation.

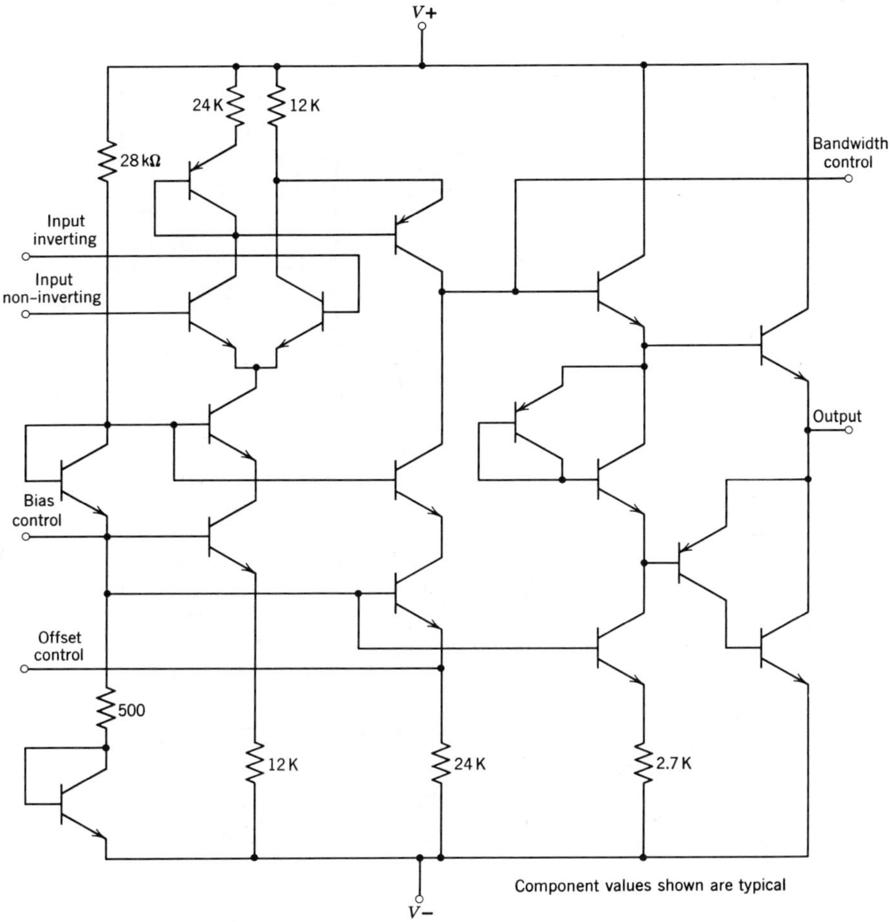

Figure 18.2. Circuit diagram for the RA909.

For a gain of 100, the preamplifier rolloff will be 70 kHz. The undistorted output voltage is 26 V peak-to-peak.

18.3 UNITY-GAIN BROADBAND OP AMP

The RA909 provides the 6 dB/octave high-frequency rolloff required for unconditioned stability without external compensation. Other

Figure 18.3. Key components in RA909 op-amp IC include vertical *pnp* transistors (two of which can be seen in top left corner of IC).

characteristics provided in the circuit shown in Figure 18.5 are the following:

$V_{in} = V_{out}$	200 mV
Risetime	40 ns
Overshoot	15%
Slew Rate	5 V/μs
Bandwidth	7 MHz
Phase Margin	60°

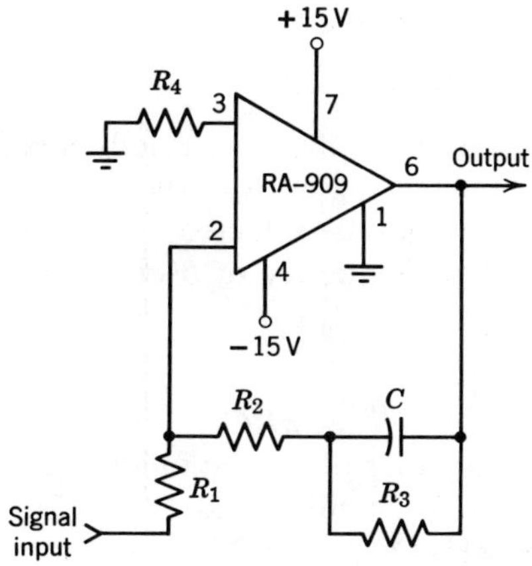

Figure 18.4. Servo preamplifier arrangement.

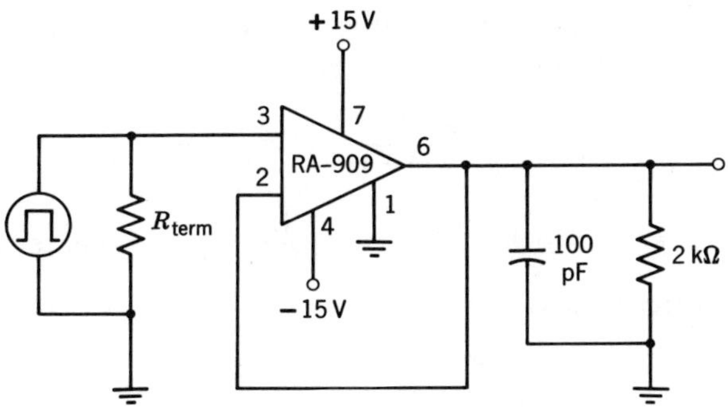

Figure 18.5. Unity-gain broadband op amp.

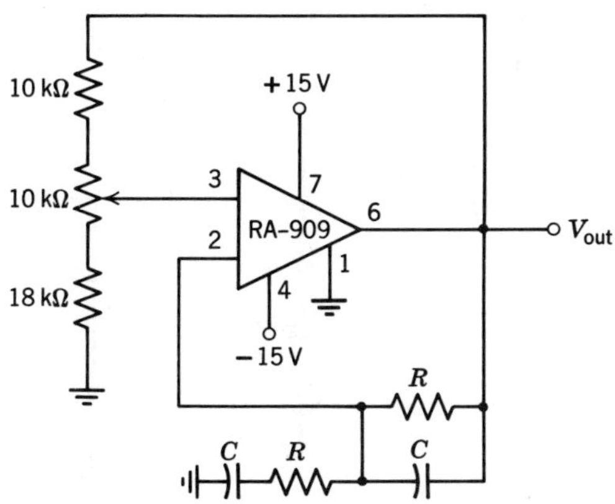

Figure 18.6. Wien bridge oscillator.

18.4 WIEN BRIDGE OSCILLATOR

A highly stable Wien bridge oscillator, virtually unaffected by temperature variations, can be built using an RA909 (see Figure 18.6). A wide range of *RC* combinations may be selected without regard to the active element of the circuit. The frequency of oscillation (up to 500 kHz) is defined by

$$f_0 = \frac{1}{2\pi RC}.$$

In this application, the amplifier offers the following typical performance characteristics with 2-kΩ R_L: undistorted E_{out} = 26 V peak-to-peak at f_0 = 10 kHz; at f_0 = 150 kHz, E_{out} = 11.2 V peak-to-peak.

18.5 HIGH-Q BANDPASS FILTER

The filter shown in Figure 18.7 features a *Q* of 100 at a center frequency of 10 kHz. The filter transfer function is

$$\frac{E_{out}}{E_{in}} = \frac{jxy}{1+y(1-x^2)+jx};\ Q = y\left(\frac{1+y}{y}\right)^{\frac{1}{2}},$$

where $x = (f/f_1)(f_0) = (1/2\pi RC)(f_0) = f_1\left(\frac{1+y}{y}\right)^{\frac{1}{2}}$, f_0 = center frequency. f_1 = lower 3 dB frequency.

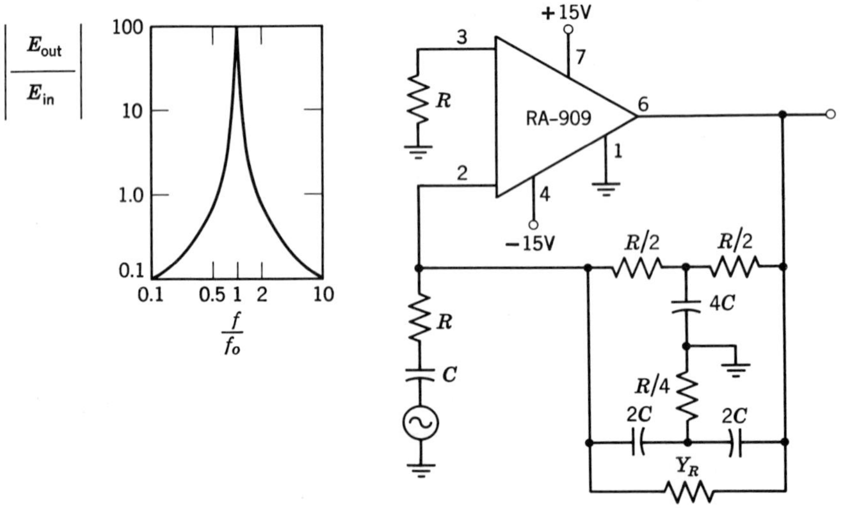

Figure 18.7. High-*Q* bandpass filter circuit and response curve.

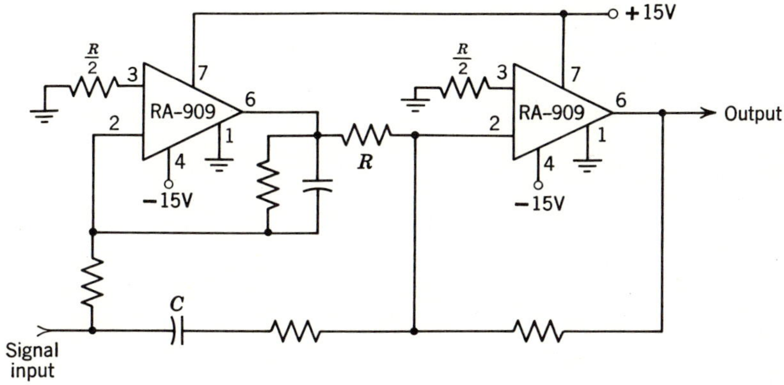

Figure 18.8. Phase-delay filter.

The typical response curve shows the high Q obtainable: $Q = 100, f_0 = 10$ kHz, $E_{out} = \pm 13$ V.

18.6 PHASE-DELAY FILTERS

The circuit shown in Figure 18.8 allows accurate determination of phase shift. Feedback components are selected to determine the 90° phase-shift frequency without regard to the active elements in the circuit configuration.

The transfer characteristic is

$$\frac{E_{out}}{E_{in}} = \frac{1 - jx}{1 + jx},$$

where $x = f/f_0$.

The 90° phase-shift frequency is $f_0 = 1/(2\pi RC)$; x is the normalized frequency ratio referenced to the 90° phase-shift frequency. For $f_0 <$ 70 kHz, the output voltage is 26.0 V peak-to-peak.

Index